A+ Operating Systems for Technicians 2004

Todd Meadors
DeKalb Technical College

Cheryl A. Schmidt
Florida Community College at Jacksonville

Scott/Jones, Inc.
P.O. Box 696
El Granada, CA 94018
Voice: 650-726-2436
Facsimile: 650-726-4693
E-mail: marketing@scottjonespub.com

A+ Operating Systems for Technicians 2004
Todd Meadors, *DeKalb Technical College*
Cheryl A. Schmidt, *Florida Community College at Jacksonville*

ZYX   543

Printed in Canada.

ISBN: 1-57676-106-1

The publisher wishes to acknowledge the memory and influence of James F. Leisy. Thanks, Jim. We miss you.

Text Design: Martie Bateson Sautter and Joshua Faigen, Folio Publishing Services, Inc.
Cover Design: Martie Bateson Sautter
Copyediting: Lois Principe
Composition: Diane DeMarco, Option C
Proofreading: Kristin Furino, Holbrook Communications
Book Manufacturing: WebCom

Scott/Jones Publishing Company
Editorial Group: Richard Jones, Denise Simon, and Patricia Miyaki
Production Management: Audrey Anderson
Marketing and Sales: Victoria Chamberlin, Richard Jones, and Leata Holloway
Business Operations: Michelle Robelet, Cathy Glenn, and Bill Overfelt

A Word About Trademarks

All product names identified in this book are trademarks or registered trademarks of their respective companies. We have used the names in an editorial fashion only, and to the benefit of the trademark owner, with no intention of infringing the trademark. IBM and Lotus are trademarks of International Business Machines, Intel, Pentium, Intel386, Celeron, Intel486, OverDrive, Xeon, Camino, 440ZX, 440EX, 450NX, 3GIO, MMX, and MMX2 are trademarks of Intel Corp. AMD, AMD Athlon, AMD-760, AMD Duron, 3DNow!, and HyperTransport are trademarks of Advanced Micro Devices, Inc. Alpha is a trademark of Digital Corp. Windows, Windows 95, Windows 98, NT Workstation, Windows NT, Windows 2000, Windows XP, DirectX, MSD, Terminal, HyperTerminal, and Word are trademarks of Microsoft Corporation. Apple, PowerPC, QuickTime, and FireWire are trademarks of Apple Computer. RapidIO is a trademark of Motorola, Inc. Cyrix is a trademark of Cyrix Corp. DEC is a trademark of Digital Electronics Corporation, Inc. Northern Telecom is a trademark of Nortel Networks, Corp. CardBay and PCMCIA are registered trademarks of the Personal Computer Memory Card International Association. ALI is a trademark of Acer Labs, Inc. VIA technologies is a trademark of VIA Technologies, Inc. AMD is a trademark of Advanced Micro Devices. Compaq is a trademark of Compaq Corp. HP is a trademark of Hewlett-Packard Company. Zenith is a trademark of Zenith Electronics. Epson is a trademark of Seiko Epson. NEC is a trademark of NEC Corp. Wyse is a trademark of Wyse Technology. AST is a trademark of AST Research, Inc. Acer is a trademark of Acer America Corp. Tandy is a trademark of Tandy Corp. Olivetti is a trademark of Olivetti S.p.A. SiS is a trademark of Silicon Integrated Systems Corp. OPTi is a trademark of OPTi, Inc. Award is a trademark of Award Software International, Inc. PhoenixBIOS is a trademark of Phoenix Technologies Ltd. AMI BIOS is a trademark of American Megatrends, Inc. Dell is a trademark of Dell Computer Corp. Norton Utilities is a trademark of Symantec Corp. CheckIt is a trademark of Touchstone Software Corp. Nut & Bolts is a trademark of Helix Software Corp. SpinRite is a trademark of Gibson Research Corp. EZ-Drive is a trademark of Western Digital Corp. Seagate is a trademark of Seagate Technology. Sound Blaster is a trademark of Creative Labs. Hayes is a trademark of Hayes Microcomputer. x2 is a trademark of 3Com Corporation. Toshiba is a trademark of Toshiba America Information Systems. Trinitron is a trademark of Sony Electronics Inc. WordPerfect is a trademark of Corel Corp. Macromedia Flash and Macromedia Shockwave are trademarks of Macromedia, Inc. Acrobat Reader is a trademark of Adobe Systems, Inc. RealPlayer is a trademark of RealNetworks, Inc. WinZip is a trademark of WinZip Computing, Inc. Download Accelerator Plus is a trademark of Speedbit Ltd. Go!Zilla is a trademark of Radiate. NetSonic is a trademark of Redmond Ventures, Inc. Eudora Light is a trademark of QUALCOMM, Inc. A+ is a trademark of The Computing Technology Industry Association, Inc. UNIX is a trademark of the The Open Group. D-Link and D-Link Air are trademarks of D-Link Systems, Inc. Red Hat is a trademark of Red Hat, Inc. Novell and NetWare are trademarks of Novell, Inc. Apple and MAC OS are trademarks of Apple Computer, Inc.

Contents

Chapter 5 Introduction to Microsoft Windows NT

Chapter 6 Introduction to Windows 2000

Chapter 7 Introduction to Microsoft Windows XP

Chapter 8 Introduction to UNIX/Linux

Note: Appendices are on the CD.

Preface

A+ Operating Systems for Technicians is written for an introductory course in operating systems software. At the beginning of each chapter is the latest CompTia A+ Operating System Technologies exam. Although this book focuses mainly on operating systems, Chapter 9 is dedicated to Networking including a section on wireless. There are numerous exercises embedded within the chapter and lab projects at the end of each chapter.

Organization of the Text

This textbook is organized in the following manner:

Chapter 1 The Operating System Environment covers basic hardware and software terminology.

Chapter 2 Basic Operating System Theory covers operating systems software in greater detail.

Chapter 3 DOS and the DOS Command Line Interface details the commands used in DOS and the DOS command line. Coverage includes the file system hierarchy, commands and batch files.

Chapter 4 Introduction to Windows 98 deals with Windows 98 from installation, managing files and folders to the boot process and troubleshooting.

Chapter 5 Introduction to Windows NT handles Windows NT installation, managing files and folders, compression, the Registry, configuring and troubleshooting.

Chapter 6 Introduction to Windows 2000 covers Windows 2000 installation, managing files and folders, compression and encryption, dual-booting, the Registry, Dr. Watson, configuring and troubleshooting.

Chapter 7 Introduction to Windows XP details Microsoft's newest desktop operating system. It covers Windows XP installation, dual-booting, the Registry, the boot process, Task Manager, Event Viewer, Performance Monitor and Dr. Watson.

Chapter 8 Introduction to UNIX/Linux covers the basic UNIX/Linux operating system commands, and concepts that are common to nearly all versions of UNIX or

Linux. Topics include understanding the tree structure, file and directory commands, redirection, filtering (piping) and wildcards (pattern matching). Also, included in a section on shell programs and security using the chmod command.

Chapter 9 Introduction to Networking highlights networking theory, protocols, topologies, access mechanisms, the OSI model, cabling, and three sections on wireless.

Appendix A Number Conversions covers how to convert numbers to and from different bases such as binary, decimal and hexadecimal. Problems are included for reinforcement.

Appendix B DOS (including DOS Command Prompt) and Linux Commands highlights the commands used by DOS and Linux discussed in this textbook.

Appendix C is the ASCII Chart with a few student problems to complete.

Appendix D includes the answers to the odd-numbered questions.

The Appendices will be on the CD included with the book.

From a pedagogical stance, if you lack the time to cover all chapters, you could go over Chapters 1, 2, 3 and then any of the Microsoft Windows chapters 4 through 7 (depending upon what you have available), Chapter 8 on UNIX/Linux and then Chapter 9, the Networking chapter.

Features
Easy-to-understand text Each chapter includes with excellent explanations of basic and advances concepts.

Hands-on exercises embedded Hands-on exercises are embedded with chapters to reinforce a concept.

A+ Operating System Technologies objectives At the beginning of each chapter is a list of the A+ Operating System Technologies objectives and the corresponding page number(s) of the topic. The UNIX/Linux chapter does not include a list because it is not on the current A+ exam.

Coverage of the major operating systems The major operating systems are discussed. These include: Windows 98, NT, 2000, XP, DOS, and UNIX/Linux.
Tech Tips Technical tips are included. A Sherlock Holmes icon precedes these tips.

Chapter Summary A chapter summary is included at the end of each chapter to highlight the chapter's main points.

Chapter Review Questions Each chapter has 20 multiple choice review questions.

Lab Projects (including Challenge) Each chapter has from 12 to 22 Lab Projects. Challenge Lab Projects are included too.

Internet Discovery Labs Each chapter has additional labs that require the student to perform Internet research as it relates to chapter content.

Soft Skills Help Desk Support This is the OTJ (On-The-Job) section of each chapter that includes real life working scenarios.

Critical Thinking Each chapter includes thought provoking questions forcing the student to stretch his or her knowledge about the subject matter.

Study Skills Each chapter includes a study skill topic followed by questions that the student should be focusing on.

Glossary A glossary of key terms.

Appendices Topics here include: number conversions, a summary of DOS and UNIX/Linux commands, the ASCII chart and answers to odd-numbered questions.

Instructor Support Ancillary materials include a Test Bank, Power Point slides and a CD with all of the answers to the Chapter Review Questions.

Acknowledgements

We would like to thank the many people who have encouraged me along the way to make this project succeed. First, we would like to thank Richard Jones of Scott Jones Publishers for giving us the opportunity to write this book. This has been the best writing experience we've ever had. We would also like to thank Audrey Anderson for managing the production process for me – she is great at what she does! We would like to thank Lois Principe, the book's proofreader and Diane DeMarco, the book's compositor.

Todd would like to thank his boss, Ernie Hensley, at DeKalb Technical College, where Todd teaches. He encouraged Todd to finish this book because no other book on the market has labs like this one. Todd also would like to thank other faculty members at DeKalb Technical College who gave him encouragement along the way: Walter Dula, William Monahan, Keith Humphrey, and Wayne Brown.

We also want to thank the real audience of this book - our students. We hope this book will enrich your working career!

Other thanks to these DeKalb Technical College colleagues: Dr. Paul Starns, President, Dr. Robin Hoffman, Vice President, Julian Wade, Dean and Fred Gibson, Assistant Dean.

Todd saves the best for last: his wife Micki for her steadfast support. She was Todd's technical reviewer and did a superb job! Todd dedicates this book to her and their two wonderful children, Zachary and Jessica.

Additionally, Todd thanks his parents, Dr. and Mrs. Lawrence H. Meadors, for the sacrifices they made for his education. "Thanks Mother for tutoring me so much in the 3rd grade – it continues to pay countless rewards".

Thanks to these reviewers for helping to make this a great textbook:

Kirk Ruby, College of Southern Idaho

Dave Bosilovatz, Bay de Noc Community College

Carol Mills, Cotton Boll Institute

Sally Douglass, Central Florida Community College

Martyns Kanu, Canada College

Gerald Sampson, Walla Walla Community College

Ross Decker, Brevard Community College

Tom Melvin, Allegany Community College

Professor M. Aghili, McNesse State College

David Oliver, Johnson Community College

Bonita A. Moyer, DeSales University

Russell Foszcz , McHenry County College

Karl Linderoth , Bay College

Roy F. Bonnett, Jr., Blue Ridge Community College

Larry Dumais, American River College

Richard Kalman, Atlantic Community College

Tom Holmes, Cotton Boll Institute

Kelly Flanagan, Brigham Young University

Kimberly A. Perez, Tidewater Community College

LaVaughn Hart, Las Positas College

Dianne Boseman, Nash Community College

Cindy Herbert, Longview Community College

Greg Stefanelli,, Carroll Community College

Connie Ivey, Robeson Community College

If you need to contact us, please feel free to point out any corrections or offer suggestions to this email address: **ltmeadors@yahoo.com**

A Note to Students
"I hear, and I forget
I see, and I remember
I do, and I understand"
 —Chinese proverb

This book is written with you in mind. We write like we teach. So, while going over this book, you will find the analogies, screen shots, tips, memory lists, projects or anything we can do to get a concept across. They are meant to help you understand.

A Note to Instructors
The real impetus for writing this book was the lack of good lab projects in any of the textbooks available. Over the years, we developed dozens of labs for different operating systems. Students were often frustrated that they were required to purchase a textbook that had virtually no lab projects with any depth. Countless times, the students have said that my labs were very useful as a resource and better than any they had seen. Now, they have a book that contains both theory and labs in one place.

Our teaching approach is to 1) discuss a topic, 2) show the topic, 3) involve the student in discussion and thought and 4) give the students lots of hands-on labs because this is where real learning occurs.

Our intent is for this book to follow that teaching style by 1) discussing a concept, 2) showing through screenshots, 3) involving through questions, critical thinking and exercises within the chapter and 4) numerous Lab Projects at the end of the chapter.

We hope you find this book a valuable resource for you and your students.

Chapter 1
The Operating System Environment

OBJECTIVES

The goal of this chapter is to introduce you to the operating system environment. This chapter will help you prepare and pass the following sections of the A+ Operating System Technologies Exam:

A+ Operating System Technologies Exam Objectives
covered in this chapter (and corresponding page numbers)

In this chapter, you will complete the following sections:
- 1.1 Understanding Software
- 1.2 Understanding Hardware
- 1.3 Understanding the Types of Users
- 1.4 Understanding the Operating System
- 1.5 Understanding Classifications of Operating Systems

1.1 Understanding Software

Software is defined as a set of instructions that are processed by a computer system. A software **program** is a collection of instructions that accomplish a task. Software programs are written in **programming languages** such as Pascal, C, C++, Java, Visual BASIC, COBOL, FORTRAN or Assembly. When a person writes a program, they write instructions to perform a certain function or task. This is called an **algorithm**. An **instruction** is a statement that performs an action. An example would be add or subtract. Combine instructions together in a set of logical steps and you have a software program. Programs are written in a specific programming language. Programming languages are like speaking languages. Words make up statements and there are strict syntax rules to follow.

Languages are divided into two broad categories:
- Low-level
- High-level

Low-level programming languages are hardware specific. The most common example is the Assembly, or Assembler, language native to each processor. These languages are extremely fast because they are written in the instruction set of each processor. An **instruction set** is the set of instruction statements the processor can understand and use. Look at the following Assembly language program below. This program simply adds two variables, X and Y, and places the result in the variable named T. Another name for an assembly language statement is **mnemonic**.

```
L 2,X
A 2, Y
ST 2,T
```

You can see this is not easily understandable. If you weren't told, you probably would not be able to tell what those statements do. Low-level languages are generally more difficult for humans to read and understand because their statements are close to the processor's language.

High-level languages are easier for us to read and interpret. However, this tends to cause more work for the computer because the language statements must be converted into a form the processor understands. For example, review the following instructions from a Visual BASIC program:

```
IF HOURS > 40 THEN
      CALL OVERTIME
ELSE
      CALL REGULAR_TIME
END-IF
```

In the example, if the hours for an employee exceed 40, then overtime will be calculated. If the hours for an employee are under or up to 40, then regular time will be calculated. You can clearly see how this sample Visual BASIC program is very similar to the English language. Examples of high-level programming languages include Pascal, COBOL, C, BASIC, Visual BASIC and FORTRAN.

Programming languages must be converted from the human-readable into machine-readable form. For low-level programming languages, this conversion is known as assembling. For high-level programming languages, this conversion is known as either compiling or interpreting. The compiler or interpreter is software that is written to perform the conversion.

Assembling converts the assembly language statements and data into machine-readable (known as binary code or executable code) form for a **linker**, or **linkage editor**. Part of the assembling process is to assign memory locations to the instructions and data. The linker is used to create a program that can be executed on the computer.

Compiling is the process of converting a completely written program into an **executable** program. An executable program is one that is compiled and syntax error free. It is capable of being run or executed. You can execute a program by clicking an icon representing the program or typing in the program name. For example, if you have ever double-clicked an icon or typed in the DOS **dir** command, that is executing a program. Pascal, C, C++, Java, COBOL, and FORTRAN are all complied languages.

Interpreting is the process of converting each line of the program into executable form. Interpreting differs from compiling because one line at a time is converted then immediately executed. Think of an interpretive language as being one where a mini-compile is done for each statement. Examples of interpretive languages are Visual BASIC, BASIC and the shell in the UNIX and Linux operating systems. Compiled languages tend to take more time to convert to an executable form, but are generally faster when executed because each line is not converted as it goes. Also, once the program is in machine-readable form, there is no need for the actual compiler software.

Compiled programs need to be converted only once. Interpretive languages require their interpreter software in order to execute. This makes interpreted language software more difficult to take to another operating system to execute than a compiled language. Interpreted programs need to be converted every time they are executed.

Because assembly language programs are hardware specific, generally you must rewrite, reassemble, and re-link the assembly program on a computer with an instruction set different from the one originally used. Thus, you can say that assembly programs are **proprietary** in nature. Programs written using compiled and interpreted languages can generally be copied to computers with differing instruction sets without much modification. Thus, compiled and interpretive languages are considered **portable** in nature.

Programs process data into information. **Data** are the raw facts and have no meaning to us alone. **Information** is processed data that has meaning to us. Consider the number 85. This is considered data. You cannot look at the number 85 and tell if it is someone's age, the grade on an exam, or the temperature outdoors. Only when a program processes the number will it be information.

The computer processes everything in **binary**. Binary is the system where bits are used to represent numbers and characters. You can refer to Appendix I for a discussion of number conversions including binary. The term **bit** stands for binary digit and can be either a zero (0) or a one (1). Each computer system has a **character set** that maps all the letters on the keyboard to its appropriate bit sequence. A character set is the set of characters on the keyboard and their binary equivalent. For example, the letter A is represented by the binary sequence 01000001. The character set common among most computer systems is the **American Standard Code for Information Interchange (ASCII)**. It is used on most computer environments, including all Windows operating systems and the UNIX and Linux operating systems. In the computer industry, the term **byte** to refers to eight bits taken in sequence. Letters, numbers and other special symbols are considered bytes. So, the letter A and its binary equivalent, 01000001, is considered a byte.

Now let's explore the major classifications of software. Software is divided into two major categories:

• System software
• Application software

System software includes the core components of the system that must be present in order for the computer to operate. Examples of system software are the operating system kernel, process management, memory management, and device drivers. System software will be discussed further in Chapter 2.

Application software is software used to assist users in performing typical office type work. Application software can be divided into these general areas:

• Word processing
• Spreadsheet
• Database

Word processing software allows you to create, modify, delete, save, and print documents that are office quality. They also have capabilities for spell checking, and include a dictionary and thesaurus. Word processing software has been around since the early 1980s. One of the first word processing software packages was WordPerfect. Now, Microsoft makes Word and Sun Microsystems has a word processing package called OpenOffice Writer. A screenshot of OpenOffice Writer appears in Figure 1.1. It operates similarly to Microsoft Word and comes free with Red Hat Linux.

Figure 1.1: A Screenshot of OpenOffice Writer by Sun Microsystems

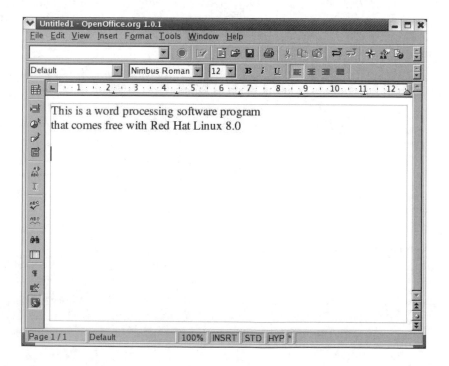

The first spreadsheet program was written in the early 1980s by college students who were taking an accounting class. Growing weary of making written changes to their accounting worksheets and having to recalculate by hand, they developed a program that automatically performed calculations and recalculated changes automatically.

Spreadsheet programs allow you to manage data in rows and columns. Spreadsheets hold data in **cells**, which is a row and column coordinate. When one cell changes, all cells referencing the original cell will change. Cells are labeled with names such as A1 or B7. Letters represent the columns and numbers represent the rows. Spreadsheet programs are extremely powerful and can include programming logic. They also include built-in functions. **Functions** are small programs where you just supply the data and the function does the work. Spreadsheet programs allow you to create your own formulas too, as shown in Figure 1.2. In Figure 1.2, you can see a spreadsheet with data and a formula.

Figure 1.2: Screenshot of the Microsoft Excel Application with Data

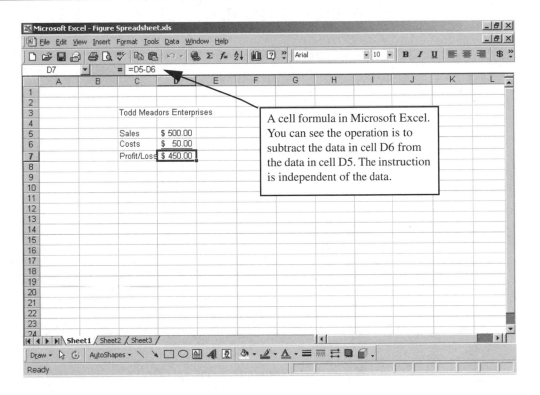

Database programs allow you to manipulate and manage data in tables. A **table** is a collection of database data stored on disk. One of the first database programs, called dBASE, was developed by the Ashton-Tate Corporation in the 1980s. Other companies such as IBM, Microsoft, Oracle and Sun Microsystems have developed database software as well. Database software allows you to create tables of data for organization. You can add, delete, or modify data in a database. You can join or combine data from multiple tables to create views or subsets of tables. Most programming languages allow you to interface with database software, letting you to create a powerful program. In Figure 1.3, an Employee database has been created. In the figure, you can see field names such as "Employee Name," "Employee Address," and so on. A **field** is a group of bytes. Multiple fields together comprise a **record**. So all of the Employee fields in Figure 1.3 make up an Employee record. The database holds the records.

Figure 1.3: A Screenshot of a Database Using Microsoft Access

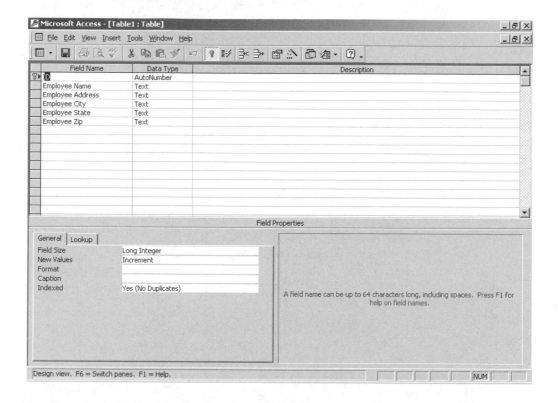

 Many software companies combine word processing, spreadsheet, and database software into a software suite. Microsoft has done this with Microsoft Office. Software that is combined in such a manner is called **integrated** software.

1.2 Understanding Hardware

If you are going to work with operating systems in the Information Technology industry, you need to understand a little about hardware. Hardware can generally be classified as follows:

- Input
- Processing
- Output

Input devices provide for data and instructions to be input into the computer system. Examples of devices that are strictly used for input are the keyboard, the mouse, a digital camera, and a scanner. After the data is input by the input device, it travels along a **bus** and is temporarily stored in **Random Access Memory** (**RAM**). The bus is a set of wires on the **motherboard** that carry data and instructions. The motherboard is the main system board in the Personal Computer that interconnects the other hardware components.

The data is processed by the **Central Processing Unit** (**CPU**). The CPU, or **processor**, is considered the "brains" of the computer system because it performs action on the data. The most important measure of a CPU is its speed. The speed of a processor is measured in **hertz**, or cycles per second. A **cycle** is made up of two stages known as the **Instruction time**, or **I-time**, and the **Execution time**, or **E-time**. During the I-time, these two events occur:

1. The instruction is fetched from RAM by the processor to the processor's own memory. Actually, the instruction is stored into a storage unit called a **register**. There are several different register types. For example, the **Accumulator** is a register that stores data. The **Instruction Register** stores the instruction.
2. The instruction is decoded and addresses for the data are generated.

During the E-time, these two events occur:

1. The instruction is actually executed by the processor.
2. The results of the execution are stored back into RAM.

Processors handle many of these instructions in a second. Common measurements are **MHz**, for mega (million) hertz, or **GHz** for giga (billion) hertz. A 3.0 GHz processor can perform 3 billion cycles per second. That is analogous to 3 billion addition operations in a second.

Let's look at general components of the CPU. The CPU is comprised of two main parts:

- Control Unit (CU)
- Arithmetic Logic Unit (ALU)

The **Control Unit** (**CU**) is the heart of the CPU. It governs all actions of the CPU.

The **Arithmetic Logic Unit** (**ALU**) function is particularly important to the topic of programming because this is where all the arithmetic and logic functions of the computer are done. Arithmetic functions include Multiply (*), Divide (/), Add (+) and Subtract (-), Exponential (**), and Parentheses. Refer to the following sample code for an arithmetic operation:

```
X = 5 * (4 + 2) / (6 + (1 / 10) - 50)
PRINT X
```

The Logic portion of the ALU performs comparisons. The types of comparisons performed by the Logic function are: AND, OR, EQUAL TO, NOT EQUAL TO, GREATER THAN, and LESS THAN. An example of a logic operation is in the calculation of the Dean's List in college. Let's assume you must have a GPA greater than 3.80 and carry 12 hours in a given semester. The logic for this operation would be an AND operation as in the following sample code:

```
IF GPA >= 3.80 AND HOURS >= 12.0 THEN
      PRINT "Dean's List"
END-IF
```

If you took 15 hours but have a GPA of 3.65, you would not be on the Dean's List. Or, if you had a GPA of 4.00 but only took a three-hour semester course, you would not be on the Dean's List.

Output devices receive the data that are processed by the CPU. Examples of output devices are the monitor, the printer, and a plotter.

Some devices are considered both input and output. Disk drives, tapes, and memory are all both input and output devices. They are called **Input/Output (I/O)** devices. The hard disk drive is a common example of an I/O device. When you save a document from within an application program, the document is output to disk for permanent storage. Later, when you need to retrieve the document, it is input from the disk into the application you are using.

Some hardware devices store data permanently and others store data temporarily. **Random Access Memory**, or **RAM**, is temporary. RAM is also shortened to "memory." Disk drive and tape storage units are permanent. However, memory is like the rough draft version of the data that will be written to the hard disk drive. Accessing memory is faster than accessing disk drives, but if you were to lose power to your computer, the contents of memory would be lost. Under the same conditions, the contents of the hard disk would remain intact.

Figure 1.4 shows a diagram of the flow of data through a computer system.

Figure 1.4: A Diagram Representing the Flow of Data

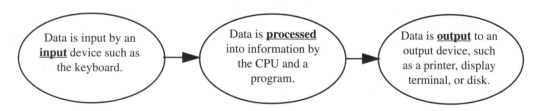

Memory and the hard drive store data in units. If you've ever heard the term megabyte or gigabyte, then you are hearing references to storage capacities. Generally, the more memory you have, the faster your computer will be. For a reference list of storage terms, see Table 1.1.

Table 1.1: Some Units of Storage in a Computer System

Term	Storage
Bit	Either a 0 or 1.
Byte	8 bits—for example the letter A (ASCII 01000001) would be considered a byte. A byte is also known as a character.
Kilobyte (KB) or thousand	1,024 bytes—typically stated as 1,000 bytes.
Megabyte (MB) or million	1,024 KB—typically stated as 1,000,000 bytes
Gigabyte (GB) or billion	1,024 MB—typically stated as 1 billion bytes.
Terabyte (TB) or trillion	1,024 GB—typically stated as 1 trillion bytes.

Processors are compared by their speed (hertz). Memory and disk drives are compared by their capacity (bytes).

1.3 Understanding the Types of Users

You need to understand the types of users that exist in an operating system environment so you can know how to best serve and interact with them. Their jobs also govern the type of access they need to the system. Here are the typical user types in an operating system setting:

• Standard User or user
• Programmer Analyst
• System Administrator
• Help Desk Support Representative

The **Standard User**, or **user**, is the person who uses the system on a daily basis in support of his or her job. They use the system to access an application. The application they need depends upon their function in the organization. For example, a Payroll Manager may need access to the Payroll application, whereas an Engineer may only need access to a Computer Aided Design/Computer Aided Manufacturing (CAD/CAM) program, and a Travel Agent may need access to the Travel Booking application. Users normally have limited access to the system.

The **Programmer Analyst** is the person responsible for analyzing the business needs of the user and writing the programs for users. This person must learn the programming language being used, as well as the business processes so they can write programs effectively for users. Normally, Programmer Analysts write programs using an editor. They need to be skilled problem-solvers and they need to be able to think logically.

The **System Administrator** is the person responsible for the operations of the computer systems and network. The System Administrator may have to work a varying work schedule depending upon the nature of the business. Many are on call 24 hours a day, seven days a week, called 24/7. Also, depending upon the size of the operation, there may be more than one System Administrator in an organization. They are completely responsible for ensuring that the computer systems, the networks, and the applications are available to the users during the hours needed by the users.

In larger organizations, the System Administrator's job may be divided between a **Network Administrator**, who is responsible for the network, and an **Operating System Specialist**, who is responsible for the operating system. They perform duties such as installs, upgrades, adding users, and so on. These people usually have complete access to the system, or at least more than the typical user. On a Windows NT and 2000 system, the user with complete access to the system is called **Administrator**. On a Novell system, this user is called **Supervisor** or **Admin**, depending upon the version of operating system. On Linux systems, this user is called **root**.

The **Help Desk Support Representative**, or Help Desk Technician, typically works in a telephone support environment where users of all types call in with problems. Help Desk people must be able to learn quickly, listen carefully, and have technical skills and patience. They learn a great deal in a short period of time. They must be able to resolve problems for users when called. One of the first questions Help Desk Support Representatives should politely ask users is, "What changed?"

1.4 Understanding the Operating System

An **operating system (OS)** is a set of software instructions that allows your computer system to operate. Operating systems are written in programming languages like application programs. For example, the Linux operating system is written in both Assembly and C. The operating system interacts with the hardware allowing the computer system to function. Nearly all computer systems require an operating system to run. This includes computer systems as small as your laptop and those as large and powerful as a supercomputer. The computer system needs both the software and the hardware to operate the computer successfully; a computer system cannot operate without either.

The operating system interfaces with the hardware for the application. For example, if you were using Microsoft Word to create a word processing document and you needed to save the document, you would click File and Save As and enter a file name to save. The operating system would intercept the request to save the document to disk for Microsoft Word. The operating system also retrieves files from disk. If you have ever clicked File, then Open and entered a file name, you have experienced the operating system retrieving a file from disk.

In Figure 1.5 you can see how an operating system fits within a computer environment. In the figure, you can see that users interact with an application that in turn interacts with the operating system. The operating system interacts with hardware on behalf of the user.

Figure 1.5: The Placement of the Operating System within a Computer Environment

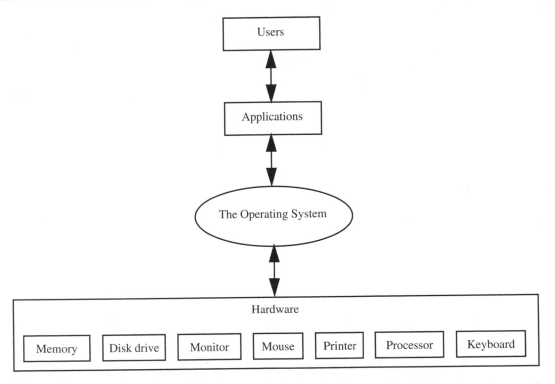

Now, let's perform a short exercise. The goal of this exercise is to help you understand how to open an application, create data in a file, and save the file. Because most all Windows operating systems have the Notepad application, this will be used. For this exercise, Windows 2000 will be assumed. Follow these steps:

1. Power on your computer. The Windows desktop should appear.
2. Click **Start**, and then click **Run**. The Run dialog box displays.
3. In the Open text box, type **notepad** and press Enter. The Notepad application opens in a separate window.
4. Refer to Figure 1.6 for a screenshot of Notepad. Enter the lines shown in Figure 1.6 replacing my name with yours.
5. In order to save the data, you will need a disk storage location. Insert a formatted floppy into the floppy drive. The floppy is referenced as A: drive.

6. You must enter a file name to save. Click **File** and then click **Save As**. The Save As window appears.
7. In the Save In drop-down box, select A: Floppy drive.
8. In the File name text box, enter **Exercise1** and click **Save**. The file will be saved with a .txt extension. File names and extensions will be discussed in Chapter 2. The point is the OS interacted with the hardware for the application and you.
9. To close Notepad, click **File** and then click **Exit**.

Figure 1.6: A Screenshot of the Notepad Application

1.5 Understanding Classifications of Operating Systems

There are quite a few operating systems on the market today. There are operating systems for mainframe computers, PCs, laptops, and even hand-held computers. Let's look at a few of the operating system (OS) types.

Mainframe operating systems are designed to handle the Input/Output (I/O), processing and storage requirements for a lot of users. Mainframe operating systems run mainframe computers and are generally used in large organizations to handle the volume of work. Mainframe computers are usually centralized in a computer room with **terminals** connecting to it from remote locations. A terminal is a device that has no computing ability and is strictly dependent upon the processing power of another, such as the mainframe. It is merely a display monitor and a keyboard with wires connecting to the mainframe. Applications are accessed on the mainframe itself. You can also use a Personal Computer (PC) with software to act like a terminal to connect to a mainframe;

this software is called terminal **emulation** software. The term emulate means imitate. An example of a mainframe operating system is IBM's Multiple Virtual Storage (MVS).

When you access the UNIX or Linux OS from a Windows PC using the **telnet** command, then your computer is acting as a terminal for that session. The telnet command allows you to remotely connect to another computer. In Figure 1.7, you can see a screenshot of the telnet command in action. The very last line tells the story. Here you can see that a Linux command, **echo $TERM**, is issued and returns a terminal type of "**ansi**" on the screen. Note that "ansi" is just a terminal type—there are many more. Don't be too concerned about this Linux command or its output for now, just understand that terminal emulation is occurring here. Linux commands will be covered in a later chapter.

Figure 1.7: A Screenshot of a Terminal Emulation Session

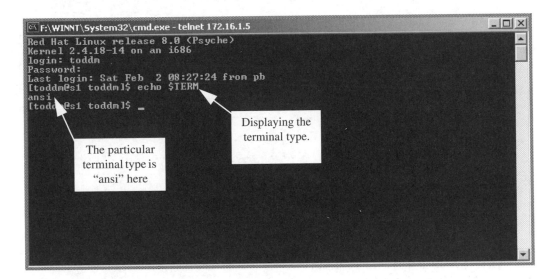

A mainframe computer has the same general components, such as processor, memory, and disk drive storage, as your own **Personal Computer**, or **PC**. They are usually physically larger and spread out over the area of a room. For example, your PC has a disk drive physically located in your **system unit**, or **chassis**. A mainframe computer may have multiple disk drive units spread out in a computer room, but these units would be connected to the mainframe computer.

Midrange operating systems have the same properties of mainframe operating systems. However, they are generally used for medium-sized organizations. They operate in a centralized manner with terminals and PCs using emulation software, accessing applications remotely. An example of a midrange operating system is IBM's OS/400, which runs on an AS/400.

A **Network Operating System (NOS)** allows computers to share resources in a network. A NOS and mainframe/midrange operating system differ in one main respect.

Where mainframe and midrange operating systems utilize centralized processing and storage capabilities, a NOS allows you to decentralize your operations. This is because instead of using terminals, a NOS relies on PCs, which have their own processing power, memory, and storage capabilities. With a NOS, you can distribute the processing load over multiple computers. For example, one computer could be a web server, another could be an e-mail server, and another could be an application server.

Which came first—the mainframe or the NOS? From an historical perspective, the use of mainframes dates back to the 1940s when they were used in military applications during World War II. Utilization of Network Operating Systems is a more recent innovation with the development of PCs in the 1980s.

A **resource** is a hardware device, a software program, or a file needed by users. A **shared resource** is a resource that is capable of being used on other computers. A printer that can be used by multiple users is an example of a shared resource. An application stored on a computer that is used by multiple users is another example of a shared resource. Let's see this in an exercise. For this exercise, Windows 2000 will be assumed. You will also use Windows Explorer to access the floppy drive. Follow these steps:

1. Power on your computer. The Windows desktop should appear.
2. Right-click **Start** then click **Explore**. A separate Explorer window appears.
3. Insert a formatted floppy into the floppy drive.
4. Scroll until you see the Floppy (A:) drive.
5. Right-click the floppy drive and click Sharing. The Properties page of the floppy appears. Refer to Figure 1.8 for a screenshot of this page.
6. Click the **Sharing** tab.
7. Click **Share this folder**. Notice the share name is "A," for the floppy.
8. Click **OK**. Now the floppy can be access from another computer remotely over the network.
9. Close the floppy's Properties page by clicking the **X** in the upper right-hand corner. This means to close the window.

Figure 1.8: A Screenshot of Sharing a Resource

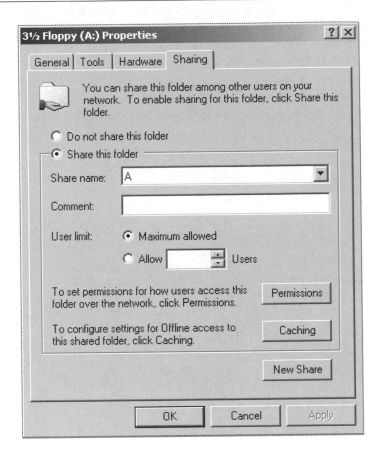

A network is a group of computers and printers connected by cables or some wireless media. Network operating systems use **protocol** software to facilitate the communication among other computers in a network. The term **Local Area Network**, or **LAN**, generally refers to a collection of computers, printers, and other networking components, such as cable, hubs, switches, and routers, which are physically close together. For example, computers interconnected in the same room or building would generally be considered a LAN.

In a LAN, **servers** are computers that allow other computers to connect to the server's shared resources. **Clients** are the computers that use the resources made available by the servers. In the **client/server** model, a server computer handles the requests made by the client computer. There are usually more clients than servers in a network. The server usually has additional memory, hard drive capacity, and processing capability because the server has to handle numerous client requests. The computers generally use a **Network Interface Card**, or **NIC**, to connect to the server in either a wire or wireless configuration.

Examples of operating systems that run on servers are Novell NetWare, Microsoft Windows NT Server, 2000 Server, and UNIX/Linux. DOS, Windows 3.x, Windows 95, Windows 98, Windows NT Workstation, Windows 2000 Professional, and Windows XP are examples of client operating systems. The UNIX/Linux operating system can also be used as a client. Servers and clients are required to use the same protocol in order to communicate with one another. Although it can be a complicated process, the use of a common protocol can even allow different operating systems to communicate with one another. In Figure 1.9, you can the client/server LAN environment.

Figure 1.9: The Client/Server LAN Environment

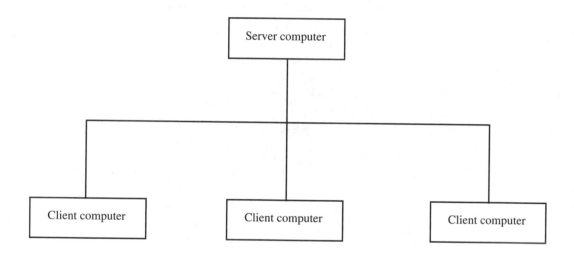

Centralized processing occurs when the processing and storage resources are performed at a central site. There may be multiple processors and storage units, however they are usually in one physical location. Access to these resources occurs locally over communication wires, or links, and remotely via terminals or terminal emulation. Mainframes and midrange operating systems are examples of centralized processing.

Distributed processing occurs when the processing and storage resources are spread out, hence distributed, over multiple computers. Instead of terminals, PCs are used to share the load. The advent of Network Operating Systems in a client/server LAN environment has brought about the use of distributed processing.

The **Transmission Control Protocol/Internet Protocol (TCP/IP)** is the most commonly used protocol today. It was developed by the United States Department of Defense. TCP/IP is what allows you to connect to the Internet and browse web sites, take college courses, perform searches, or buy products. Actually, the first use of the Internet was an electronic mail (e-mail) message sent from one user to another on different coasts of the United States. Each computer, called a **host**, must have an IP (Internet Protocol) address to communicate with other computers on the network. Note that each host must also have an operating system that supports TCP/IP. The host may also have a **Domain Name Service (DNS)** name such as www.meadors.org. This is sometimes called the host's "user friendly name." The DNS name keeps us from having to remember the IP address. Networking technologies will be discussed in more detail in a later chapter.

DNS is what allows you to enter a name such as http://www.scottjonespub.com or http://www.meadors.org in the address line of your web browser software and view a web page on your computer.

Let's view your IP address settings in an exercise. This exercise assumes Windows 2000 running on a PC. Follow these steps:

1. Connect to the Internet from home, work, or school.
2. Click **Start**, then click **Run**. The Run dialog box displays.
3. In the Open text box, type **command** and press Enter. The DOS command line interface displays. (On Windows 2000, you could also have entered **cmd** to display the command line interface.)
4. In order to display your IP configuration, enter **ipconfig** and press Enter. (If you happen to be on Windows 95, the command is **winipcfg**.) The IP configuration of your PC displays.
5. Refer to Figure 1.10 for a sample screenshot. The Internet connection for the PC shown in Figure 1.10 has an IP address of 67.30.192.18.
6. Close the window. The command line interface window closes.

Figure 1.10: Displaying the TCP/IP Properties of a Computer

```
F:\WINNT\System32\command.com                                    _□×
(C)Copyright Microsoft Corp 1990-1999.

F:\>ipconfig

Windows 2000 IP Configuration

Ethernet adapter Local Area Connection:

        Connection-specific DNS Suffix  . :
        Autoconfiguration IP Address. . . : 169.254.254.251
        Subnet Mask . . . . . . . . . . . : 255.255.0.0
        Default Gateway . . . . . . . . . :

Ethernet adapter Local Area Connection 2:

        Media State . . . . . . . . . . . : Cable Disconnected

PPP adapter NetZero:

        Connection-specific DNS Suffix  . :
        IP Address. . . . . . . . . . . . : 67.30.192.18
        Subnet Mask . . . . . . . . . . . : 255.255.255.255
        Default Gateway . . . . . . . . . : 67.30.192.18

F:\>_
```

Traditionally, mainframe computers have not supported the TCP/IP. This is because mainframe computers were around before TCP/IP was developed. Mainframe manufacturers such as IBM had to develop their own communications software, such as IBM's System Network Architecture (SNA), so the terminals could connect with the mainframe. However, with the advent of the Internet, many mainframe manufacturers began to provide products allowing Internet connectivity. Thus, mainframe computers and midrange computers now allow connectivity using TCP/IP.

The use of the Internet is so pervasive that with the release of Windows 95, Microsoft incorporated its own web browser, Internet Explorer, with its operating system. This was an issue in litigation during the 1990s. Now, products released by Microsoft come with Internet Explorer for Internet connectivity. Other operating systems, such as Red Hat's Linux, also come with Internet browser software ready to use. Connecting your computer to the Internet involves **Wide Area Network (WAN)** technologies. A WAN refers to computers connected remotely. While LANs usually connect many devices in physical proximity, WANs generally connect devices in remote locations.

A telecommunications carrier such as AT&T, Sprint, or Bell South provides WAN technologies, such as Dial-up, DSL, T1, or ISDN. WANs sometimes connect separate LANs that are spread over geographic distances. For example, if a company had a LAN in Jacksonville, Florida, and another LAN in Lawrenceville, Georgia, a WAN could be configured to allow these two separate LANs to communicate.

To help you understand this concept, Figure 1.11 represents the Internet as "the cloud." Typically you connect to the Internet using your **Internet Service Provider (ISP)**, such as AT&T, Sprint, or Bell South, and they provide remote access. The "lightning bolts" refer to a WAN type connection.

Figure 1.11: Your Home Connection Along with LAN Connections to the Internet Cloud

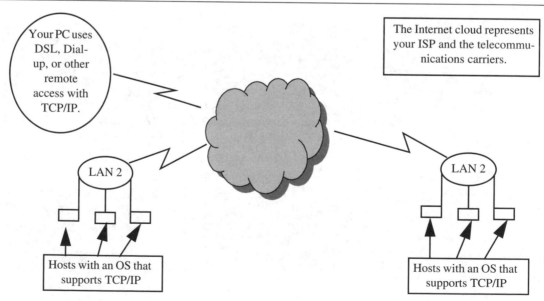

Table 1.2 provides a summary of the more common operating systems used on PCs. Note that each of these operating systems supports TCP/IP for Internet access.

Table 1.2: A Few of the More Common Operating Systems Used on PCs Today

Operating System	Description
DOS	One of the first operating systems developed commercially by Microsoft in the early 1980s. It is a text-based, single-tasking operating system.
Windows 3.1	Microsoft Windows 3.1 was not really an operating system but is a Graphical User Interface (GUI) for managing DOS. Windows 3.1 was developed after DOS and is executed after DOS is up and running.
Windows 9x	The Windows 9x class of operating systems include Windows 95 and 98. Windows 95 is the first true GUI operating system developed by Microsoft. Windows 95 and 98 are multi-tasking operating systems with a DOS shell component allowing you to run DOS commands. These operating systems have many GUI tools as well as Internet Explorer.
Windows NT	Microsoft developed Windows NT Workstation and Server to compete in the server-based networking market. Users typically use the Workstation product and the Server product performs network functionality for the Workstations. NT is a GUI operating system with DOS command line access.
Windows 2000	Microsoft developed Windows 2000 Professional and Server as upgrade paths to Windows NT Workstation and Server. Windows 2000 includes more administrative tools and functionality than NT. These are also GUI operating systems with DOS command line access.
UNIX	AT&T developed the UNIX operating system in the 1960s. There were two main original versions of UNIX: AT&T System V (Five) and Berkeley Source Distribution (BSD). UNIX is a multi-tasking, multi-user text-based operating system. DOS has many similar characteristics to UNIX.
Linux	Linux is a version of UNIX that runs on PCs and includes both a GUI and text interface to the operating system. It usually runs as a server-based product and was developed in the 1990s.

| NetWare | Novell developed its network operating system, called NetWare, in the 1980s as a server-based product. NetWare runs in a DOS-based partition. |
| Mac OS | Apple developed the Mac operating system. |

A **platform** is comprised of <u>both</u> the hardware and software that a given system runs on. For example, Linux version 8.0 is the software running on an Intel-based hardware system. The two combined are called the platform. Another example would be Windows 2000 on an Intel-based system.

Chapter Summary

- Software is a set of instructions processed by the computer. Software programs are written in programming languages. Programs are either compiled or interpreted. Programs are converted from programming languages to the processor's machine language. Data is defined as raw facts that are processed into information. Software is divided into two categories: system software and application software.
- Hardware is the tangible part of the computer system. Hardware devices are classified as input, processing, or output. Input devices facilitate data coming into the computer system. The Central Processing Unit (CPU) performs the processing activity for the computer. The CPU is comprised of the Arithmetic Logic Unit and the Control Unit. Output devices receive data processed by the CPU. Some devices are considered I/O devices. They perform either input or output functions. Examples of I/O devices are the hard disk, tape, and memory.
- The Standard User utilizes the operating system to perform their daily duties. The Programmer Analyst is responsible for analyzing business needs of users and writes the programs for them. The System Administrator is responsible for all operating systems functions. The Network Administrator ensures the network is up and running. The Help Desk Support Representative provides telephone support for customer problems.
- The operating system is a set of software instructions allowing your computer system to operate. It works with the hardware, users, and the applications.
- Mainframe operating systems handle I/O, processing and storage requirements for many users. Midrange operating systems have similar properties as mainframes except they are typically used in medium-sized organizations. A Network Operating System (NOS) allows you to share resources in a network environment.

Review Questions

1. Which of the following is defined as a set of instructions that are processed by a computer system?
   a) Firmware
   b) Software
   c) ASCII
   d) Low-level programming language

2. Which of the following is a collection of instructions that accomplish a purpose?
   a) Program
   b) Hardware
   c) ASCII
   d) I-time

3. A (n) _____ is a statement that performs an action.
   a) Program
   b) Instruction
   c) CPU
   d) Compiler

4. A (n) _____ is hardware specific.
   a) Application software program
   b) High-level programming language
   c) Low-level programming language
   d) Compiler

5. A (n) _____ is the set of instruction statements the processor can understand and use.
   a) Application software program
   b) High-level programming language
   c) Low-level programming language
   d) Instruction set

6. A (n) _____ is a program that is compiled and syntax error-free.
   a) Executable program
   b) Record
   c) Hardware device
   d) Piece of data

7. What is data processed into?
   a) I-time
   b) ASCII
   c) Information
   d) Compiled language

8. The character set common to most PCs is called _____.
   a) I-time
   b) ASCII
   c) Information
   d) System software

9. With _____ software, you can create and modify office-quality documents.
   a) Spreadsheet
   b) ASCII
   c) Word processing
   d) Database

10. With _____ software, you can manage data in cells and perform calculations on the data.
    a) Spreadsheet
    b) ASCII
    c) Word processing
    d) Database

11. With _____ software, you can manipulate and manage data in tables.
    a) Spreadsheet
    b) ASCII
    c) Word processing
    d) Database

12. A _____ is a group of bytes.
    a) RAM
    b) Table
    c) File
    d) Field

13. During the _____, an instruction is fetched from RAM.
    a) MHz time
    b) E-time
    c) CPU time
    d) I-time

14. A _____ is either a 0 or a 1.
    a) MHz
    b) MB
    c) Bit
    d) Byte

15. A _____ utilizes the computer system on a daily basis in support of his or her job function.
    a) System Administrator
    b) Programmer Analyst
    c) Standard User
    d) Help Desk Support Representative

16. A _____ writes instructions to accomplish a task.
    a) System Administrator
    b) Programmer Analyst
    c) Standard User
    d) Help Desk Support Representative

17. A _____ is responsible for operations of the computer systems.
    a) System Administrator
    b) Programmer Analyst
    c) Standard User
    d) Help Desk Support Representative

18. A _____ mainly provides telephone support for customer problems.
    a) System Administrator
    b) Programmer Analyst
    c) Standard User
    d) Help Desk Support Representative

19. Which of the following is defined as software that facilitates communication among computers within a LAN?
    a) Protocol
    b) Server
    c) Client
    d) NIC

20. Which of the following best describes when processing and storage resources are spread over multiple computers?
    a) Centralized
    b) Distributed
    c) DNS
    d) TCP/IP

Lab Projects

Lab Project 1

In this project you will need to visit an organization that uses a computer system and discuss the type of hardware and software it uses. You can use your school, work, library, or church.

1. Locate an organization that uses a computer system.

2. Record the name of the organization.

   _____

3. Interview someone with System Administrator responsibility and record the classification of computer system, the operating systems, and versions used.

   _____

   _____

   _____

4. Interview a user and record the type of applications used.

   _____

   _____

   _____

5. If they use the Internet, record how it is used.

   _____

   _____

   _____

6. Politely thank the users and leave.

Lab Project 2

In this project you will go through the process of purchasing a computer system to help you understand the terminology of an operating system environment. You are to identify the computer make and model, vendor, the total price, and your source of information.

1. With a fictitious budget of $2,500, you will purchase an operating system and hardware for a computer meeting these requirements:

Hardware:

    256 MB RAM
    800 MHz or higher CPU
    60+ GB Hard disk drive
    Floppy drive
    DVD/CD-ROM (50 X speed)
    100 Mbps Network Interface Card
    Video card
    17" monitor
    ink jet printer
    modem card

2. Record the operating system you chose.

   _____

3. Record the reason for choosing this operating system.

   _____

4. Present your findings to the class.

5. Turn in all documentation to your instructor.

Lab Project 3

The goal of this project is to help you understand how to view additional TCP/IP settings on your computer. This exercise assumes Windows 2000 Professional as the operating system on a running computer. Follow these steps:

1. Connect to your Internet Service Provider (ISP).

2. While connected, click **Start**, and then click **Run**. The Run dialog box displays.

3. In the **Open** text box, type **command** and press Enter. The DOS command line interface displays.

4. Type **ipconfig /all** to display your TCP/IP properties with information about your remote connection.

5. Record the settings for the following:
   IP address: _____
   Subnet mask: _____
   Default gateway: _____
   DNS server: _____

6. Record two additional settings.

   _____

   _____

7. Refer to Figure 1.12 for a sample screenshot. In the figure, you'll notice the "PPP adapter NetZero" connection. This is the WAN connection going to the Internet. Notice the IP address information. This was sent to the computer from the ISP.

8. Close your window. The command line interface window closes.

Figure 1.12: Internet Connectivity Showing IP Address Settings Using the WAN Protocol (Point-to-Point, or PPP)

```
F:\WINNT\System32\command.com                                      _|□|x|
Ethernet adapter Local Area Connection:

        Connection-specific DNS Suffix  . :
        Description . . . . . . . . . . . : SMC EZ Card 10/100 (SMC1211TX)
        Physical Address. . . . . . . . . : 00-E0-29-9E-E6-06
        DHCP Enabled. . . . . . . . . . . : Yes
        Autoconfiguration Enabled . . . . : Yes
        Autoconfiguration IP Address. . . : 169.254.254.251
        Subnet Mask . . . . . . . . . . . : 255.255.0.0
        Default Gateway . . . . . . . . . :
        DNS Servers . . . . . . . . . . . :

Ethernet adapter Local Area Connection 2:

        Media State . . . . . . . . . . . : Cable Disconnected
        Description . . . . . . . . . . . : SMC EZ Card 10/100 (SMC1211TX) #2
        Physical Address. . . . . . . . . : 00-E0-29-9E-E5-9B

PPP adapter NetZero:

        Connection-specific DNS Suffix  . :
        Description . . . . . . . . . . . : WAN (PPP/SLIP) Interface
        Physical Address. . . . . . . . . : 00-53-45-00-00-00
        DHCP Enabled. . . . . . . . . . . : No
        IP Address. . . . . . . . . . . . : 65.58.21.222
        Subnet Mask . . . . . . . . . . . : 255.255.255.255
        Default Gateway . . . . . . . . . : 65.58.21.222
        DNS Servers . . . . . . . . . . . : 209.247.164.35
                                            209.247.164.36
F:\>_
```

Lab Project 4

The goal of this project is to help you understand how to convert storage capacities.

1. Convert 10 MB to KB. Record your answer.

   _____

2. Convert 5 GB to MB. Record your answer.

   _____

3. Convert 1,700 KB to GB. Record your answer.

   _____

4. Convert 1 KB to bits. Record your answer.

   _____

5. Record how many 1.5 KB files will fit on a 1.44 MB floppy.

   _____

Lab Project 5—Challenge

The goal of this project is to help you understand how to create a file, share it, and view it from another computer in a network. This project assumes at least two computers are connected. The project also assumes Microsoft 2000. If you don't have two computers, you can simulate a second one just by viewing the network icon on the same computer you create the file on.

1. Create a folder on your hard drive, C:.

2. Share the folder.

3. List the steps you took to share the folder.

   _____

4. From a second computer (or the same one if you don't have a second computer), go to the **My Network Places** icon until you see the share name of the folder on the other computer. Refer to Figure 1.13 for a sample screenshot. In the figure, the folder that is shared is named Shared Folder.

5. List the steps you took to view the folder on the second computer.

   _____

6. Close all windows of all computers and logout.

Figure 1.13: A Screenshot of Viewing a Shared Folder from Another Computer

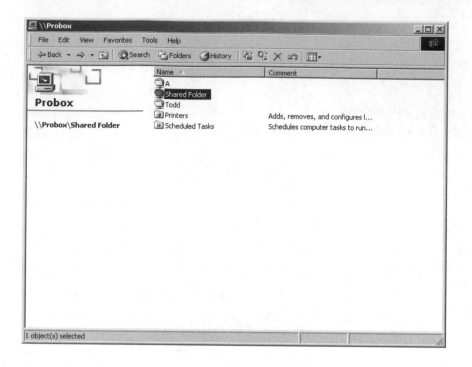

Internet Discovery

Internet Discovery Lab 1

The goal of this Internet Discovery Lab is to help further your knowledge of chapter terms by accessing the Webopedia web site.

1. Connect to the Internet.

2. Open your web browser.

3. In the address box, type **http://www.webopedia.com** and press Enter.

4. Locate the search box, enter the following terms, search for them, and write a brief synopsis of each term. Close your browser when complete.

    a. Firmware

    _____

    _____

    _____

    b. Compiler

    _____

    _____

    _____

    c. Instruction

    _____

    _____

    _____

Internet Discovery Lab 2

The goal of this Internet Discovery Lab is to have you determine the minimum hardware needed for installing Red Hat Linux 8.0.

1. Connect to the Internet.

2. Open your web browser.

3. In the address box, type **http://www.redhat.com** and press Enter.

4. Locate the search box and type **minimum** and press Enter.

5. Record the CPU type, amount of memory, disk drive capacity, and any other necessary hardware requirements for installing Red Hat Linux 8.0.

   _____

   _____

   _____

6. Close your browser.

Internet Discovery Lab 3

The goal of this Internet Discovery Lab is to have you determine the minimum hardware needed for installing Microsoft operating systems.

1. Connect to the Internet.

2. Open your web browser.

3. In the address box, type **http://www.microsoft.com** and press Enter.

4. Browse the web site until you find the minimum hardware requirements for these operating systems:

   a. Windows 98

   _____

   _____

   b. Windows 2000

   _____

   _____

   c. Windows XP

   _____

   _____

5. Close your browser.

Internet Discovery Lab 4

The goal of this Internet Discovery Lab is to help further your knowledge of chapter concepts by accessing the Howstuffworks web site.

1. Connect to the Internet.

2. Open your web browser.

3. In the address box, type **http://www.howstuffworks.com** and press Enter.

4. Search for "operating system."

5. Review any related articles.

6. Record your findings. Find an example of an operating system not discussed in this chapter and record the name and manufacturer.

   _____

   _____

   _____

7. Close your browser.

Soft Skills: Help Desk Support

1. A user calls and tells you they are having trouble saving a file to disk. What type of questions do you ask? What steps do you tell the user to take to successfully save the file?

2. You are working the help desk for your company and a user calls and says they would like to delete a payroll file on the system. What types of issues should you consider?

Critical Thinking

1. Discuss how an operating system environment is like a human body.

2. Oppose or defend the statement: All computer systems require some type of operating system to operate.

3. Defend what you think is more important—software or hardware. Which do you think came first?

Study Skills

Obtaining Success in Your Technical Studies

Technical courses can be quite challenging. Not only are you learning new skills but also you are exposed to a wide range of terms and acronyms. However, attaining these new skills will help you on your first job. Listed below are the Study Skills for Chapter 1. In future chapters, you will be introduced to additional Study Skills to help you to a rewarding and exciting computer career. Following these will put you on the right path to improving your success in class and on the job!

- Attend class **on time every day!** Attending all class sessions is one of the best habits you can get into towards your goal of **obtaining success**. Missing class will put you behind. Get to class 10 minutes early. At the very least, you can review the chapter material or work on a Lab Project. If you have to be out, contact your instructor via telephone or e-mail as soon as you can.
- Take complete and accurate notes—Taking notes will help reinforce your knowledge when preparing for a test.
- Listen to your instructor and ask questions when you don't understand. You instructor has a wealth of knowledge and experience. Listen to his or her directions and pay attention to their rules! Don't be afraid to ask questions. There are probably others that have a similar question so you'll help them too.
- Bring all necessary materials. Many students go to class with the idea that the instructor will provide everything. If your instructor asks you to bring floppy disks, bring them each class session. Always bring your textbook, notebook paper, pen or pencil and a positive attitude!
- Get a Study Skills Partner. Contact a classmate who you can count on when you get into trouble. This is particularly useful for Lab Projects where one mistake can cost you time.

Study Skills: Self-Study Question(s)
1. Identify at least one Study Skill you did today to "Obtain Success in Your Technical Studies".
2. Who is your Study Skills Partner?

2

Chapter 2
Basic Operating
System Theory

OBJECTIVES

The goal of this chapter is to introduce you to basic operating system theory. This chapter will help you prepare and pass the following sections of the A+ Operating System Technologies Exam:

A+ Operating System Technologies Exam Objectives
covered in this chapter (and corresponding page numbers)

Domain 1 Operating System Fundamentals

1.1 Identify the major desktop components and interfaces, and their functions. Differentiate the characteristics of Windows 9x/Me, Windows NT 4.0 Workstation, Windows 2000 Professional, and Windows XP.

1.4 Identify basic concepts and procedures for creating, viewing, and managing disks, directories and files. This includes procedures for changing file attributes and the ramifications of those changes (for example, security issues).

Domain 2 Installation, Configuration, and Upgrading

2.4 Identify procedures for installing/adding a device, including loading, adding, and configuration device drivers, and required software.

In this chapter, you will complete the following sections:
- 2.1 Understanding the User Interface
- 2.2 Understanding Application Management
- 2.3 Understanding File System Management
- 2.4 Understanding Hardware Management
- 2.5 Understanding the Operating System Kernel

2.1 Understanding the User Interface

One goal of the operating system is to interface with the user. There are two major classifications of user interfaces that are available with operating systems. They are:
- Command Line Interface (sometimes generically called CLI)
- Graphical User Interface (GUI)

The most common operating systems using a **Command Line Interface (CLI)** are Microsoft DOS (MS-DOS or just DOS) and UNIX/Linux. To use a command line interface, you must be able to type and learn the command line syntax and structure. Figure 2.1 shows a screenshot of the DOS command line displaying a list of files on the floppy drive. The command being executed is the DOS **dir** command. It includes an option, **/a**, to show all files. This command and its options will be covered in detail in Chapter 3 on DOS and the DOS command line. For now, just understand that the command line exists and it is not graphical.

Figure 2.1: The MS-DOS Command Line Interface

```
A:\>dir /a

 Volume in drive A has no label
 Volume Serial Number is 2156-18E8
 Directory of A:\

IO       SYS      222,390  05-11-98  8:01p  IO.SYS
MSDOS    SYS            9  05-11-98  8:01p  MSDOS.SYS
COMMAND  COM       93,880  05-11-98  8:01p  COMMAND.COM
DRVSPACE BIN       68,871  05-11-98  8:01p  DRVSPACE.BIN
         4 file(s)          385,150 bytes
         0 dir(s)         1,071,104 bytes free

A:\>_
```

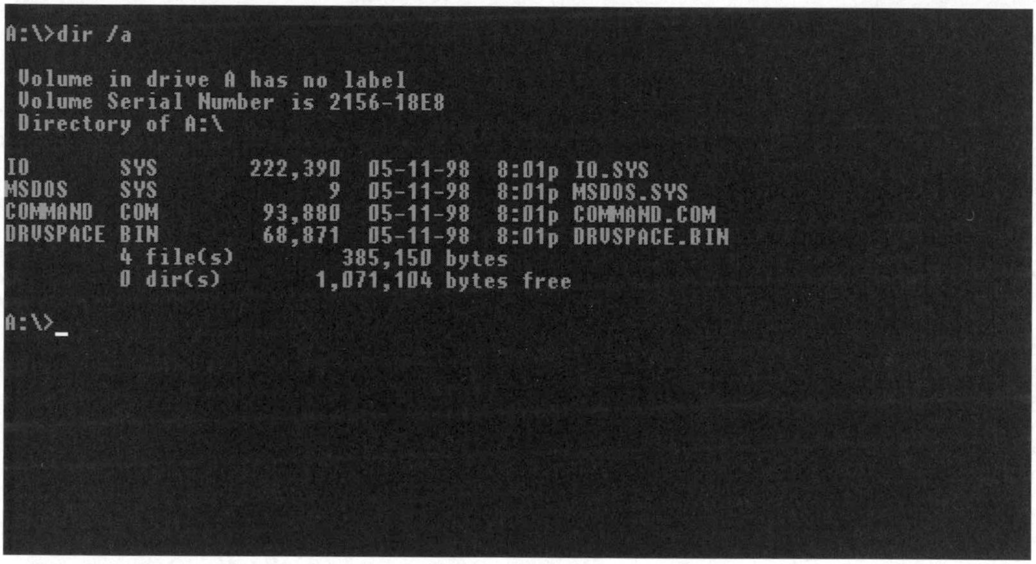

In UNIX/Linux terminology, the command line interface is called a **shell**. Figure 2.2 contains a screenshot of the Red Hat Linux shell. In this figure, two commands are executed to show you how the shell works. The **who** command is executed first and displays users currently logged into the system. Next, the **pwd** displays the current directory location. These commands will be covered in detail in Chapter 8 on the UNIX/Linux operating system.

Figure 2.2: The Red Hat Linux Command Line Interface (Shell)

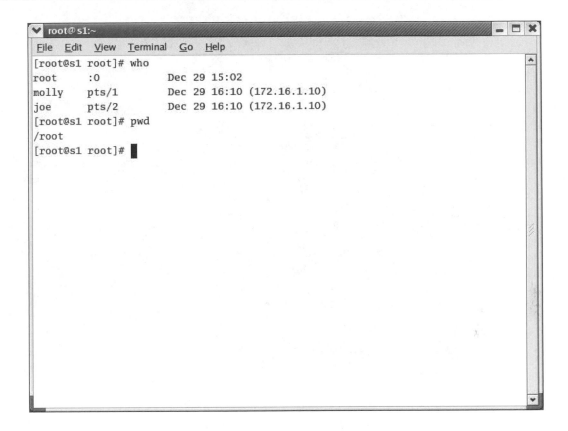

To help you understand the user command line interaction, you will perform an exercise in the DOS command mode of a Windows operating system. You will display the version of the operating system and the memory statistics. Understanding the version of the operating system and the amount of memory available is important because you may have to install an application that requires a specific operating system level and a minimum amount of memory. These commands will assist you in determining if the application you plan to install will run. This exercise assumes a PC is up and running with the Windows 98 (or higher) operating system. Follow these steps:

1. Click **Start**; then click Run. The Run dialog box displays.

2. In the Open text box, type **command** and press Enter. The DOS command line interface displays. Note that you could enter **cmd** instead of **command** for Windows NT, Windows 2000, and Windows XP. Either command will work on those operating systems.

3. In order to display the current version of the operating system, type **ver** and press Enter. The operating system version displays.

4. In order to display memory statistics, type **mem** and press Enter. The memory statistics display.
5. Refer to Figure 2.3 for a screenshot of this exercise.
6. To close the window, type **exit** and press Enter. The command line interface window closes.

Figure 2.3: Working with the DOS Command Line Interface

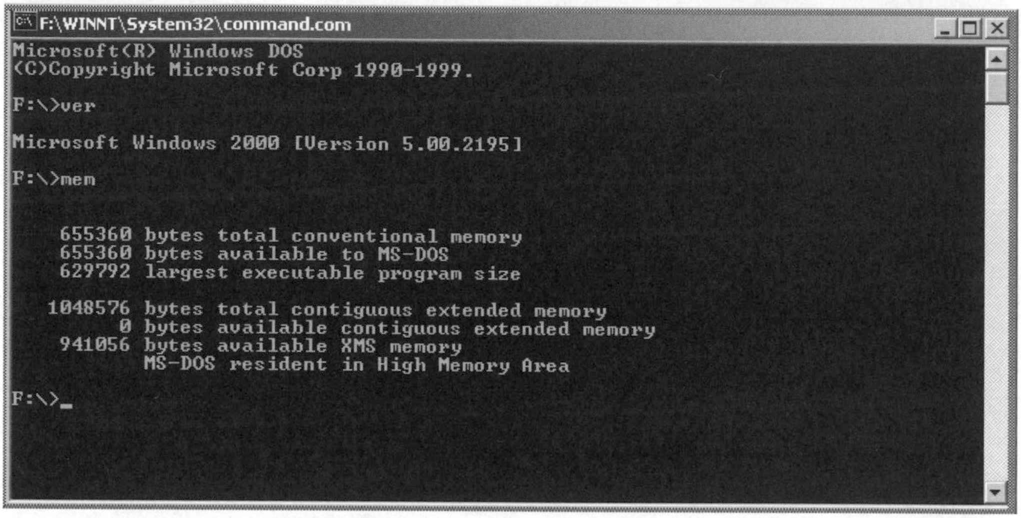

Today most operating systems have a **Graphical User Interface (GUI)** where a pointing device, such as mouse, is used to navigate, or **point and click**, around the operating system. The area on the screen where the user interacts with the operating system is called the **desktop**. Users click on **icons**, which are pictures that represent items such as programs and folders. Both Windows and Linux operating systems have a GUI interface.

Refer to Figure 2.4 for a screenshot of the Windows 98 Explorer program displaying the same files that are shown in Figure 2.1. On the left hand side of the screen you see folders and on the right you see the contents of the folders. Notice the horizontal bar at the top of a window. This is called the **Title Bar** and indicates the name of the program that is running. In the upper right hand corner of the Title Bar are three buttons. These are the **Minimize**, **Maximize,** and **Close** buttons. Notice how the same files from Figure 2.1 show up as icons here instead of simply file names. You can view most of the same information using the command line or the GUI with a given system.

Figure 2.4: Microsoft Windows 98 GUI Operating System Explorer Program

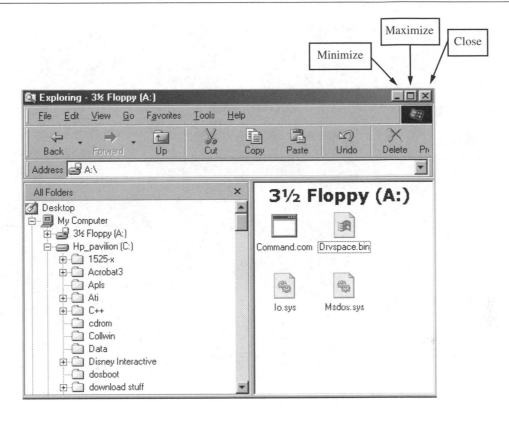

Because you have most likely seen a GUI desktop in Windows but not in Linux, Figure 2.5 shows a screenshot of the Red Hat Linux 8.0 operating system GUI. You can see the icons on the desktop, the Panel at the bottom of the screen, and the Internet browser software.

Figure 2.5: A Screenshot of a Red Hat Linux 8.0 Operating System Graphical Desktop

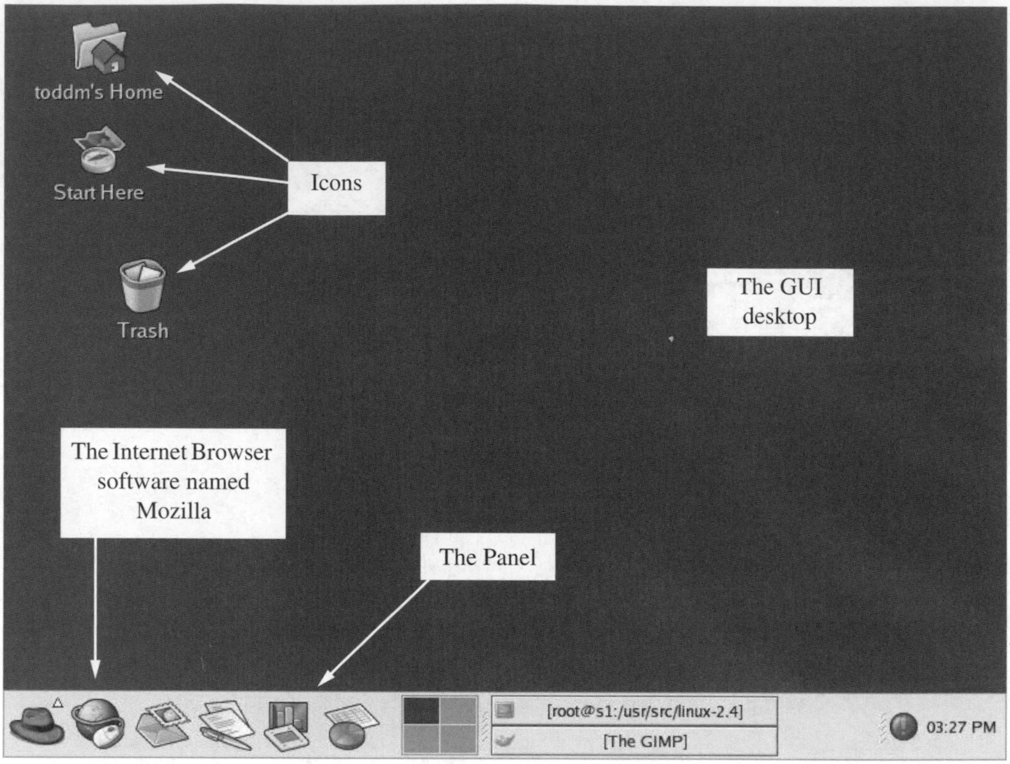

2.2 Understanding Application Management

Operating systems provide application management in the following ways:

- Access the hardware for applications
- Facilitate installing and uninstalling of applications
- Provide a consistent interface among applications through an Application Program Interface (API)

One role of the operating system is to perform hardware **Input/Output (I/O)** for applications. Some examples of I/O operations are reading from the hard drive, writing to the hard drive, printing to the printer, storing data in memory, or displaying data on the screen. In some older operating systems, applications were able to access hardware directly. When this occurred, a **General Protection Fault (GPF)** could occur causing the computer to halt, or crash. A GPF occurs when a program tries to use the same area of memory as another program. GPFs occur less often in more current operating systems such as Windows 2000 and Windows XP because they don't allow applications direct access to the hardware. The operating system calls upon device drivers when performing I/O operations. **Device drivers** are software programs that tell the device how to operate. They will be discussed further in a later section of this chapter.

Another role the operating system performs is to allow users to install and uninstall applications on their computer system. This allows the user to customize his or her system for maximum personal benefit. On many Windows operating systems, this is performed using the Add/Remove Programs software tool. On Red Hat Linux you can run a command called **rpm**, Red Hat Package Manager, to install and uninstall software programs.

Additionally, the operating system provides consistency among applications. For example, in both Microsoft Word and Excel, you save a file by clicking File and then Save, or Save As. This similarity makes learning a new application quicker. Application consistency is provided through a set of software routines and tools for building software called an **Application Program Interface (API)**. APIs are designed for programmers and guarantee all programs, sharing a common API, have similar interfaces.

Many operating system manufacturers encourage programs be written to conform to the **Common User Access (CUA)** standard developed by IBM and Microsoft. The CUA makes sure Windows programs are consistent. Programmers use an API to implement the CUA. Having a common interface makes it easier for the user to navigate Windows applications. Microsoft operating system tools and applications support CUA. This standard is also supported on X Windows, which is the GUI for UNIX/Linux operating systems.

An API is a programming standard. The CUA is a Windows standard facilitated by the use of a standard API.

Microsoft Windows applications consist of executable files and **Dynamic Link Libraries (DLLs)**. A DLL is an executable file that contains smaller programs, called functions, and data. They allow applications to be loaded, unloaded, and reused easily. An application that is currently running is known as a **process** or job. A process has system resources allocated to it, such as the processor, memory, and environment variables. A process will be given a **Process Identification** number, called a **PID**, which is used by the operating system to reference the process. You can change the amount of processor time a process has or terminate a process by its PID. Figure 2.6 shows a screenshot of Task Manager in Windows 2000. Notice the columns for each process such as Image Name, PID, CPU (percentage used), CPU Time (used), and Mem Usage (Memory Usage).

Figure 2.6: A Display of Process Information Using Task Manager in Windows 2000 Professional

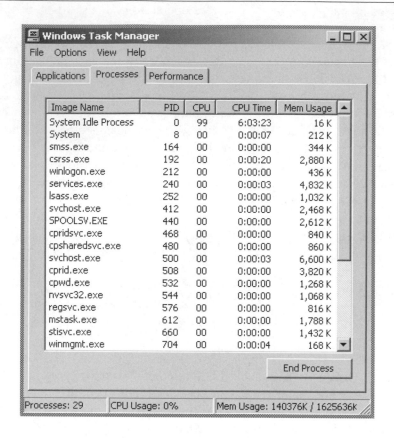

Figure 2.7 is a screenshot of the Linux System Monitor. To access this menu under Linux Red Hat 8.0, you would follow these steps:

1. Log in as a user.
2. Click the Red Hat icon on the taskbar. A list of tools appears.
3. Click **System Tools**. Additional tools appear.
4. Click **System Monitor**. System monitor appears.
5. Click **Process Listing**.
6. Once you are through reviewing System Monitor, click **x** to close the window.

This program is the equivalent of Windows 2000 Task Manager. It shows that Linux displays a column for the user running the process in addition to the ID, %CPU, and memory used. This is because Linux is a multi-user OS. The far right column, labeled "ID," is the PID.

Figure 2.7: System Monitor in Linux Displaying Process Information

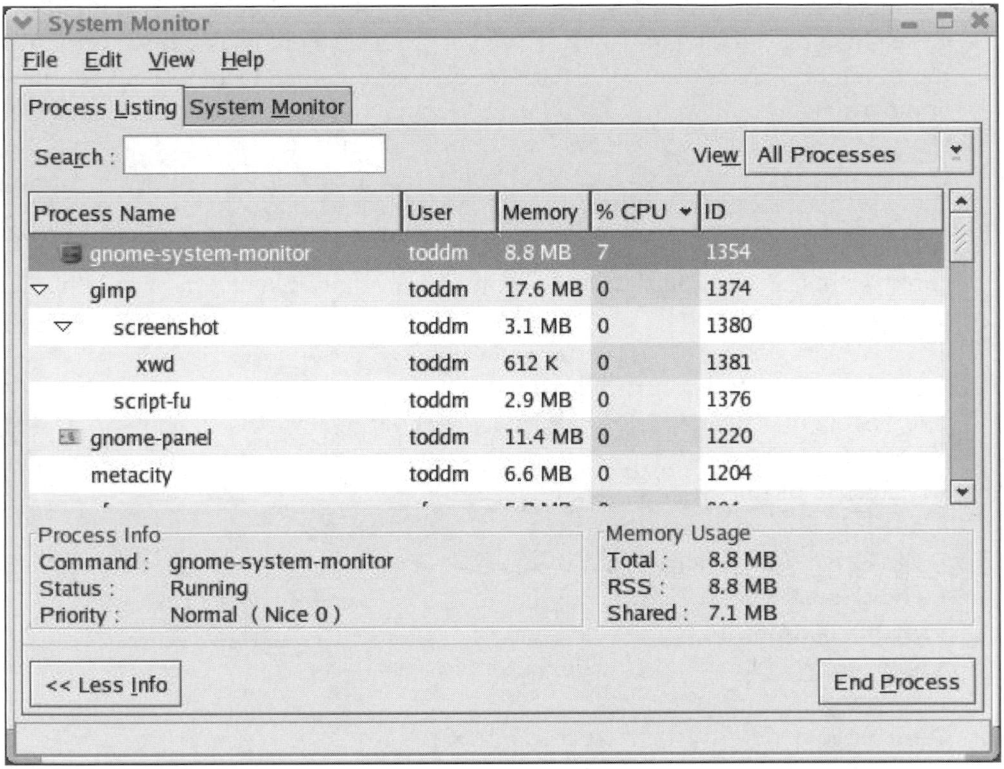

A **thread** is a basic unit of instruction that is allocated processor time by the operating system. Think of a process as being made up of many threads. The threads of a process execute the process code. A thread can execute any part of the code, which includes portions that are currently run by another thread. Threads synchronize with each other to coordinate resource access. This prevents one thread from interrupting another. Operating systems supporting **multithreaded** applications allow threads to be executed concurrently. Most Windows and Linux operating systems support multithreaded applications. For example, multithreading occurs anytime you are using products like Microsoft Excel or OpenOffice.org Writer and save a large file while printing another file concurrently.

On computers with multiple processors and an operating system that supports multiple processors, you can set which processor actually runs a process. Although more costly, having multiple processors speeds up your computer system because the processes can be divided among available processors. Microsoft server products support multiple processors.

Let's look at an analogy to help you understand the concepts of a process and threads. Everyone has purchased a product at one time or another that has been made using an assembly line production process. Suppose TM Industries, Inc., produces a product known as Product Z. Product Z has to be created on the assembly line. When an order comes in from a customer, a Production Number is assigned to the order so the product can be made for the customer. The Production Number is like the <u>PID</u> of a process. Once the Production Number is created for Z, then resources are allocated to it. In the computer system, resources are processor time and memory, but resources in the company will be time, parts, and labor. Manufacturing product Z is divided into several steps. When an employee makes Part A for Product Z, this step is analogous to a <u>thread</u>. If this step takes more than one shift, it may involve a shift turnover meeting where employees meet to discuss the status of Product Z. The next shift knows to continue where the employee on the previous shift left off. Shift turnover is like the thread <u>syn-chronization</u>. If TM Industries had multiple assembly plants, that would be like having multiple processors on the system.

2.3 Understanding File System Management

File system management is a complex issue covering a variety of major topics. The topics covered here are:
- Overview of File System Access and Storage
- The File System Hierarchy
- The Drive Specification and Files
- The File Allocation Table
- Fragmentation
- Specific Operating System File System Formats

Overview of File System Access and Storage

The operating system also manages the **file system** for users and applications. The file system defines how files and folders are stored on disk. The file system is created when you **format** a **partition**. Formatting is the process of preparing the disk for use. A partition is a logical division of a hard disk drive. In DOS, you use the **fdisk** command, short for Fixed Disk, to create partitions. Linux has an fdisk program, which accomplishes the same thing.

You can install multiple operating systems on your PC. This is called **multibooting**. However, you can only boot one operating system at a time. You can install multiple operating systems on the same partition but it is recommended to place them on separate partitions.

The **primary partition** contains your operating system and is the first partition on a drive. In order to boot your operating system, you must have a primary partition and it must be marked as the **active partition**. Microsoft assigns letters to partitions. In DOS, your primary partition is the C: partition. An **extended partition** type can be further divided into multiple **logical** partitions, called logical drives. Here, you could have a drive D:, E:, all the way up to Z:. Note that A: and B: both represent your floppy drive and C: typically represents your hard drive. A primary partition cannot be further divided.

Under DOS and Windows 95/98, you can have only two partitions: one primary and one extended. Windows NT, 2000, and XP use the terms **system partition** and **boot partition**. The system partition is where the load files are located. The system partition must be marked as active in order for these operating systems to load. The boot partition is where the operating system itself is located. The system and boot partitions can be on the same or different partitions in NT/2000/XP. On Windows NT, 2000, and XP, you can have up to

• four primary partitions, or

• three primary partitions and one extended (with multiple logical) partitions.

The **Master Boot Record (MBR)** is a record that tells the operating system about the partitions. The MBR looks for the active partition and boots it. If the MBR becomes corrupt, you can run the **fdisk /mbr** command to fix it.

You can see the partitions, labeled as C:, E:, F:, and so on, in Windows XP by using the Disk Management tool. Disk Management is located in Administrative Tools under Control Panel. For a screenshot of the partitions on a computer, refer to Figure 2.8. To access this menu, follow these steps:

1. Log on as Administrator.
2. Click **Start**, point to **All Programs**, point to **Administrative Tools** and click **Computer Management**. The **Computer Management** window appears.
3. In the left window pane, click **Disk Management**.
4. View and then close the window by clicking **x**.

Figure 2.8: Displaying the Partitions Using the Disk Management Tool

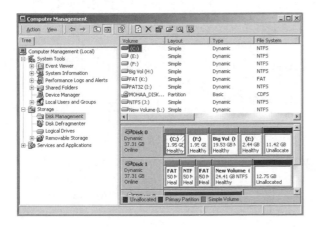

In Windows products, the file system stores a file in units called **clusters**, or **allocation units**. A cluster is equal to a specific number of sectors on disk. A **sector** is equal to 512 bytes. A hexadecimal number references a cluster location. Refer to Appendix A for a discussion of hexadecimal numbers. A cluster is the smallest amount of space the operating system reserves for a file. The size of the partition determines the cluster size. If you have a file that increases in size, it may need additional clusters. If the file does not occupy all of the space in a cluster, the excess is wasted and cannot be used for additional file storage.

Let's compare large and small cluster sizes. While a large cluster size is typically more efficient than a small cluster size, it generally wastes more disk space. Large cluster sizes are more efficient because the file system has a fewer number of clusters to access when compared to small cluster sizes. Large cluster sizes potentially waste more space because more sectors would be allocated for each file. On the other hand, a small cluster size tends to waste less space because fewer sectors are allocated for each file. However, a small cluster size is generally less efficient than a large cluster size because of the greater number of clusters the operating system must manage.

File systems with large cluster sizes are best for data intensive applications like database systems, while small cluster size file systems are best for e-mail systems, which typically contain small file sizes.

Data is magnetically stored on disk in **sectors** and **tracks**. As you can see in Figure 2.9, a sector is a slice, or section, of the disk while a track is a concentric circle running around the center of the disk.

Figure 2.9: Sectors and Tracks on a Disk Surface

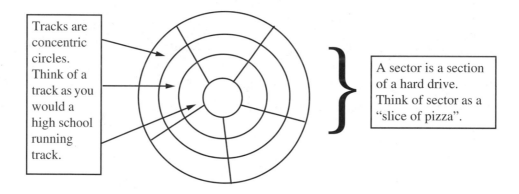

Tracks are concentric circles. Think of a track as you would a high school running track.

A sector is a section of a hard drive. Think of sector as a "slice of pizza".

Figure 2.10 shows the effect of waste in a cluster. Let's suppose you had a file named **filename.dat** that completely occupied the first two clusters, 1 and 2, but only a portion of cluster 3 (about two-thirds of cluster 3). The remaining one-third of space has already been allocated to the file and cannot be used. It is considered waste in file system terminology. This waste occurs naturally when storing data on disk because cluster sizes are specific sizes and not all files are the same size as the cluster size. Later in the chapter, you will see the cluster sizes for the file system types. These sizes vary depending upon the size of the partition.

Figure 2.10: Wasted Cluster Space

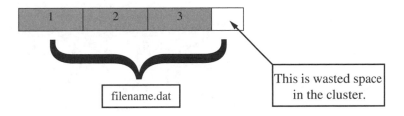

Linux use the term **block** instead of cluster.

The File System Hierarchy

A **file** is the lowest unit of storage within the file system. A **folder**, or **directory**, can contain other folders or a folder can contain files. Placing files within folders is useful for organizational purposes. It is easier to locate a file if it is stored in a related folder. For example, you could create a folder called Personal for documents such as your resume. Or, you could use the system created folder called My Documents. The file system is also hierarchically represented in most GUI tools, such as Microsoft Explorer. The hierarchical design makes referring to file and folders easier.

The term **folder** is used when in the GUI and the term **directory** is used when referring to the command line of an OS. The terms are synonymous.

Figure 2.11 shows a hierarchical representation of the file system. Note in the figure that we are referring to a Windows operating system. The drive is C:, or the hard drive, and the forward slash, \, is the root, or top level directory.

Figure 2.11: Hierarchical Representation of the File System

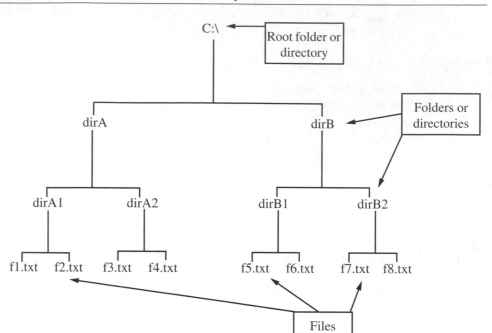

 In Windows, the root symbol is \. In Linux, the root symbol is /. So in Figure 2.11, if you changed C:\ to /, you would have a Linux hierarchy.

Let's explore this hierarchy concept by using an analogy. The file system hierarchy is analogous to an actual tree you might buy at a garden store that has its root system bundled up in a burlap bag. If you turn that tree upside down, you have the concept of the file system hierarchy. The root system is now on top, just like the <u>root</u> directory is at the top of the file system. The branches and leaves stem from the root. A <u>directory</u> is analogous to a branch on the garden tree. A leaf is analogous to a <u>file</u>. Just as a leaf on the tree cannot contain branches or other leaves, a file in the file system hierarchy cannot contain directories or other files. Just as a branch on the tree can contain other branches and leaves, a directory in the file system hierarchy can contain other subdirectories and files.

Let's follow up with an exercise. The goal of this exercise is to help you understand the hierarchical relationship that exists on Windows operating systems by creating a few folders and files. This exercise assumes the Windows 2000 operating system is up and running. Follow these steps:

1. Right-click **Start**; then click **Explore**. A separate Explorer window appears.
2. Insert a formatted floppy disk in the floppy drive.

3. Scroll until you see the Floppy (A:) drive.

4. Right-click the floppy drive and click **Open**. Another window with the floppy's contents appears.

5. To create a folder, click **File**, then point to **New**, and then click **Folder**. A folder icon appears prompting you to enter a folder name.

6. Name the folder **Chapter1** and press Enter. The folder is created.

7. Double-click the folder named **Chapter1**. The folder's contents appear.

8. To create a subfolder within Chapter1, click **File**, then point to **New**, and then click **Folder**.

9. Press Enter to accept the name, **New Folder**.

10. To create a file within Chapter1, click the Chapter 1 folder, then click **File**, point to **New**, and click **Text Document**.

11. Press Enter to accept the name, **New Text Document.txt**. Notice the extension. Notepad uses the .txt extension for files. So, this is a Notepad compatible file.

12. To create another file within Chapter1, repeat the previous step. Notice the file name is different from the one in the previous step.

13. Close all windows.

14. Go back into Windows Explorer and open up all the folders on the floppy until you see the hierarchical relationship similar to the one shown in Figure 2.12.

15. Close Windows Explorer.

Figure 2.12: Display of Windows Explorer's Hierarchical View

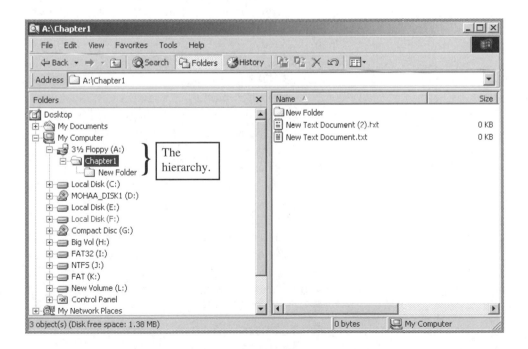

The Drive Specification and Files

When referencing a drive in DOS or the DOS command line in Windows, you <u>must</u> include the drive letter and the colon. Collectively, this is known as the **drive specification**, or **drive spec** for short. If you leave out the colon, DOS will display an error message. UNIX and Linux also have a hierarchical view of their file system; however, they do not refer to drives using drive letters. They refer to a file system by its **mount point**. A mount point is a directory name that is associated with a partition.

Remember that the file system is logical not physical. The view of the file system that we see in Windows Explorer is hierarchical and for our use only. The operating system does not store files and folders in the same hierarchical fashion that we see in Explorer.

Data and programs are stored in files within folders on disk. There are two major types of data that a file can contain. They are:
• Executable or binary
• Non-executable (including text, pictures, graphics, and motion picture files)

Executable or **binary** programs are ones that have been **compiled** and perform some type of operation or task. They are written using text editors following the strict syntax of the language, and then they are compiled. The compilation process creates the executable code.

Non-executable, or data, files are files that are used by the executable programs for a specific activity. For example, a payroll program will use a payroll data file to process checks. A game program will use where you last saved in the game as the starting point data for the next round. Finally, your digital camera software will use the picture you took as data when displaying the image to you on the screen.

Files have names so you can easily identify them. The Linux operating system uses an index node, or **inode**, number when referencing the file. It is a unique number Linux uses to locate the file.

Let's look at a quick analogy to help you understand. Suppose you are a <u>file</u> in this analogy, the <u>inode</u> is your Social Security Number (SSN), and the government is the <u>OS</u>. You may go by your name with your friends, but the government (the OS) knows you by your SSN (inode). So just as your SSN is the number that the government uses to reference you, the inode is the number the operating system uses to reference a file.

In DOS file names follow the **8.3** rule. There can be up to eight characters in a file name, followed by a dot, and a three-character extension. The **extension** is a set of characters used to associate the file to an application. You can run most DOS-based commands by entering the file name and leaving off the extension. DOS will see files ending in .com, .exe, and .bat, as programs it can execute. These are also known as **file types**.

In Windows, the extension is what allows you to open the application and the file, as data, by double-clicking on the file name. This is also known as **file type association.**

Not all characters are allowed in a DOS file name. For example, the space is not allowed and you can only use uppercase characters. DOS converts all characters to uppercase. Chapter 3 discusses illegal DOS file name characters in detail. For now be aware that a DOS file name must conform to the 8.3 standard. Microsoft Windows operating systems, from Windows 95 and up, allow you to store file names up to 255 characters and allow you to use a space in a file name. The Linux operating system allows up to 255 characters in a file name and does not require an extension.

To help you remember the 8.3 rule for DOS file names, just think "filename.ext." There are 8 characters in the word "filename" (with no space), followed by a dot, and the 3-character extension, "ext" in this case. In other words, "filename.ext" contains the maximum letters that are allowed for the name of the file and the extension.

One other aspect about files and directories is the concept of a **path**. The path is the location of the file, or directory. The two types of paths are **full** and **partial**. A full path always begins from the root directory of a partition and a partial path is simply the file name without any reference to the root directory. Think of the full path as your full name and the partial path as the name you go by. Your full name can usually be used to uniquely identify you. But a first name, such as Joe, does not uniquely identify you because there could be several people with the same first name.

The term "full path" is also called the **fully qualified path** or **absolute path** (the path is absolutely defined). The "partial path" is also called a **relative path** (the path is relative to your current position in the directory tree structure).

The File Allocation Table

Files and folders are accessed through the use of a table. A Windows operating system keeps track of files and folders on disk by using a file system format called the **File Allocation Table**, or **FAT**. Table 2.1 shows the operating systems and the specific file system formats they support on their hard drives. The specific formats will be discussed later in this section.

Table 2.1: File System Formats Supported by Various Operating Systems

Operating System	File System Format
Windows XP	FAT16, FAT32 and NTFS
Windows 2000	FAT16, FAT32 and NTFS
Windows NT	FAT16 and NTFS
Windows 98	FAT16 and FAT32
Windows 95 OSR2	FAT16 and FAT32
Windows 95 (before OSR2)	FAT16
DOS	FAT16

One file system format not mentioned in Table 2.1 is VFAT. Microsoft developed VFAT (Virtual FAT) with Windows 95 to allow file names to exceed the 8.3 rule. VFAT allows up to 255 characters in a file name, including spaces. These file names are called **Long File Names**, or **LFNs**. These FAT types will be discussed further in a later section.

If you have ever seen a file name on a Windows 95 operating system with a tilde (~) in the name, you are looking at an LFN. Figure 2.13 is a listing of the **dir** command displaying file names with their 8.3 and LFN equivalents. (On Windows 2000, you can use the **dir /x** command to display the Long File Name of a file name.) The two rightmost columns display the file names. The first file name column shows the 8.3 file name, while the last column shows the LFN equivalent. Because DOS can have only 8.3 uppercase characters, how does it fit a file name of potentially 255 characters into 8.3? The 8.3 format uses the same first six legal characters as the LFN, but reserves the last two for uniqueness. If you have several LFN file names with the same first eight characters, then the last two file name characters of 8.3 notation would be ~1 for the first occurrence, ~2 for the second, ~3 for the third, and so on as shown in Figure 2.13.

Figure 2.13: A Directory Listing of 8.3 and LFN File Names

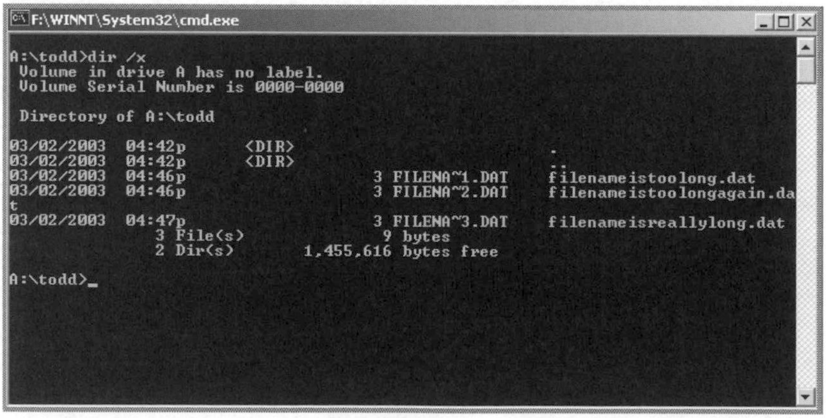

The operating system does not store all of the files you have on disk in memory. The operating system only stores the disk address locations of the files. The FAT is usually **cached**, or stored, in memory for speedy access. Caching is the concept of storing frequently used files or tables in memory. Accessing memory is faster because memory access is electronic and disk access is mechanical. But not everything is cached. If a file or program is not heavily used, it does not need to be cached. It can still be accessed on disk when needed.

It would be impossible for the operating system to store all files from disk in memory anyway. Computer disk capacity far exceeds the memory capacity. For example, it is common to have a computer system with an 80 GB (gigabytes) hard drive and only 512 MB (megabytes) of RAM.

Let's see how the operating system uses the FAT to access a file. Suppose you have a file named Payroll.dat located on the C:\ drive in the Apps directory. The full path of the file is C:\Apps\ Payroll.dat. A partial path would be simply Payroll.dat. You need to perform a File Open to open the file. How does the operating system retrieve the file? Refer to Figure 2.14 while you read through this section to help your understanding. Although these steps assume a Microsoft operating system, the same general theoretical concept applies to how many other operating systems retrieve their files and folders (directories).

Because a file is stored in a directory, the operating system first looks to the directory for the location of the file. The location is specified as a cluster number. The directory has the starting cluster location of the actual file. In Figure 2.14, this is cluster 4. The operating system then goes into the FAT at cluster location 4 to retrieve the data. Once the data from this cluster is retrieved, the operating system goes to the next cluster and retrieves the data there. You can see in the figure that the next cluster, after cluster 4, is cluster 5. The data located in cluster 5 is retrieved, but the next cluster listed is cluster 10. How did it skip clusters 6 through 9? Disk access is random as shown in this figure. Another file could be located in clusters 6 though 9. Eventually, cluster 10's data is retrieved, and it points to cluster 11 as the next location of data. Cluster 11's data is retrieved and a special set of characters representing an **End Of File (EOF)** tells the operating system file retrieval is complete.

Think of End Of File (EOF) as "The End" in a movie.

Figure 2.14: The Logical Representation of the File System Table

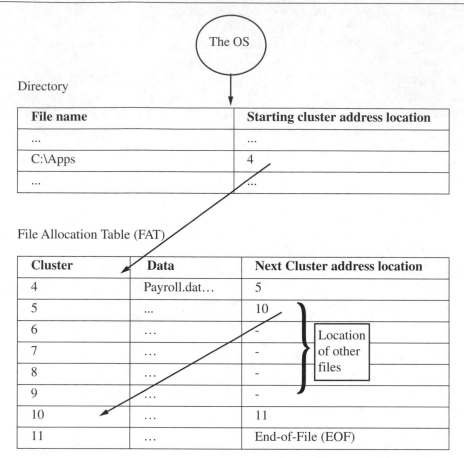

Directory

File name	Starting cluster address location
...	...
C:\Apps	4
...	...

File Allocation Table (FAT)

Cluster	Data	Next Cluster address location
4	Payroll.dat…	5
5	...	10
6	...	-
7	...	-
8	...	-
9	...	-
10	...	11
11	...	End-of-File (EOF)

Location of other files

Let's look at another analogy. When trying to understand the FAT, it may help to think of it as an index in the back of a book. When you want to locate a term, you go to the index. To the right of the term is the page number for the associated term. You then turn to the page for an explanation of the term. The page number is a location for the term in the book, just like a <u>cluster address</u> is a location for the file or directory on disk. The explanation of the term is analogous to the data within the file. Without the book's index, you would have to sequentially read the entire book starting from page 1. Without the FAT, the operating system would have to sequentially start reading the entire disk starting from the first cluster until it found the file. That would take far too long.

There are two important overall concepts to remember about an OS. They are:
- The operating system stores files and directories logically in a hierarchical fashion for organization.
- The operating system relies on tables for quick access and lookup capabilities. When you think of a table, think about an index in the back of a book. You use the index to locate the page reference for a term you don't understand.

Fragmentation

Because the file C:\Apps\Payroll.dat shown in Figure 2.14 is not located in **contiguous**, or adjacent, cluster locations, it is considered **fragmented**. A fragmented file is one that is physically spread over non-contiguous areas of the disk. The disadvantage of fragmentation is that it increases the time to locate a file because the file is spread out on different parts of the disk.

Files become fragmented because they change over time. For example, assume you created a document named Resume.doc and you store your resume information in it. At the time it takes up several contiguous clusters. Sometime later, you gain new experience and decide to update your resume to reflect this fact, but another file named File2.doc occupies the clusters immediately after Resume.doc. The operating system cannot place the changes to Resume.doc immediately after the original because File2.doc is stored there. It uses cluster locations after File2.doc. Resume.doc now occupies clusters that are not contiguous. The file is spread out on the disk in fragments, which increases the time it takes to retrieve it.

In Figure 2.15, you can see fragmentation. Notice cluster 3 is about two-thirds full. The document did not take up the entire cluster, but the rest of it cannot be used so it is considered wasted disk space.

Fragmentation on a system increases over time. If the files on your system have been modified over a period of time, it has probably become fragmented. Before the days of defragmentation tools, System Administrators would back up all of the data to tape, reformat the hard drive, then restore the hard drive from the tape backup. By backing the data up to tape, the file's fragments were adjacent to one another. It could take hours to perform this operation, but it sped up system performance for the users in the long run.

Figure 2.15: Diagram of Fragmentation

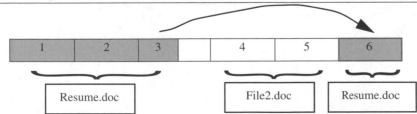

Defragmenting is the process of placing a file into contiguous areas of the disk. This speeds access to the file because there is less mechanical movement of the read/write mechanism in the drive. Microsoft includes a Disk Defragmenter tool to help you analyze and defragment your disk. Microsoft recommends you click Analyze to analyze your disk before you actually try to defragment it. It is possible that you would not save enough disk space compared to the amount of time disk defragmentation takes.

Notice what happens to Resume.doc after it has been defragmented by referring to Figure 2.16. There is still wasted space in cluster 3. However, the pieces of the file are contiguous. This translates to quicker access by the hard drive because the pieces of the file are in close proximity.

Figure 2.16: The Result of Defragmenting a Disk

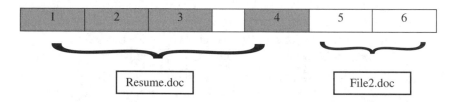

Figure 2.17 shows a screenshot of the Disk Defragmenter in Windows 2000. You can see that the C: drive has been analyzed and has fragmented files. The fragmented files will appear color-coded on an actual screen.

Figure 2.17: The Disk Defragmenter Tool Available in Windows 2000 Professional

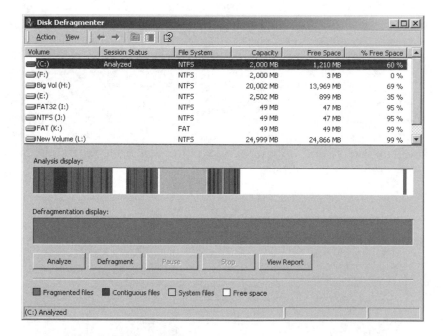

Depending upon the version of Microsoft's defragmentation tool, you should try to analyze the disk for defragmentation purposes prior to actually attempting to defragment it. There may be little to gain from spending resources performing the actual defragmentation.

Specific Operating System File System Formats

Some of the most common file system formats are FAT12, FAT16, and FAT32. The numerals in the names refer to the number of bits required for a file allocation table entry.

- **FAT12**—This is used on floppies and is supported by all operating systems allowing storage up to 1.44 MB. A floppy has 2,847 clusters of 512 bytes each. If you multiply these two numbers, you get 1,457,664 total bytes, which is where we get 1.44 MB. In Windows XP Professional, FAT12 is used only on floppy disks and on volumes smaller than 16 MB.
- **FAT16**—This is the original File Allocation Table file system used on DOS, Windows 95, 98, and NT. It is limited to 8.3 file names. One disadvantage of FAT16 is that the root directory can manage a maximum of 512 entries. FAT16 has no built-in file system security or compression scheme. In FAT16, there is a primary FAT and a copy of the primary FAT. The use of a backup copy of the FAT is for **fault tolerance**. Fault tolerance is used to describe the degree to which a system can tolerate a problem, or fault. It is much like storing a backup copy of an important document, such as your resume, taxes, or a term paper on floppy. You may have a primary copy on the hard drive, but if something happens to it, you can go to your backup on the floppy.

The size of the cluster is dependent upon the size of the partition and the file system format.

- **FAT32**—This offers improvements over FAT16. The main benefit being it supports larger partitions than FAT16.
- **VFAT**—Virtual FAT came out with Windows 95 to support file names longer than 8.3. It supports file names up to 255, lowercase characters, and spaces in the file names.
- **NTFS**—New Technology File System, provides performance and reliability not present in FAT. Some of the features provided by NTFS are:
  - Disk quotas, which allow you to limit the amount of storage a user can utilize on a partition.
  - Encryption, which allows you to encrypt the contents of a file for security purposes. **Encryption** is the process of disguising data to hide its true value.
  - Compression, which allows you to compress a whole partition and saves disk space. **Compression** is the process of removing redundant data thereby saving disk space.
  - Dynamic disks, which allow you to extend a partition from unallocated free space on the disk.

- Auditing, which allows you to track changes to files and folders by users in order to track usage of files.
- Bad-cluster remapping, which allows the system to mark physical locations on disk that are bad and not use them in the future.
- File level security, which allows you to set permissions for a user on a file. Although FAT only allows you to set permissions on a folder, NTFS allows you to set permissions on both folders and files.

NTFS uses a **Master File Table** (**MFT**), which is analogous to the table used by FAT16 and FAT32. The MFT is the first file on an NTFS volume that contains information about each file and folder on each volume.

Table 2.2 compares the default cluster sizes for FAT16, FAT32, and NTFS on varying partition sizes. You can see that NTFS is more efficient in terms of disk space usage due to its smaller cluster sizes. For example, if you have a 1,000-byte partition, FAT16 would allocate a cluster size of 16,000 (16 KB), FAT32 would allocate a cluster size of 4,000 (4 KB), and NTFS would allocate a cluster size of 1,000 (1 KB). A 1-byte file would occupy 16 KB on the 1,000-byte FAT16 partition, 4 KB on the 1,000-byte FAT32 partition, and 1 KB on the 1,000-byte NTFS partition.

Typically, the larger the partition size, the larger the cluster size. For example, on FAT16, a 50 MB partition uses a cluster size of 1 KB, a 100 MB partition uses a cluster size of 2 KB, and a 150 MB partition uses a cluster size of 4 KB.

Table 2.2: Cluster Sizes for Various Partition Sizes

Partition Size	FAT16 Cluster Size	FAT32 Cluster Size	NTFS Cluster Size
7 MB–16 MB	2 KB	Not supported	Not supported
17 MB–32 MB	512 bytes	Not supported	512 bytes
33 MB–64 MB	1 KB	512 bytes	512 bytes
65 MB–128 MB	2 KB	1 KB	512 bytes
129 MB–256 MB	4 KB	2 KB	512 bytes
257 MB–512 MB	8 KB	4 KB	512 bytes
513 MB–1,024 MB	16 KB	4 KB	1 KB
1,025 MB–2 GB	32 KB	4 KB	2 KB
2 GB–4 GB	64 KB	4 KB	4 KB
4 GB–8 GB	Not supported	4 KB	4 KB
8 GB–16 GB	Not supported	8 KB	4 KB
16 GB–32 GB	Not supported	16 KB	4 KB

- **CDFS**—Used on read-only CD-ROM drives.
- **UDFS**—Universal Disk Format, is primarily used for read-only DVD/CD-ROM media.

- **ext2**—The file system supported by the Linux operating system.
- **ext3**—A newer version of ext2 that uses an on-disk journal to keep track of changes thereby keeping the file system in a consistent state. The ext3 file system allows for a power outage without having to clean the file system.
- **NFS**—Network File System, developed by Sun Microsystems, is a file system accessible from a network connection. With NFS, you don't have to have the disk physically attached to your computer.

You should go through the shutdown process for your operating system. This will write any necessary system information from memory to disk safely. If you have ever experienced a power outage while using your system, the operating system may need to fix the file system. This is because the table in memory may become corrupt due to the abrupt loss of power. Many versions of Linux are particularly susceptible to this problem. With the ext3 file system, this problem is less severe. The tool to fix the disk is CHKDSK or SCANDISK on most Microsoft operating systems. The Linux command is **fsck** (file system check).

Let's perform a short exercise to help you understand file systems. You will run a file system check on a floppy disk. This exercise assumes Windows 2000 is up and running. Follow these steps:

1. Right-click **Start** then click **Explore**. A separate Explorer window appears.
2. Insert a formatted floppy disk in the floppy drive.
3. Scroll until you see the Floppy (A:) drive.
4. Right-click the floppy drive and click **Properties**. Another window with the floppy's properties appears.
5. Click the **Tools** tab and click **Check Now**. The **Check Disk** window appears. A window such the one shown in Figure 2.18 appears. Notice you can elect to automatically fix file system errors or scan for bad sectors. Otherwise, the Check Disk procedure simply displays the status of the disk.
6. Click **Start** to check the disk.
7. Close all windows when complete.

Figure 2.18: Running the Check Disk Command

Check Disk	? X

Check disk options

☐ Automatically fix file system errors
☐ Scan for and attempt recovery of bad sectors

[Start]   [Cancel]

2.4 Understanding Hardware Management

Another function of the operating system is hardware management. Hardware devices connect to the computer either through **expansion slots** located inside the computer or **ports** located externally. For example, a Network Interface Card (NIC) usually fits into an expansion slot that is housed on the motherboard itself. Devices such as your printer, mouse, keyboard, display monitor, CD-Writer, speakers, Digital Camera, and scanner use a cable to connect to the computer system. Figure 2.19 shows a view of external ports.

Figure 2.19: External Ports Available on PCs

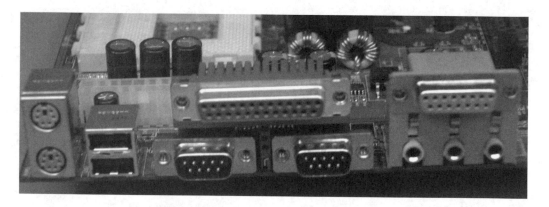

The **BIOS (Basic Input/Output System)** is an important function for the computer system from the standpoint of most operating systems. The **BIOS** is a set of instructions permanently stored on a chip that directs basic input and output operations such as the keyboard. This is an example of **firmware**. Firmware is a set of instructions permanently stored on **Read-Only Memory (ROM)** chips. You cannot simply delete the contents of the ROM firmware by shutting down the computer. The BIOS has basic functionality for your keyboard, display adapters, serial ports, and other functions.

Firmware is "software on hardware." The software is "firmly" (permanently) coded on the chip.

The operating system uses the ROM firmware for booting, or starting, the computer. When you power on your computer system, it goes through a Power On Self Test (POST) to ensure the basic components such as memory, keyboard, mouse, and the hard drive are available. Once this occurs, the Master Boot Record (MBR) is used to locate the address of the operating system stored on disk. Next, the operating system is loaded into RAM for your use.

The term booting comes from the phrase "pull yourself up by your bootstraps", as in the boots that go on your feet.

Devices need software in order to communicate. If the software is not included in the BIOS, then you will need a device driver for the device. You must configure the correct driver for the type of hardware you install. It's not uncommon to install a device driver incorrectly. For example, if you install the wrong printer driver for a given printer, output on the printer may be garbled and unreadable.

Let's look at memory addressing before we delve into additional hardware management topics. Random-Access Memory (RAM) is a physical piece of hardware. However, data and programs are stored in RAM and accessed logically using memory addresses. The necessary portions of the operating system are loaded into RAM during the boot process and stored in memory address locations. These addresses are given in hexadecimal notation like disk addresses. The operating system, other programs, and data are referenced using their memory address location. RAM is considered volatile because its contents are constantly changing as programs are loaded and unloaded and data is modified. When you double-click to run a program or you key in the program's name in the CLI of an operating system, the command will be loaded in vacant memory. Any data you enter will also be placed into memory. The contents of memory are lost during a shutdown—this is normal. So the data must be saved from memory to disk in order for you to keep a permanent copy of it.

In Figure 2.20, you can see an example of how memory address locations contain programs and data. Memory is comparable to the rows and columns of Post Office (P.O.) boxes. The P.O. boxes have an address. The contents of the P.O. boxes are constantly changing just like memory.

Figure 2.20: The Contents of Memory Address Locations

Memory Address Location	Contents
A100	OS
A101	OS
A102	Software instructions for other programs.
...	Software instructions for other programs.
FFFE	Data

When thinking of data and instructions being stored in memory, just visualize a spreadsheet. The cells are like memory address locations. These cells store data and instructions to be performed.

System resources are features that control how devices on a computer system work. System resources are set either manually or within software when installing the device. The manual method is the older form of setting system resources. Typically, there was a series of little switches that could be turned on or off. Their combination created a numeric setting.

A new, improved method of setting system resources is **Plug and Play (PnP)**. Plug and Play, developed by Intel, is a standard that allows a computer to automatically detect and configure the installed device. Both the operating system and the device must be Plug and Play compliant for this method to work. Windows 95/98/2000/XP are PnP compliant, but Windows NT is not. During configuration, the PnP operating system will assign a unique set of system resources to a device. If you have ever plugged or unplugged a device on your system that is PnP compliant, the operating system will typically display a message indicating the change. In the case of a new device, the PnP operating system will install the associated driver and allocate necessary system resources.

On PnP systems, you should not manually change the system resources. If you change them improperly, it can disable your hardware and cause your computer to malfunction. Resource settings should only be changed if you are sure the new settings won't conflict with other hardware. On many older systems called "legacy" computers, you have to manually set system resources by setting jumpers or switches on the particular board itself. However, some legacy boards came with configuration software allowing you to accomplish the same task.

The system resources used by devices are as follows:
• Interrupt Request (IRQ)

- Input/Output ports
- Base memory address range
- Direct Memory Access (DMA)

The operating system and devices communicate using **interrupts**. An **Interrupt Request (IRQ)** occurs when a signal is sent by a device to the processor when it needs attention. For example, each time you press the Enter key on the keyboard, you are interrupting the processor.

There are two types of interrupts. A **hardware interrupt** is an interrupt requested by a hardware device such as a disk drive or printer. A **software interrupt** occurs when application software requests use of a device. The IRQ numbers do not change for either type of interrupt. For example, when you want to print or save a file from within an application such as MS-Excel, a software interrupt occurs. Interrupts are sent over hardware lines, or wires, that interconnect the components. There are 16 such lines that are labeled 0 through 15. Table 2.3 lists the IRQs and the devices that use them.

Table 2.3: TIRQ and Device Assignments

IRQ	Device
0	System timer
1	Keyboard
2	Cascade, or bridge, to IRQs 8–15
3	COM2 and COM4 (serial ports)
4	COM1 and COM3 (serial ports)
5	LPT2 (parallel port)
6	Floppy
7	LPT1 (parallel port)
8	Real-time clock
9	Available
10	Available
11	Available
12	Mouse
13	Math coprocessor
14	Primary hard drive
15	Secondary hard drive

When an interrupt occurs, an **interrupt handler** goes to work. The interrupt handler is software that takes care of the processor interruption. The interrupt handler uses an **Interrupt Vector Table** to locate the software, whether it is the BIOS or a device driver stored in memory.

Think of the Interrupt Vector Table as a table that has pointers to memory locations for the driver needed. This table is also known as the **dispatch table**.

Let's look at an example. Suppose you have a printer connected to your parallel port, LPT1. You have your resume opened within an application on your computer and you want to print it. This will trigger an interrupt to occur. Referring back to Table 2.3, LPT1 uses IRQ 7. Once the interrupt occurs, the interrupt handler refers to the Interrupt Vector Table to locate the address of the device driver for the printer. You can see this in Figure 2.21. In the first table of Figure 2.21, you can see that memory address A7 is the location of the driver for IRQ 7. In the second table of the figure, the actual memory location is referenced and the appropriate driver is executed for the device.

Figure 2.21: How the Interrupt Handler Uses Interrupt Vector Table to Locate Device Instruction

IRQ	Memory Address Location
0	A0
1	A1
...	...
7	A7
...	...
15	AF

Interrupt Vector Table

Memory Address Locations	Contents
A0	...
A1	...
...	...
A7	Software instructions for printer.
...	...
AF	...

Snapshot of Memory

Let's look at an analogy to help you understand IRQs. Imagine the President of the United States with 16 telephones next to his desk. Each phone number is like an <u>IRQ</u> number, the President is like the <u>processor</u> and the heads of state of other countries are like <u>devices</u>. Assume the heads of state have a particular number they use. For example, the Prime Minister of Great Britain uses phone number 1, which is like IRQ 1. When the Prime Minister calls the President, phone 1 rings, interrupting the President. The President knows who it is because phone number 1 is associated with the Prime Minister of Great Britain. The President stops his business and handles the call. Once over, the President resumes his previous duties. This is similar to how devices use IRQs and how the processor responds.

Devices also use **Input/Output port** (**I/O ports**) to transfer data. An I/O port is really a memory address through which data is transferred between a device and the microprocessor. Think of the I/O port as a channel. The I/O port addresses are reserved for each device and are given in hexadecimal values. Table 2.4 contains the most common I/O ports.

Table 2.4: Common I/O Port Addresses for Devices

I/O Port Address	Device
040–043	System Timer
060 and 064	Keyboard
03F8–03FF	COM1
02F8–02FF	COM2
03E8–03EF	COM3
02E8–02EF	COM4
0378–037F	LPT1
0278–027F	LPT2
01F0–01F7	Primary Disk Controller
0170–0177	Secondary Disk Controller
03F0–03F7	Floppy Disk Controller

A device driver needs to be loaded into memory in order to be utilized. Each device is allotted an area of memory known as a **memory address range**. These memory addresses are given in hexadecimal. Figure 2.22 shows a screenshot of the System Information tool containing memory address ranges for devices.

Figure 2.22: Memory Address Ranges for Devices

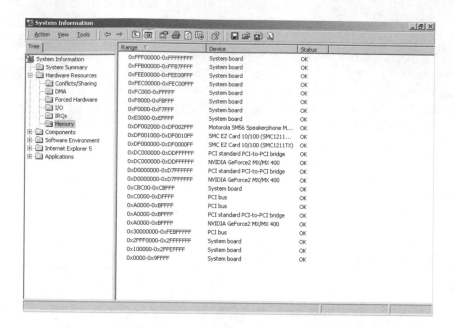

The CPU takes charge and directs the flow of traffic in and out of memory for most devices on Windows platforms. But some devices can access memory directly. This is called **Direct Memory Access** (**DMA**). DMA is most often used for data transfer directly between memory and a device such as a hard disk drive. The use of DMA improves the performance of a device.

Let's perform an exercise in to help you understand system resources. You will use the Device Manager tool in Windows 2000 to view system resources for your keyboard. The Device Manager tool is used to manage devices in Windows. This exercise assumes Windows 2000 is up and running. Follow these steps:

1. Right-click the My Computer icon on the Windows desktop and then click **Properties**. The System Properties page appears.
2. Click the **Hardware** tab and click **Device Manager**. The Device Manager page appears with your computer name.
3. Double-click your computer name until you see the keyboard icon listed.
4. Double-click the keyboard icon until you see a specific keyboard beneath the main keyboard icon. For example, the one used in this exercise is the **Standard 101/102-Key or Microsoft Natural PS/2 Keybaord**.
5. Double-click the icon for the specific keyboard type. The properties page appears.
6. Click the **Driver** tab and then click **Driver Details**. A Driver Details window appears.
7. Review the details and click **OK** to close the Details window.

8. Click the **Resources** tab.
9. Review the resource settings. Figure 2.23 contains a sample screenshot of system resources.
10. Compare the Interrupt Request value to the one in Table 2.3. It should be the same.
11. Compare the I/O Port Address to the one in Table 2.4. It should be the same.
12. Close all windows when complete.

Figure 2.23: Viewing System Resources of the Keyboard Using Device Manager

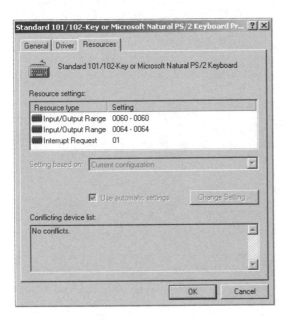

2.5 Understanding the Operating System Kernel

The kernel is the "heart" of the operating system. When any computer system boots, a set of instructions is loaded from the hard drive and kept in memory. Many of these instructions are discarded from memory after being used. The **kernel** however is the core of any operating system, and it occupies memory as long as the computer remains on. The kernel controls all other software activity. Only the most important and widely used programs are part of the kernel. The kernel calls upon programs that are held on disk or in memory by interfacing with other operating system programs and the hardware of the computer system. Most operating systems have a kernel that remains in memory. Examples include Linux, Novell NetWare, Windows NT, Windows 2000, and Windows XP.

As the kernel boots up, it looks in certain locations on disk to determine what to load. On Windows platforms, the kernel looks to the Registry. The **Registry** is a hierarchical database that contains values for specific computer settings. Figure 2.24 shows a screenshot of the Registry on a Windows 2000 computer using the **regedt32** command. The folders in the left hand pane are called **keys**. The items in the right hand pane are the

value entries with data. For example, the "Identifier" entry in Figure 2.24 has a value of "AT/AT COMPATIBLE." The term **hive** is used to represent a set of keys and values. So, in Figure 2.24, HKEY_LOCAL_MACHINE is considered a hive.

Figure 2.24: A Screenshot of the Registry Editor, regedt32

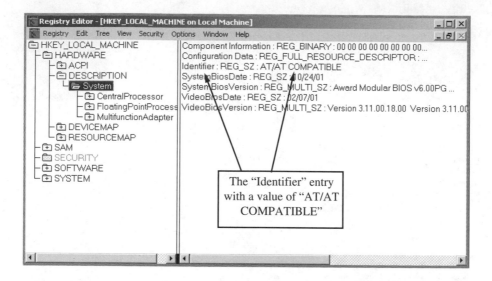

 It is not advisable to change the Registry unless you are sure about the change. Setting a key entry to an incorrect value could make your system malfunction.

The Linux operating system uses configuration files stored in various system directories to inform the kernel how and what to boot. Some of these are located in the Linux system directory named /etc. Figure 2.25 shows the Linux kernel configuration screen for changing kernel values.

Figure 2.25: The Kernel Configuration Screen for Red Hat Linux 8.0

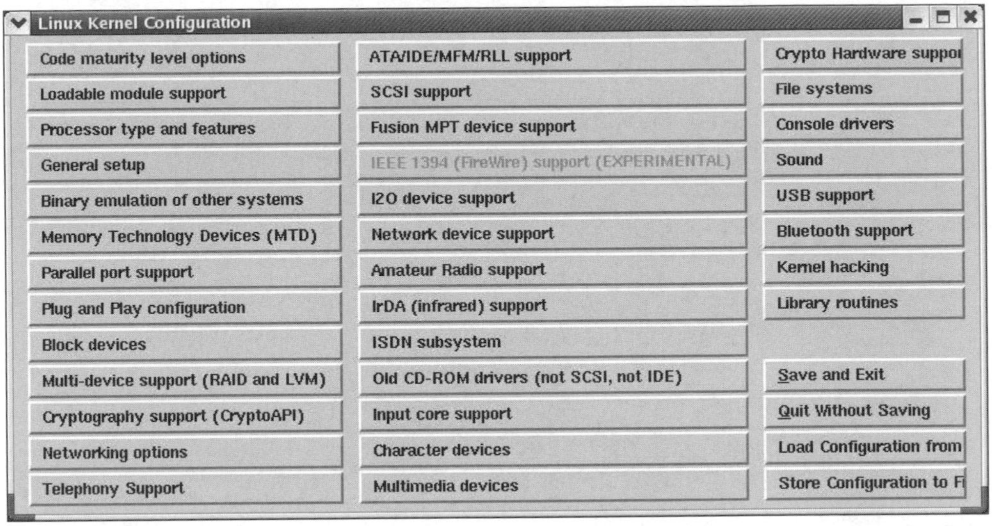

Linux Kernel Configuration		
Code maturity level options	ATA/IDE/MFM/RLL support	Crypto Hardware suppor
Loadable module support	SCSI support	File systems
Processor type and features	Fusion MPT device support	Console drivers
General setup	IEEE 1394 (FireWire) support (EXPERIMENTAL)	Sound
Binary emulation of other systems	I2O device support	USB support
Memory Technology Devices (MTD)	Network device support	Bluetooth support
Parallel port support	Amateur Radio support	Kernel hacking
Plug and Play configuration	IrDA (infrared) support	Library routines
Block devices	ISDN subsystem	
Multi-device support (RAID and LVM)	Old CD-ROM drivers (not SCSI, not IDE)	Save and Exit
Cryptography support (CryptoAPI)	Input core support	Quit Without Saving
Networking options	Character devices	Load Configuration from
Telephony Support	Multimedia devices	Store Configuration to F

An operating system kernel that can utilize several processors concurrently is called a **multi-processor** operating system. Many UNIX and Linux operating systems versions, and Windows NT/2000/XP, can utilize multiple processors. Use of multiple processors improves the performance of processor intensive applications. For example, programs that perform mathematical computations on large amounts of data for the government or engineering firms would benefit from multiple processors. The ability of a computer system to use multiple processors that can execute different portions of a program is called **Symmetrical Multi-Processing** or **SMP**. Windows and Linux operating systems support SMP.

If the operating system can support multiple users at the same time, then the operating system is a **multi-user** operating system. Mainframe, midrange, UNIX, and Linux operating systems are multi-user operating systems.

A **multitasking** operating system kernel is one that appears to handle multiple tasks at the same time. A **task** is a program that is running. A task is also known as a **job** or **process**. The operating system actually does one task at a time, but using something called time slicing the operating system gives each task a little bit of CPU time. The tasks take turns being processed in a round-robin fashion. A **time slice** is a unit of time allocated to a task.

For example, assume an operating system gives each task two minutes of CPU time—which is actually quite a bit. If a job exceeds the two-minute time slice, the job and its data are placed into the virtual memory page file. **Virtual memory** uses RAM and a section of the hard disk to accommodate multitasking and multiple users. The **page file**, or **swap file**, is the section of the hard disk used for virtual memory. Figure 2.26 shows the Virtual Memory settings on Windows 2000.

Figure 2.26: A Screenshot of the Windows 2000 Virtual Memory Settings

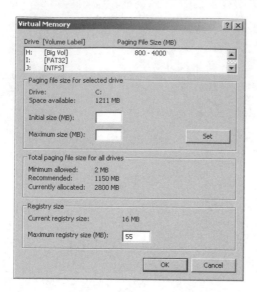

Once the job is moved to the page file, the next task in line will be processed. If this job finishes in less than two minutes, then the system can retrieve the first task from the swap area to continue processing. This is done for all jobs until there is no more work to be done. The advantage of multitasking is one large task won't bottleneck jobs requiring less time.

On Windows 2000, the page file is named Pagefile.sys. The minimum size of the file is 2 MB and the recommended page file size is 1.5 times memory.

There are two variations of multitasking—preemptive and cooperative. With **preemptive multitasking**, the operating system has the ability to take control of the computer system from an application. With **cooperative multitasking**, the application takes control of the system resources. Preemptive multitasking is much more effective because if an application fails, the system won't necessarily crash. In cooperative operating systems, a failed application can cause the whole computer to freeze or crash.

People multitask using time slicing all the time. For example, you may be busy studying at home when the phone rings. While you are on the phone, the doorbell rings and the dryer shuts off with a beep. You manage each of these processes by putting the person on the phone on hold, going to the door, checking the dryer, etc., and continuing to give little slices of time to each process until they are all handled.

Let's briefly discuss real and protected mode operating systems and software. An operating system can operate its application software in one of two modes: **real** (also called 16-bit mode) or **protected** (also called 32-bit mode). The term 16-bit means the software was developed using a 16-bit processor such as the Intel 80286. The term 32-bit means the software was developed using faster 32-bit processor such as an Intel Pentium processor. The features of real mode are as follows:
- The single application can access hardware directly—"real" access.
- It supports only single-task (one application) execution.
- The processor can only access up to 1 MB of memory (it can access more than 1 MB with Extended Memory Specification or XMS (this will be discussed in Chapter 3.)
- The failure of a single application can cause the whole system to fail.
- It is slower than 32-bit processors and operating systems.
- It is written for 16-bit processors (processors with 16 physical bus lines), such as the Intel 8088.

Some of the features of protected mode are as follows:
- The failure of a single application does not necessarily cause the other applications (or system) to fail—they are "protected" from one another.
- It supports multi-tasking (multiple applications).
- The processor can access up to 4 GB of memory.
- It is faster than 16-bit processors and operating systems.
- It is written for 32-bit processors (processors with 32 physical bus lines), such as the Intel 80286, 80386, 80486, and Pentium class of processors.
- The applications cannot access hardware directly.

In real mode, the application software has direct (hence the term "real") access to the hardware (such as memory). In real mode, the failure of a single application can cause the whole computer to fail. The reason is the application can access hardware directly and could potentially overwrite the memory address space occupied by operating system programs. Native DOS runs in real mode.

In protected mode, the application software does not have direct access to the hardware. Also, multiple applications are assigned their own memory address space and are "protected" from one another. Thus, the failure of one application does not necessarily mean another will fail. So, other applications are protected and so is the hardware. Windows 98, NT, 2000, and XP boot into real mode and then run in protected mode. This is why you can end one task without affecting another while in those operating systems.

Application software and device drivers are written for either a 16-bit or a 32-bit operating system. The 16-bit applications and device drivers run on DOS and Windows 3.1 (16-bit operating systems). These applications operate in real mode and could potentially fail causing complete system failure. The 32-bit applications and device drivers run under Windows 98, NT, 2000, and XP (32-bit operating systems). These applications operate in protected mode. A 16-bit operating system cannot run 32-bit software. The 32-bit operating systems can run a 16-bit application in an **emulated** environment, which means that the 16-bit application "thinks" it is running on a 16-bit operating system. However, a 32-bit program actually runs the 16-bit application.

Chapter Summary

- Operating systems facilitate user interaction with a Command Line Interface (CLI) or a Graphical User Interface (GUI). For most users, the GUI is user-friendlier than the CLI. You must learn the command syntax when executing commands in the CLI. In the GUI, you point and click on programs you want to execute.
- Operating systems provide application management by accessing hardware for applications, allowing applications to be installed or uninstalled, and providing a consistent interface among applications with an Application Programmer Interface (API). Microsoft and IBM developed the Common User Access (CUA) standard, which makes applications consistent for users.
- The operating system manages the file system for both users and applications. The file system defines the organization of the files and folders on disk and is created when you format a partition. The file system on Windows and Linux operating systems are hierarchical.
- The operating system also manages the hardware. The Basic Input/Output System (BIOS) is firmware that guides basic input and output functions of the PC. Booting the operating system means loading the operating system into memory so you have a functional system. System resources control how the devices on a computer work. Examples of system resources are: Interrupt Requests (IRQs), I/O ports, base memory address, and Direct Memory Access (DMA).
- The operating system kernel is the core set of instructions that remain in memory once the system is booted. On Windows, the kernel looks at the Registry for information on how and what to load. On Linux, the kernel looks to configuration files for this information.

Review Questions

1. What is the CLI called in the Linux operating system?
   a) DOS
   b) Shell
   c) ASCII
   d) Compiler

2. What is the name of the program to get to the command line on Windows 98/NT/2000/XP?
   a) Command
   b) Firmware
   c) COBOL
   d) Desktop

3. What is the area on the screen where a user interacts with the GUI of an operating system?
   a) Thread
   b) Firmware
   c) Desktop
   d) ASCII

4. What occurs when a program tries to occupy the same area of memory as another program?
   a) Thread
   b) Firmware
   c) BIOS
   d) GPF

5. Which of the following is an executable file that consists of functions and data?
   a) Kernel
   b) JAVA
   c) BIOS
   d) DLL

6. Which of the following is a standard written by Microsoft and IBM to make Windows programs consistent?
   a) CUA
   b) Firmware
   c) BIOS
   d) API

7. Which of the following is used to guarantee similar interfaces among programs?
   a) CUA
   b) Firmware
   c) BIOS
   d) API

8. Which of the following is a number used to reference a process?
   a) PID
   b) CUA
   c) BIOS
   d) API

9. Which of the following is a basic unit of instruction allocated processor time by the operating system?
   a) CUA
   b) Thread
   c) BIOS
   d) API

10. Operating systems supporting _____ allow threads to be executed at the same time.
    a) Virtual memory
    b) Multitasking
    c) Multithreading
    d) APIs

11. The _____ defines the organization of the files and folders stored on disk.
    a) File system
    b) Active partition
    c) Boot partition
    d) System partition

12. The _____ contains your operating system and is the very first partition on the drive.
    a) File system
    b) Active partition
    c) Primary partition
    d) System partition

13. On Windows NT/2000/XP, you can have up to _____ primary partitions.
    a) 1
    b) 2
    c) 3
    d) 4

14. What is the name of the record that tells the operating system about the partitions?
    a) System partition
    b) MBR
    c) CUA
    d) API

15. On Windows operating systems, the file system stores a file in _____.
    a) Clusters
    b) Other files
    c) CUA
    d) API

16. _____ are the concentric circles on disk.
    a) Clusters
    b) Tracks
    c) Sectors
    d) Allocation units

17. The drive letter and the colon are known as the _____.
    a) Drive specification
    b) Extension
    c) Folder
    d) Cluster

18. What is used to associate a file to an application?
    a) Extension
    b) Tracks
    c) Folder
    d) CUA

19. In DOS, file names must follow the _____ rule.
    a) Extension
    b) 255
    c) 8.3
    d) Fragment

20. A file that is stored in non-adjacent clusters is _____.
    a) Extended
    b) Multithreaded
    c) Fragmented
    d) Paged

Lab Projects

Note that these projects assume the PC is powered on. Although you can perform these projects in other versions of Microsoft Windows with minor changes, the assumed operating system is Windows 2000 unless otherwise specified. You are logged in as the Administrator user.

Lab Project 1

The goal of this project is to help you understand how to view system information. In this project, you will view both hardware and software information.

1. Click **Start** and point to **Programs**, **Accessories**, and **System Tools**. Then click **System Information**. The System Information window opens.

2. In the left hand windowpane of System Information, click the **System Summary** folder. Note: This folder may already be opened. In the right hand windowpane Item names and their Values appear in columns.

3. Record the operating system Name, Version and System Manufacturer.

   _____

4. Choose two other Items and Values and record them.

   _____

5. In the left hand windowpane of System Information, double-click the **Hardware Resources** folder. Additional folders appear.

6. Click the **IRQs** folder. The list of IRQs on the computer appear.

7. Click the **IRQ Number** column to sort the IRQ numbers from lower to higher.

8. Record the IRQs for COM1, COM3, and the Primary IDE controller.

   _____

9. Choose two other devices and record their IRQs.

   _____

10. Double-click the **Hardware Resources** folder to close it.

11. Double-click the **Software Environment** folder. Additional folders appear beneath the Software Environment folder.

12. Click **Loaded Modules**. A list of modules appears. Notice some of the modules are DLLs due to the .dll extension. Refer to Figure 2.27 for a screenshot.

Figure 2.27: Viewing Loaded Modules in System Information

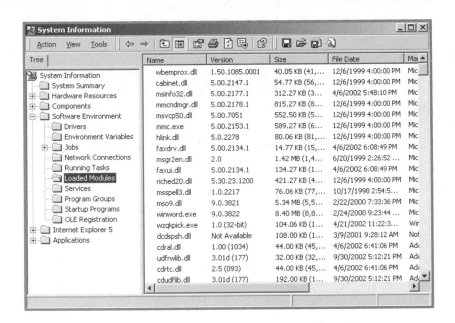

13. Record the name, version, size, file date, and path of three modules. Note you may have to scroll to the right to locate the path.

_____

_____

_____

14. Close System Information.

Lab Project 2

The goal of this project is to help you understand how a task impacts hardware resources. In this project you will use Task Manager on Windows 2000 to view process information of the Calculator program. Because the Calculator program is processor intensive, you will perform calculations causing the CPU usage to increase.

1. Click **Start** and then click **Run**. The Run dialog box appears.

2. Type **calc** and press Enter (or click **OK**).

3. The Calculator program window appears. Keep the window open.

4. Execute Task Manager by pressing Ctrl + Alt simultaneously and then Del. The Windows Security dialog box appears.

5. Click the **Task Manager** button. The Windows Task Manager program appears.

6. Click the **Processes** tab. Scroll until you locate the calc program in the Image Name column. It should be named calc.exe.

7. Record the PID, the CPU, CPU Time, and Mem Usage values for calc.

8. Click the **Performance** tab in Task Manager.

9. Go back to the Calculator program by clicking its name in the Title Bar of the Window. Click **View** and the click **Scientific**. To view both windows on the screen, you may have to move them around.

10. In order watch the CPU Usage increase, click the number **9** and then click the **x^3** key 10 times.

11. Notice the CPU Usage in Task Manager reaches 100% as shown in Figure 2.28.

12. To end the calculator program within Task Manager, click **Applications**.

13. Locate the Calculator and click **End Task**.

14. Close Task Manager.

Figure 2.28: Viewing CPU Usage Reaching 100%

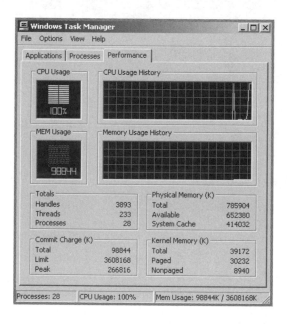

Lab Project 3

The goal of this project is to help you understand how to analyze and defragment a disk partition. In this project you will use the Disk Defragmenter in Windows 2000 to analyze and defragment a disk. An analysis is performed to determine if defragmenting the drive is cost-effective.

1. Click **Start** and point to **Programs, Accessories**, and **System Tools**. Then click **Disk Defragmenter**. The Disk Defragmenter program opens.

2. Click the C: drive and click **Analyze**. The analysis occurs. When the analysis is done an Analysis Complete window appears.

3. Click **View Report**. Analysis Report opens.

4. Record whether your drive needs to be defragmented.

   _____

5. In the Volume Information section of the report, record the total fragmentation and file fragmentation percentages. You may need to scroll up and down to locate these.

   _____

6. In the Most fragmented files section, record the file name with the most fragments, its size, and total number of fragments. You can click the Fragments column name to sort the listing from higher to lower.

   _____

7. Close the Analysis Report.

8. Whether recommended or not, defragment your drive.

   _____

9. Record the total fragmentation now.

   _____

10. Record the difference in fragmentation before and after.

    _____

11. Close all windows.

Lab Project 4

The goal of this project is to further your understanding of the file system hierarchy. In this project you will create folders and files. Although the figure in this project is taken from a Windows 2000 PC, you could perform this project on Windows 95 or higher.

1. Insert a formatted floppy in the floppy drive.

2. Create the file system hierarchy shown in Figure 2.29. You will have to use the **File**, **New** command to create the folders and files. The MIS, Payroll, and Sales folders are immediately beneath the floppy (the root folder). MIS contains two files: Project1.txt and Project2.txt. Payroll contains two folders: Paychecks and Managers, which are empty. Sales contains these folders: Eastern and Western. The Eastern folder contains these files: Jan.txt, Feb.txt, and Mar.txt. The Western folder is empty.

3. Close all windows.

Figure 2.29: A Screenshot of the Partial Hierarchy for Project 4

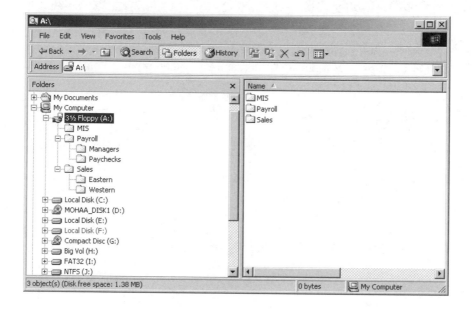

Lab Project 5

The goal of this project is to help you understand file system formats. In this project, you will view disk property information and convert the FAT file system to NTFS. The drive letter used in this project is K: but another drive letter will suffice as long as it is FAT. Once you change the drive to NTFS, you cannot change it back to FAT.

1. Double-click the **My Computer** icon on your desktop. The My Computer window appears with various devices and folders on the system.

2. Right-click on the K: drive or another driver spec as long as it is FAT.

3. Record the file system format (although you know it is FAT) and the used and free space on the partition.

---

4. To access the command line in order to convert the drive using the **convert** command, click **Start**, and then click **Run**. In the Run dialog box, type **cmd** and press Enter or click **OK**. The Windows command line prompt appears.

5. At the prompt, enter **convert *drive_spec:* /fs:ntfs** and press Enter. Note that you will supply your own specific drive letter in the place of the drive specification. The option **/fs:ntfs** indicates to convert to the NTFS file system.

6. Figure 2.30 shows a sample screenshot. In the figure, drive k: is converted.

Figure 2.30: A Screenshot of Converting FAT to NTFS

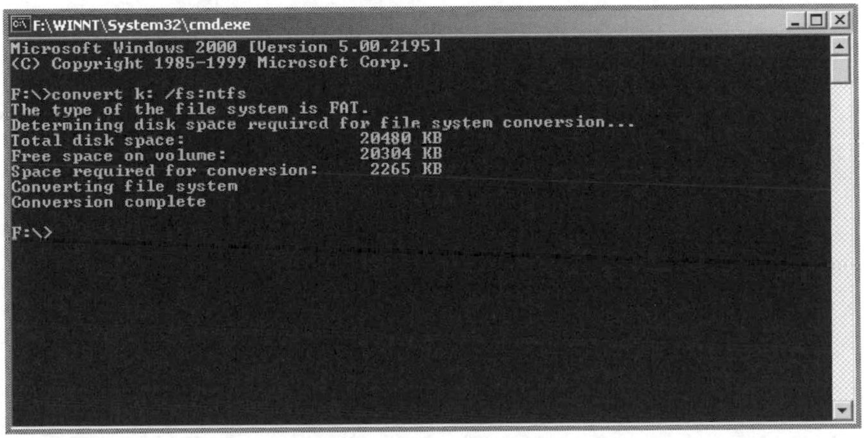

7. View the file system format again.

8. Now record the file system format and the used and free space on the drive.

---

9. Close all windows.

Lab Project 6

The goal of this project is to help you understand Device Manager in Windows 2000. In this project, you will scan for Plug and Play (PnP) devices and view device information.

1. Right-click the **My Computer** icon on your desktop and click **Properties**. The **System Properties** window appears with various tabs. The **General** tab is selected by default.

2. Record the System information.

   _____

3. Record the Computer information.

   _____

4. Click the **Hardware** tab. Additional command buttons appear.

5. Click **Device Manager**. The Device Manager window appears.

6. To scan for Plug and Play devices, right-click the computer's name icon and click **Scan** for hardware changes. Another box appears indicating it is scanning.

7. Record whether or not it found any new PnP hardware.

   _____

8. View three devices other than the keyboard (because it was viewed in a chapter exercise).

9. Record whether the device is working properly, driver details, and any resource information.

   _____

   _____

   _____

10. Close all windows.

Lab Project 7

The goal of this project is to help you understand how to view and modify performance settings on your computer.

1. Right-click the **My Computer** icon on your desktop and click **Properties**. The System Properties window appears with various tabs.

2. Click the **Advanced** tab. Additional command buttons appear.

3. Click **Performance Options**. The Performance Options screen appears.

4. Record the Application Response setting.

   _____

5. Click **Change**. The Virtual Memory window opens.

6. Scroll in the top portion of the window until you see a drive letter with a paging file.

7. To highlight the drive spec with a page file, click the driver letter for a drive that has a page file. If you only have one partition, it may already be highlighted. The initial and maximum size entry values appear in the middle of the screen.

8. Record the initial and maximum size of the paging file.

   _____

9. Record the Registry's current size and maximum.

   _____

10. Now, let's increase the paging file size. Add 5 to the maximum page file size.

11. In the Maximum size, enter the new value.

12. To set the new maximum, click Set.

13. Record the new maximum.

   _____

14. Click OK three times. Restart your computer, if prompted to do so. You have increased your page file size.

Lab Project 8

The goal of this project is to help you understand the Registry on Windows 2000. In this project, you will see what happens when you change a Registry setting.

1. Click **Start** and then click **Run**. The Run dialog box appears.

2. To open the Registry Editor, type **regedt32** and press Enter (or click **OK**). The Registry Editor opens.

3. Double-click **HARDWARE**. The HARDWARE key opens.

4. Double-click **DESCRIPTION**. The DESCRIPTION key opens.

5. Double-click **System**. The System key opens with values in the right hand windowpane.

6. Double-click the **Identifier** item. A box appears with the value.

7. In order to change the value entry, type **your_name** and press **OK**, where *your_name* is your own name.

8. Refer to Figure 2.31 for a sample screenshot.

Figure 2.31: A Screenshot of the Registry After Changing the Identifier Key

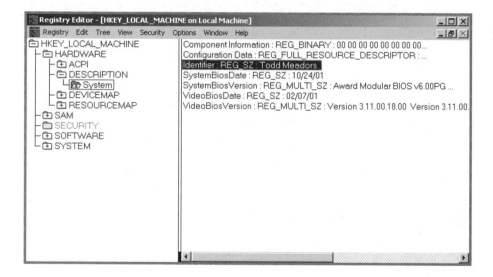

9. Close the Registry Editor.

10. To verify the change, right-click the **My Computer** icon on your desktop and click **Properties**. The **System Properties** window appears with various tabs. The **General** tab is selected by default.

11. Record the Computer information.

_____

12. Notice how this differs from the Computer information you recorded in Step 3 of Project 6.

13. Close all windows.

Lab Project 9 Challenge

The goal of this project is to further your understanding of the file system hierarchy. In this project, you will implement a file and folder hierarchy in Linux.

1. Repeat Project 4 using the Linux operating system GUI.

2. Close all windows when complete.

Lab Project 10 Challenge

The goal of this project is to help you understand how cluster sizes vary on partitions for various file system types. You will need Windows 2000 and between 4 and 6 GB of unallocated free disk space.

1. Open Disk Management located in Administrative Tools under Control Panel.

2. Create three 1,000-byte partitions and format them as FAT (for FAT16), FAT32, and NTFS.

3. Create a 1-byte file in each partition named a.dat.

4. View the properties of the file on the FAT16 partition.

5. Refer to Figure 2.32 for a sample screenshot of the Properties page for the file named a.dat.

Figure 2.32: A Screenshot Showing File Size and Size on Disk

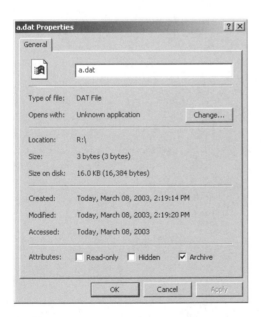

6. Notice the Size and Size on Disk amounts in Figure 2.32. The Size amount for the FAT16 partition is 3 bytes. Although there is only 1 byte in the file, the extra 2 bytes are overhead. Included in this overhead is the EOF character sequence. The Size on Disk amount

for the FAT16 partition is 16 KB. Note the Size on Disk amount is the cluster size. There is quite a bit of wasted disk space (16, 384 − 3 = 16,381 bytes wasted).

7. Record the Size and Size on Disk amounts for a.dat on the FAT32 partition.

_____

8. Record the Size and Size on Disk amounts for a.dat on the NTFS partition.

_____

9. Compare the Size on Disk values to Table 2.2.

10. Record how they relate.

_____

_____

11. Delete the partitions.

12. Create three 1,025-byte partitions and format them as FAT (for FAT16), FAT32, and NTFS.

13. Create a 1-byte file in each partition named a.dat.

14. Record the Size and Size on Disk amounts for a.dat on the FAT16 partition.

_____

15. Record the Size and Size on Disk amounts for a.dat on the FAT32 partition.

_____

16. Record the Size and Size on Disk amounts for a.dat on the NTFS partition.

_____

17. Compare the Size on Disk values to Table 2.2.

18. Record how they relate.

_____

_____

19. Close all windows.

Internet Discovery

Internet Discovery Lab 1

The goal of this Internet Discovery Lab is to help further your knowledge of chapter terms by accessing the Webopedia web site.

1. Connect to the Internet.
2. Open your web browser.
3. In the address box, type **http://www.webopedia.com** and press Enter.
4. Locate the search box, enter the following terms, and write a brief synopsis of each term.

    a. Multitasking

    _____

    _____

    _____

    b. Partition

    _____

    _____

    _____

    c. Fragmentation

    _____

    _____

    _____

    d. Cluster

    _____

    _____

    _____

Internet Discovery Lab 2

The goal of this Internet Discovery Lab is to have you obtain a copy of the Linux operating system that boots on a CD. This version is called Knoppix.

1. Connect to the Internet.
2. Open your web browser.
3. In the address box, type **http://www.knoppix.com** and press Enter.
4. Download the Knoppix version or order a copy. Ordering a copy costs under $30.00. Downloading is free but you will need a CD Writer and a 700 MB CD.
5. Once you obtain the operating system, boot the CD and try out the GUI. Notice the differences and similarities between Linux and Windows.

Soft Skills: Help Desk Support

1. A user calls you and indicates that opening files takes longer than it used to take. What suggestions do you offer them as to what could be happening?
2. You receive an e-mail from a user asking you to compare the FAT and NTFS file systems because his or her manager wants to know the benefits. Summarize the differences.
3. A user wants to add a new device to his or her system and calls you for assistance. What solution do you suggest?
4. You get paged in the middle of the night from a user because the MBR on his or her computer has become corrupt. What do you tell the user to do?

Critical Thinking

1. If a file can become fragmented, do you think a folder can become fragmented? Why or why not?
2. Assuming the file system table was not present yet everything else still functioned normally, how do you think the operating system would go about locating a file you wanted to open?
3. Referring to Table 2.2, why do you think the minimum cluster size is equal to 512 bytes?
4. Explain the pros and cons of a fictitious cluster size of 1 byte.

 Study Skills

Reinforcements are on the way!

- Read your textbook—again! This provides wonderful reinforcement of the content of a chapter.
- Summarize the chapter in your own words. By doing this, you will go over the material at least twice. This also provides great reinforcement of the material.
- Do all homework assignments. Doing your homework will improve your test score results.
- Do all Lab Projects because that is where learning really occurs. Plus, you'll have fun doing them.
- Repeating Lab Projects provides invaluable reinforcement.
- Understand the analogies. Your instructor may come up with good everyday analogies that help "hook" the material in your brain. An example of an analogy is the garden store tree analogy discussed in this chapter. You can also come up with your own analogies to help you understand a topic.
- Review the tips embedded within a chapter to further your understanding of the finer points of a topic.
- Search the Internet for study questions on the A+ test. Again, reinforcement!
- Ask the instructor questions about material that you don't understand. If you are afraid to ask in class, ask your question in private. Your instructor is there to help!

Study Skills: Self-Study Question(s)
1. Identify at least one reinforcement technique you performed today!
2. Identify at least one analogy that reinforced your knowledge. (You **cannot** use the garden store tree analogy.)

3

Chapter 3
DOS and the DOS
Command Line
Interface

OBJECTIVES

The goal of this chapter is twofold:
- To introduce you to DOS and the DOS Command Line Interface (CLI)
- To help you prepare and pass the following sections of the A+ Operating System Technologies Exam:

A+ Operating System Technologies Exam Objectives
covered in this chapter (and corresponding page numbers)

In this chapter, you will complete the following sections:
- 3.1 Understanding DOS and the DOS Command Prompt in Windows
- 3.2 Understanding the DOS Tree Structure
- 3.3 Understanding the Types of DOS Commands
- 3.4 Understanding Drive Related Commands
- 3.5 Learning about Wildcard Characters
- 3.6 Learning about Redirection and Filtering
- 3.7 Learning about Batch Files
- 3.8 Learning about Additional Commands

3.1 Understanding DOS and the DOS Command Prompt in Windows

Microsoft developed the Microsoft Disk Operating System or MS-DOS in the early 1980s for the IBM PC. **MS-DOS**, sometimes called DOS, is a single-user, single-tasking operating system. This means only one user can do only one task at a time. If you printed a document in DOS, you can't do anything else until the document is through printing. You cannot load several applications at once like you can on Windows-based platforms.

DOS is an operating system comprised of various programs that help it control the overall operation of the PC. The basic categories of DOS software are:

1. **System programs and files**—used to start DOS during the boot process. Examples include IO.SYS, MSDOS.SYS, and COMMAND.COM. These are sometimes called the **boot files**.
2. **File system management programs**—used to manage the directories and files. Examples include COPY, DIR, and CD.
3. **Utilities**—used to augment the operating system. Examples include FDISK and FORMAT.

Let's look at system programs in more detail. File system management programs and utilities will be discussed in later sections.

As the PC is powered on and passes its Power-On Self Test (POST), it begins to load the operating system into memory. MS-DOS uses three system files for booting. The files are:

IO.SYS
MSDOS.SYS
COMMAND.COM

The first two, IO.SYS and MSDOS.SYS, are hidden to prevent accidental deletion. Without these files, DOS won't boot. These two files provide generic input/output and file management functionality. You cannot even see the files without doing a little extra work. However, if you run the DOS ATTRIB command you will be able to see that they do exist. Figure 3.1 shows a screenshot of the ATTRIB command displaying these files. The ATTRIB command is used to view and set file attributes. In DOS, a file can have these four attributes: Read-Only, Archive, System, and Hidden. You will learn about more attributes and the ATTRIB command later in this chapter. In Figure 3.1, you also see another hidden file named DRVSPACE.BIN commonly used for disk drive compression to save disk space.

The system program named **COMMAND.COM** is known as the **command interpreter**. It is the last boot file listed in Figure 3.1. It gives you the DOS prompt and is responsible for processing every command typed into the computer. In Figure 3.1 the prompt **A:\>** appears to the left of the ATTRIB command. Remember from Chapter 2, this references the floppy.

Figure 3.1: Viewing the System Program Files on a DOS Disk

```
A:\>attrib
A   SHR       A:\IO.SYS
A   SHR       A:\MSDOS.SYS
A   SHR       A:\DRVSPACE.BIN
A             A:\COMMAND.COM

A:\>_
```

You can create a DOS boot disk by running the SYS command. The SYS command transfers the DOS system program files and command interpreter to the drive specified. So SYS A: makes a floppy bootable. You can also create a DOS boot disk by running the FORMAT command using the /S option. The command FORMAT A: /S will format the floppy and make it bootable. The difference between SYS and FORMAT /S is the SYS command does not format the disk. So, if you have a floppy that has files you want to keep on it, and you want to make it bootable, use the SYS command.

If COMMAND.COM is missing from your disk, you will receive an error on the screen. The error displays a message such as "**Bad or missing Command Interpreter**" during the boot process. This indicates that COMMAND.COM is either corrupt or indeed missing. To fix this problem, you need to get a copy of COMMAND.COM on the boot disk. You can do this by typing the SYS command or using the /S option on the FORMAT command.

There are two other optional system files. They are called **AUTOEXEC.BAT** and **CONFIG.SYS**. The AUTOEXEC.BAT file is a batch file. A batch file contains DOS commands and its last three characters are BAT for "batch." When creating batch files, you are creating a program that is interpreted. This file is used to customize your boot process. Because it is a batch file, it can be executed anytime. The following is a partial AUTOEXEC.BAT file:

PROMPT PG
PATH C:\DOS;C:\WINDOWS

The first line sets the prompt and the second line allows DOS commands to be executed anywhere in the tree structure. If you had Windows 3.1 installed, you could load it into memory by placing the statement WIN (to load WIN.COM) as the last statement in the AUTOEXEC.BAT file.

The CONFIG.SYS file is a configuration file that has entries for device drivers such as memory, the mouse, and the CD-ROM. The CONFIG.SYS file is read only during the boot process. If you change it, you must reboot for the changes to take effect. A partial CONFIG.SYS file looks like this:

DEVICE=C:\DOS\HIMEM.SYS
DOS=HIGH
FILES=30

The first line allows use of Extended Memory, the second line loads DOS into High Memory and the last line sets the number of concurrent open files to 30. These types of memory will be discussed next.

If the statement has an equal sign, it generally goes in CONFIG.SYS. Device driver and other variable references go in CONFIG.SYS. Commands go into AUTOEXEC.BAT. You must reboot your system in order to use the changes made to CONFIG.SYS. You can simply enter AUTOEXEC to implement changes made to AUTOEXEC.BAT.

You need to be aware of the DOS system file load order for troubleshooting purposes and to help you with the A+ OS Technologies exam. The MS-DOS system files are loaded in the following order:

1. IO.SYS
2. MSDOS.SYS
3. CONFIG.SYS
4. COMMAND.COM
5. AUTOEXEC.BAT

During the boot sequence, you can press the **F5** key to bypass executing CONFIG.SYS and AUTOEXEC.BAT. You can press the **F8** key and DOS will prompt you to confirm executing each line of CONFIG.SYS and AUTOEXEC.BAT. The use of these keys allows you to test different configurations on your PC when running DOS.

For example, pressing **F8** would show a confirmation line such as the following:

MS-DOS will prompt you to confirm each CONFIG.SYS command.

DOS=HIGH [Y,N]?

You would then enter Y to load DOS high. If you entered N, DOS would not be loaded high.

During the boot process, DOS is loaded into **conventional memory**. Conventional memory, or base memory, is memory in the range of 0 to 640 KB. This is where MS-DOS and other applications reside. Figure 3.2 shows a diagram of the types of memory used by DOS.

Figure 3.2: The Memory Ranges Used by MS-DOS

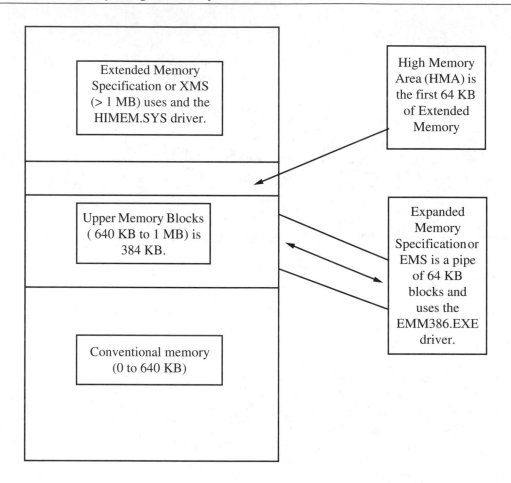

DOS uses **upper memory** for the system BIOS, video BIOS, and other functions. Upper memory is addressed in blocks known as **Upper Memory Blocks** or UMBs and is sometimes called reserved memory. This is the 384 KB of memory from 640 KB to 1 MB. You can make UMBs available to DOS by placing this line in CONFIG.SYS:

DOS=UMB

If you ever get the following message, then check your floppy drive. You may have placed a non-system disk in the drive. Remove or replace it and press any key to continue booting.

Invalid system disk

Replace the disk, and then press any key

Over time, software demanded that hardware run more effectively. With the release of MS-DOS Version 5.0, the use of **Extended Memory Specification** or XMS was developed to break the 640 KB barrier. (You can determine the version of your operating system by running the VER command.) With XMS, DOS uses additional memory in the PC. Applications and data can now be loaded in the memory range made available by XMS. XMS requires a device driver named HIMEM.SYS to use it. You place the following in the CONFIG.SYS file:

DEVICE=C:\DOS\HIMEM.SYS

Lotus, Intel, and Microsoft developed a standard called **Expanded Memory Specification** or EMS. This standard is called the LIM (the initials of Lotus, Intel, and Microsoft) EMS. EMS uses a technique where 16 KB blocks of data are transferred in and out of a reserved 64 KB section of upper memory. If you think back to Chapter 2, this is similar to how virtual memory swaps programs and data between memory and the page file on disk. In order to use EMS, you have to use a device driver named EMM386.EXE and place the following in the CONFIG.SYS file:

DEVICE=C:\DOS\EMM386.EXE ON

The option **ON** following the device driver name activates it.

Another type of memory called the **High Memory Area** or HMA is the first 64 KB block of memory in extended memory. HMA actually came about as a result of a software bug. Once it was discovered, DOS was retrofitted to utilize it. DOS can be loaded into HMA thus freeing conventional memory for use by applications and programs. Extended memory will generally activate HMA, but you could do it by placing the following in CONFIG.SYS:

DOS=HIGH

You can combine the HIGH and UMB option on the DOS= statement by adding this entry to CONFIG.SYS to load DOS high and create UMBs:

DOS=HIGH, UMB

Programs such as HIMEMS.SYS, EMM386.EXE, COMMAND.COM, and IO.SYS are called **Terminate and Stay Resident** or TSR programs. This means they execute, terminate normally, and then stay resident in memory until needed by the operating system.

You can think of a TSR as executing and then "going to sleep" until needed. When an activity they perform is required, they "wake up" and go to work. Once the activity is completed, they "go back to sleep" until the next time they are needed. Other programs such as DIR or ATTRIB execute, terminate, and do not stay resident.

Originally Microsoft did not think users would need more than 1 MB of memory. Nowadays, many GUI operating systems need a minimum of 32 MB just to run properly.

Table 3.1 summarizes the types of memory used by MS-DOS as discussed in this section.

Table 3.1: Types of Memory Used by MS-DOS

Type of Memory	Description
Conventional Memory	The range of memory from 0 to 640 KB used by DOS and applications.
Upper Memory Blocks (UMB)	The range of memory is reserved for video memory and other needs.
High Memory Area (HMA)	The first 64 KB of extended memory used by DOS.
Extended Memory Specification (XMS)	Extended memory is memory above 1 MB used by applications.
Expanded Memory Specification (EMS)	Blocks of data are swapped in and out of a 64 KB section of memory between UMBs and EMS.

There are two different modes where you can execute DOS commands at a command line prompt. They are:
- Within the DOS operating system—the PC is booted using DOS.
- Within the DOS command prompt of a Windows operating system—the PC is booted using a Microsoft Windows operating system such as Windows 95/98/2000/XP.

Using the first method, DOS is booted and there is no Graphical User Interface. With the second method, you enter the name of the command interpreter to go into DOS mode. It is called COMMAND.COM on Windows 95/98. On Windows NT/2000/XP, it is called COMMAND.COM or CMD.EXE.

You must learn the DOS commands and command line syntax regardless of version used. Most commands, such as CD and DIR, work across all operating systems revisions. But not all commands exist in all revisions of DOS. Likewise, not all commands exist in all versions of the DOS command prompt mode provided by each Windows operating system revision. For example, the TREE command exists in DOS 6.22, but not in Windows 98. Microsoft included it in Windows XP.

You may be wondering why you need to spend time learning MS-DOS. After all, using the GUI of an operating system is so much easier than learning DOS commands. Besides laying the fundamental groundwork, an actual experience may offer you a good reason. Recently, one of the author's PCs would not boot. A backup had not been performed in about a week—not a good situation. Pressing F8 during the boot process of Windows 2000 to go to Safe Mode did not work. Finally, the author booted Windows 2000 to "Safe Mode with Command Prompt" and used the DOS COPY command to copy important files to the floppy, A:, and then to another system. The PC had to be completely reinstalled. Although the data was intact after reinstalling, without understanding how the DOS commands and command line worked, the important data files on the PC might have been lost.

Note: The remaining sections in this chapter will cover commonly used DOS-related commands available to most of the Microsoft operating systems. Discussion includes use of commands actually within DOS or at the DOS command prompt mode of Windows 98/2000/XP. It does not really matter how you get to the command prompt; what matters is understanding the purpose of the commands and topics discussed. Additionally, because the command prompt can be changed, it is assumed that the prompt is the default prompt in all exercises and lab projects throughout this chapter. Thus, the current drive and directory is assumed for the command prompt.

Let's perform an exercise to get you started with DOS. This exercise assumes a PC is up and running a Microsoft operating system with access to DOS or the DOS command prompt.

Because there are several methods you can use to get to the DOS command prompt, choose an appropriate method for the operating system you are using.

A. DOS: You are already at the command prompt.

B. Windows 98:

Method 1: Click Start, point to Programs, and then click the MS-DOS Prompt option.

Method 2: Click Start, click the Run option, and type in COMMAND at the dialog box and press Enter.

Method 3: During start-up, hold down either the Ctrl key or press the F8 key to access the Command-Prompt-Only mode.

C. NT Workstation:

Method 1: Click Start, point to Programs, and then click on the Command Prompt option.

Method 2: Click Start, click the Run option, type in either CMD or COMMAND (either command works), and click on the OK button.

D. Windows 2000 or Windows XP:

Click Start, click the Run option, type in CMD or COMMAND (both commands work) in the dialog box, and click the OK button.

1. Notice your prompt. Your prompt should look something like this: C:\WINDOWS> or C:\>. It will have a drive letter specification followed by the root, \, symbol.
2. At the command prompt, type **FORMAT A:**. A message appears indicating you need to insert a new diskette for drive A: and press Enter.
3. Insert your floppy in the drive.
4. The disk is formatted. If you receive a message indicating the disk has an unusable track, replace the disk and begin again with step 1.
5. Once the disk is formatted, you can enter a volume label. It is a good practice to include your last name as the volume label. However, you are limited to 11 characters. The volume label is like a "software version" of a sticker on the front of your disk with your name on it.
6. When prompted for formatting another disk, enter **N**.
7. Refer to Figure 3.3 for a sample screenshot taken using the FORMAT command under Windows XP.
8. Shut down your PC and operating system by following one of these methods:

A. DOS: Press **Ctrl+Alt** at the same time and while you are holding these down, press **Del**. This is known as the Ctrl+Alt Del key sequence.

B. Windows 98, **NT Workstation,** or **Windows 2000:** Click Start and then click Shut down. When the **Shut Down Windows** dialog box appears, click **Shut down** and then click **OK**. Your PC turns off.

C. Windows XP: Click Start then point to **Turn Off Computer**. When the **Turn Off Computer** dialog box appears, click **Turn Off**. Your PC turns off.

9. Insert the newly formatted floppy in the drive.
10. Power up your PC.
11. During the boot process, press the appropriate key to enter the BIOS Setup. You will need to watch the screen as the PC goes through its POST to determine the appropriate key to press to enter the Setup. Typically it is the Del key, but it depends upon the BIOS manufacturer.
12. In the BIOS Setup, locate the option to change the boot sequence.
13. Change it to boot the floppy (A:) if it is not already set. Make sure that the hard drive (C:) is the next boot device.
14. Save the changes. The system boots to the floppy.
15. You receive an error message indicating that the disk is not a system disk.
16. In order to fix this problem, remove the disk and press Enter.
17. The system boots normally.

Figure 3.3: A screenshot of Formatting a Floppy Under Windows XP

```
C:\>format a:
Insert new disk for drive A:
and press ENTER when ready...
The type of the file system is FAT.
Verifying 1.44M
Initializing the File Allocation Table (FAT)...
Volume label (11 characters, ENTER for none)? Meadors
Format complete.

    1,457,664 bytes total disk space.
        5,120 bytes in bad sectors.
    1,452,544 bytes available on disk.

      512 bytes in each allocation unit.
    2,837 allocation units available on disk.

       12 bits in each FAT entry.

Volume Serial Number is 285D-A969

Format another (Y/N)? n

C:\>_
```

3.2 Understanding the DOS Tree Structure

DOS files can be organized like chapters in a book. However, DOS files are grouped into **directories**. The starting point for all directories is the **root directory**. From the root directory, other directories can be made. The garden tree analogy discussed in Chapter 2 applies to the DOS tree. Figure 3.4 shows how a hard disk's file structure might be organized.

Figure 3.4: The DOS Tree Hierarchy

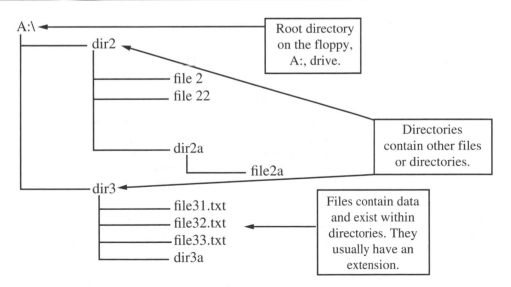

Figure 3.5 shows the TREE command output of the same structure. The TREE command normally shows only directories, but with the /F option, it shows files too.

Figure 3.5: The DOS TREE Command of the Tree Structure in Figure 3.4.

```
A:\>tree /f
Folder PATH listing
Volume serial number is 71FAE346 0000:0000
A:.
├───dir2
│       file2.txt
│       file22.txt
│
│   └───dir2a
│           file2a.txt
│
└───dir3
        file31.txt
        file32.txt
        file33.txt

    └───dir3a
```

A hierarchy is useful because it helps with organization and that translates to quick access for the user.

Notice in Figure 3.4 how each directory has a unique name. A **file name** in DOS can be a maximum of eight characters followed by a dot, then a three-character extension. However, the DOS command mode that comes with most Windows products can support additional characters in the file name.

The general format is as follows:

filename.ext

Note *filename* is the name of the file, the dot is literally a period, and *ext* is the extension. Extensions are used to associate files with software. For example, .DOC are MS-Word documents and .XLS are MS-Excel spreadsheets.

In DOS, a file name *cannot* contain the following characters:

. (period)	**; (semicolon)**	**/ (forward slash)**
, (comma)	**" (quotation marks)**	**\ (backslash)**
| (vertical bar)	**? (question mark)**	**[(left bracket)**
: (colon)	**= (equal sign)**	**] (right bracket)**
* (asterisk)	**(space)**	**< (less than)**
> (greater than)		

The general rule of thumb is if you use the character elsewhere, you cannot use it for a file name character. You need to be aware of these characters for the A+ exam.

Let's discuss navigating the tree structure. Assume the tree structure shown in Figure 3.4 has already been created and the current directory is root, or \. In DOS, there is a command prompt that is set to your current directory path location by default. This is useful because you can just look on your screen and tell where you are located.

In order to change directory locations to dir2, you would issue this command: **CD dir2** and press Enter. Now let's navigate to the directory beneath dir2 named dir2a. By the way, any directory beneath another directory is called a subdirectory, but they'll just be referred to as directories in this chapter.

In Figure 3.4, the directories named dir2 and dir3 are children of root. Or said another way, root is their parent. A **parent directory** is the directory that holds a given file or directory. It is up one level from the current directory. A **child directory** is a directory contained within its parent. It is down one level. The parent directory is represented by two dots, **..**, commonly called dot-dot. The current directory is represented by a single dot.

To help you remember that two dots represent the parent directory in the tree, just remember that you have two parents. So, one dot represents mom and one dot represents dad and collectively, they are called **..** in DOS.

You can issue the **CD ** to always take you to the root directory of the current drive. You can use the **CD ..** command to take you to the parent directory of your current directory. Notice in Figure 3.4 that dir2a is a child of dir2 and dir2 is a child of the root directory. The root directory is a grandparent of dir2a. If you wanted to change to root from dir2a, you would issue either one **CD ** command or two **CD ..** commands.

To change to root by name, you would issue the single command: **CD **. However, to change to any parent directory, you would enter the **CD ..** command. To change to a parent directory going up one level, you must use two dots, or dot-dot. To change to a parent's parent (grandparent), you would enter **CD ..** twice. Or, you can combine levels on one command. For example, you could issue the **CD ..\..** command to go back up two levels.

Each directory and file in the tree has a location called its **path**. There are two types of paths: full path and partial path. The **full path** identifies the directory or file location beginning from root. It is called an **absolute path** because it <u>absolutely</u> refers to the file or directory from root. A command employing the full path <u>always</u> begins at root.

A **partial path** identifies the directory or file location relative to your current location. A partial path does <u>not</u> begin at root. The partial path is also called the **relative path** because you are referring to a file or directory <u>relative</u> to where you are. Refer to Table 3.2 for a few examples. Notice how each full path has the root symbol as the very first character. Each of these paths can also be used with most any command.

Table 3.2: Examples of Full and Partial Paths

Path	Path Type
\	Full
\dir2	Full
\dir2\dir2a	Full
\dir3\file33.txt	Full
..	Partial
..\..	Partial
dir2	Partial
..\dir3	Partial

3.3 Understanding the Types of DOS Commands

In the past, many manufacturers created their own DOS version. IBM developed a version known as PC-DOS and Novell has licensed DR-DOS. Microsoft controls the operating system market today. Microsoft's DOS is known as MS-DOS. While there are many commands, only some of them will be discussed in this chapter. You can type the HELP command in the Windows 2000 and XP command mode to find the available commands for those operating systems.

DOS commands are classified as either internal or external. **Internal commands** are not visible when viewing files on a disk or hard drive but after you enter the commands, they will execute. Internal commands are built into the COMMAND.COM file and execute much faster than external. This is because COMMAND.COM is loaded into memory, which is faster than disk. Table 3.3 lists some of the more common internal commands. DOS and the command mode in Windows 98, Windows 2000, and Windows XP support all of these commands.

Table 3.3: DOS Internal Commands

Command	Description
CD or **CHDIR**	Change directory locations in the tree structure.
MD or **MKDIR**	Create a directory.
DIR	Display a listing of both files and directories.
TYPE	Display the contents of a file.
RD or **RMDIR**	Remove a directory.
COPY	Copy a file to another file.

CLS	Clears your screen.
DATE	Displays the date and allows you to change it.
PATH	Display the current search path.
PROMPT	Display or set your prompt.
REN	Change the name of a file.
ECHO	Displays text on the screen.
TIME	Displays the time and allows you to change it.
DEL	Delete a file.
VER	Displays the current DOS version.

External commands can be seen when viewing files on a disk or a hard drive. External commands execute slower than internal commands because the external commands must be retrieved and loaded from the disk or hard drive. External commands are usually stored in one of the following directories, depending upon the version of the operating system you are using:

• C:\WINDOWS—used by Windows 98 and Windows XP
• C:\WINDOWS\SYSTEM—used by Windows 98
• C:\WINDOWS\COMMAND—used by Windows 98
• C:\WINDOWS\SYSTEM32—used by Windows NT, 2000, and XP
• C:\WINNT\SYSTEM32—used by Windows NT and Windows 2000
• C:\DOS—used by MS-DOS

External commands have an extension of .EXE or .COM. Table 3.4 lists some of the more common external commands. The third column indicates which command is supported under which operating system for DOS, Windows 98, Windows 2000, and Windows XP. Although the FORMAT command is supported among these operating systems, the /S on the FORMAT command is not supported under Windows 2000 and Windows XP.

Table 3.4: Some Common DOS External Commands and Their Operating System Availability

Command	*Description*	*OS Version Availability*
ATTRIB	View or change the attributes of a file or directory.	DOS, Windows 98, Windows 2000 and XP command mode.
CHKDSK	Check the status of a disk.	DOS, Windows 98, Windows 2000 and XP command mode.
DELTREE	Remove a tree, which includes files and directories.	DOS and Windows 98 command mode. In Windows 2000 and XP, you use the **rd** **/s** command.

DOSKEY	Allows you to scroll up and down through the list of commands that you've entered. Think of DOSKEY as keeping a history of the commands you've entered.	DOS, Windows 98, Windows 2000 and XP command mode
EDIT	The DOS editor.	DOS, Windows 98, Windows 2000 and XP command mode
FDISK	Partition a hard disk.	DOS and Windows 98 command mode. Windows 2000 and XP have GUI tools.
FIND	Search for text within a file.	DOS, Windows 98, Windows 2000 and XP command mode.
FORMAT	Prepare a drive for use by DOS.	DOS, Windows 98, Windows 2000 and XP command mode. **Note:** The /S option is not available in Windows 2000 and XP.
LABEL	Change the label on a disk. Think of the label as a software version of a sticker you physically place on your floppy identifying it as being yours.	DOS, Windows 98, Windows 2000 and XP command mode.
MOVE	Used to move a file from one directory location to another.	DOS, Windows 98, Windows 2000 and XP command mode.
MEM	Display a listing of commands and processes in memory.	DOS, Windows 98, Windows 2000 and XP command mode.
SYS	Transfer system files to a device.	DOS and Windows 98 command mode.
SORT	Sort data in a file.	DOS, Windows 98, Windows 2000 and XP command mode.
TREE	Display a tree listing of a directory.	DOS and Windows 2000 and XP command mode. Not available in Windows 98.
XCOPY	Copy a whole tree, which includes files and directories.	DOS, Windows 98, Windows 2000 and XP command mode.

Commands are further classified into two broad categories:
- Directory commands
- File commands

Some commands deal only with directories and some commands deal only with files. Internal or external commands can deal with either files or directories. There are a few exceptions. For example, the DOS DIR command displays both files and directories. Most commands use the following syntax:

command options

Note *command* is the name of the command and *options* are either file and directory names or actual options that alter the command in some way. You can enter the command in upper- or lower-case. Almost all commands use options to enhance the way they operate. The symbol for using options is the forward slash, /.

Let's discuss some of the basic commands you would use in DOS. Let's say you wanted to create a directory. In order to create a directory, you would use the MD, or MKDIR, command. Assuming your current directory is the root directory on the floppy (A:\), if you wanted to create a directory named DirTM you would enter **MD DirTM**. You could issue the DIR command to see that the directory was created. If you wanted to put a file or directory within DirTM, you would first change to that directory using the **CD DirTM** command. Figure 3.6 shows a screenshot of these steps. Because the default prompt is used, your prompt changes as you change directory locations.

Figure 3.6: A Screenshot of Creating a Directory, Taking a Directory Listing, and Changing to a Directory

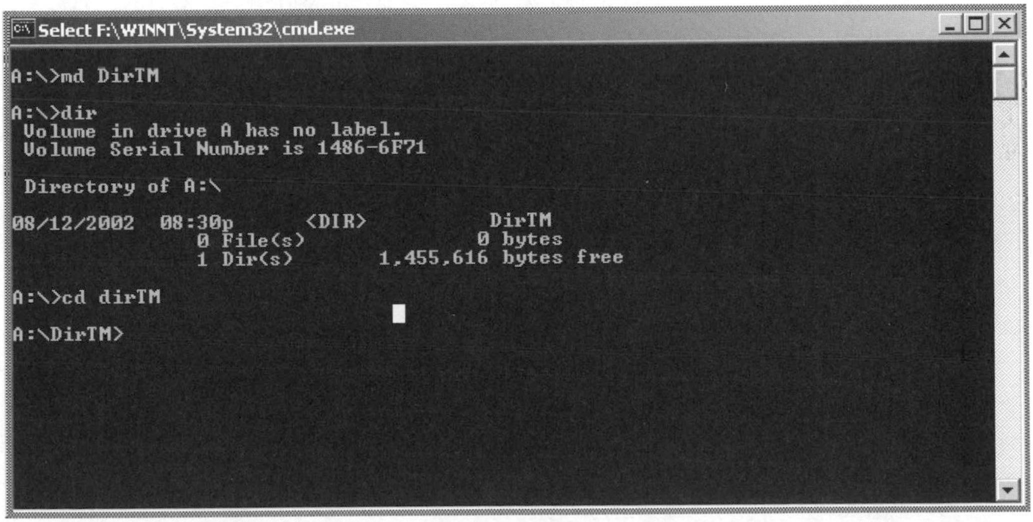

Command options modify the command in some way. For example, DIR /W displays the files and directories in column form. DIR /P pauses the listing. The most useful option for all commands is the question mark symbol, ?. It allows you to question DOS as to what a command does. Think of it as "help" on the command. Figure 3.7 shows a screenshot of DIR /?. You could use this option with most any command. Try CLS /?, CD /? and MD /? at the prompt.

Figure 3.7: A Screenshot of the DIR /? Command

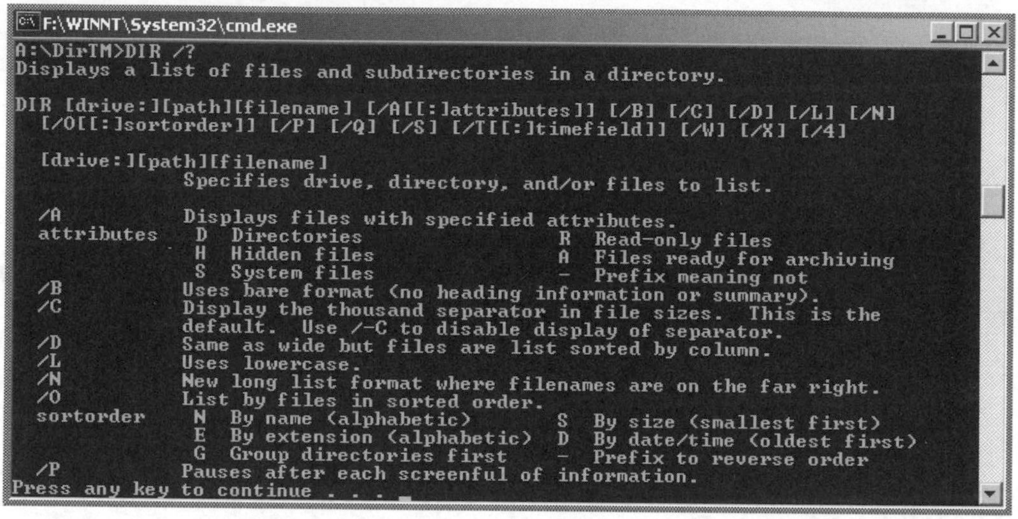

Another very useful command is the TREE command. It works on Windows 2000 but not on Windows 98. The TREE command gives you a directory listing of all subdirectories in a directory. You can use the /F option to include all files in the listing. So, if you wanted to see a tree listing of a directory named A:\PAYROLL, you would enter TREE A:\PAYROLL. If you wanted to see the files as well as directories, you would add the /F option as in TREE /F A:\PAYROLL.

Figure 3.8 shows a sample tree directory for a fictitious organization named The Firm.

Figure 3.8: A Screenshot of Using the TREE /F Command

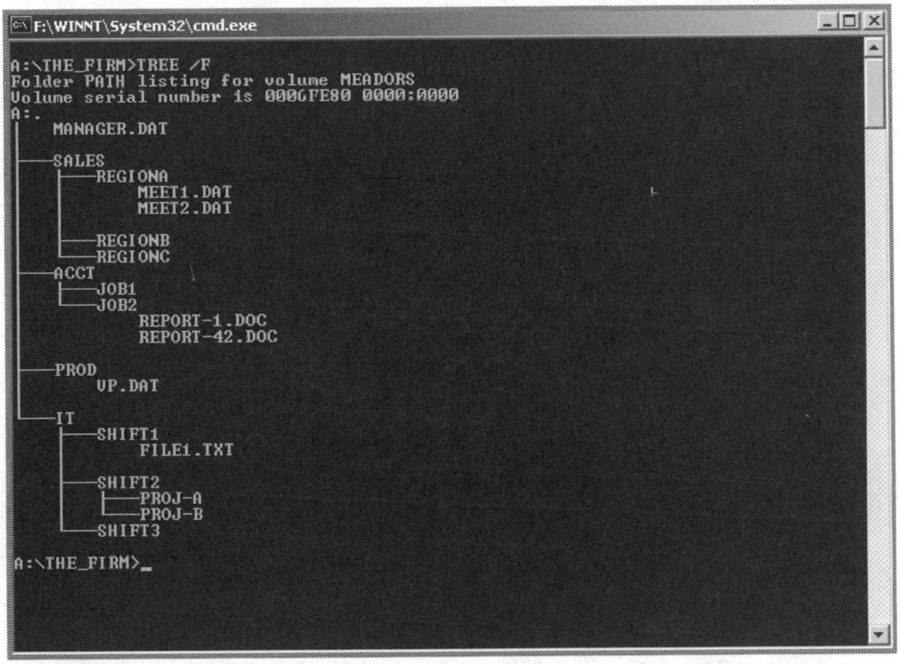

```
F:\WINNT\System32\cmd.exe                                    _ □ ×

A:\THE_FIRM>TREE /F
Folder PATH listing for volume MEADORS
Volume serial number is 000GFE80 0000:0000
A:.
│   MANAGER.DAT
│
├───SALES
│   ├───REGIONA
│   │       MEET1.DAT
│   │       MEET2.DAT
│   │
│   ├───REGIONB
│   └───REGIONC
├───ACCT
│   ├───JOB1
│   └───JOB2
│           REPORT-1.DOC
│           REPORT-42.DOC
│
├───PROD
│       UP.DAT
│
└───IT
    ├───SHIFT1
    │       FILE1.TXT
    │
    ├───SHIFT2
    │   ├───PROJ-A
    │   └───PROJ-B
    └───SHIFT3

A:\THE_FIRM>_
```

Creating and Removing Directories

In order to create a directory, you would use the MD or MKDIR command. Although it has been discussed, here is the format for the MD command:

> MD *directory-name*
> MKDIR *directory-name*

To create a directory named Payroll, you would enter **MD Payroll**. Notice that the directory Payroll is a partial path. You really don't know what parent directory it is located in. You can use the MD command with either the partial path or full path name of a directory. For example, you could enter **MD \Acct\General** to create a directory using a full path. Or, you could use a partial path, as in these examples: **MD ..\dir5, MD dir6\dir6a\dir6b**, or **MD dirTM**. Also, you could create a directory on a drive other than your current drive. For example, if your current drive and directory is A:\ (the floppy) and you wanted to make a directory on the C: (hard disk) drive, named Sales, you could enter: **MD C:\Sales**.

Next, you need to understand how to remove directories. The command to remove a directory is RD or RMDIR. Here is the format:

> RD *directory-name*

When using the RD or RMDIR command there are two things to remember:

1. The directory must be empty.
2. Your current directory cannot be the directory you want to remove.

If you get an error indicating you cannot remove the directory, try looking at your prompt or running the CD command. Typing in the CD command with no options will display your current directory. Looking at your prompt will also show your current directory as long as it has not been changed. If your current directory is the one you are trying to delete, then go to the parent using the **CD ..** command. Retry the failed RMDIR command and it should work.

Figure 3.9 shows a sample session of attempting to remove a directory while in the directory. Only when you change directory locations to somewhere higher in the tree will you be able to delete the directory.

Figure 3.9: A Screenshot of the Using the RD (or RMDIR) Command

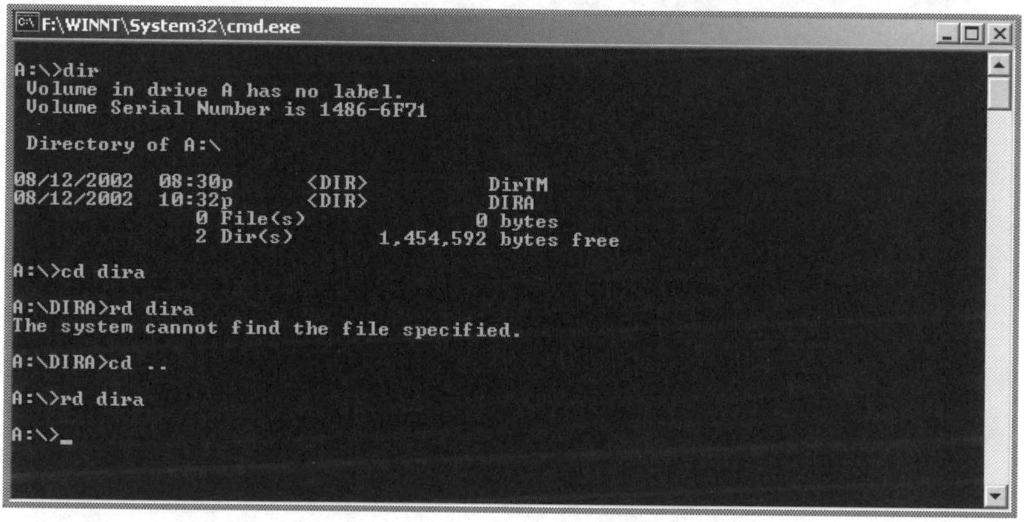

Let's perform an exercise. This exercise assumes a PC is up and running with a Microsoft operating system with access to DOS or the DOS command prompt. A formatted floppy is in the floppy drive. Choose the operating system you are using to access DOS command mode.

1. To change your prompt to the floppy drive, type **A:**.
2. To create a directory, type **MD dira** and press Enter.
3. To change directory locations, type **CD dira** and press Enter.
4. Let's make a directory within the directory named "dira." Type **MKDIR diraa** and press Enter. Remember either MD or MKDIR will work.
5. To change locations to "diraa," type **CD diraa** and press Enter.

6. To change directory locations to the parent directory, type **CD ..** at the command prompt.

7. To remove the "diraa," type **RD diraa**. Remember either RD or RMDIR will work.

8. To change directory locations to the root directory, type **CD ** at the command prompt.

9. To create a directory, type **MD DIRB**.

10. To create a directory, type **MD DIRC**.

11. To verify, type **DIR**.

12. To change to DIRC, type **CD DIRC**.

13. To create a directory, type **MD DIRC1** and press Enter.

14. To create a directory, type **MD DIRC2** and press Enter.

15. To verify, type **DIR.**

16. To change to DIRC2, type **CD DIRC2**.

17. To create a directory, type **MD DIRC2-1** and press Enter.

18. To create a directory, type **MD DIRC2-2** and press Enter.

19. To change directory locations to the root directory, type **CD ** at the command prompt.

20. If your operating system supports the TREE command, type **TREE** and press Enter. It is <u>not</u> supported by Windows 98. To verify whether your operating system supports a command, simply type the command. If it works, it is supported; otherwise, you will see a message indicating the command was not found.

21. Exit the system appropriately.

Managing Files

You need to know how to create, delete, modify, copy, and move files. In this section, we will learn how to do that. In terms of creating files, there are several ways. We are not concerned about all the methods, nor are we really concerned about the data that is in the files. We just need to create files so we can manipulate them within the tree structure.

Let's explore the methods to create files in greater detail. To create a file you could use the COPY command. Although this is not the typical use of the COPY command, it can be used to quickly make files.

The COPY command allows you to copy from a source file (the file name immediately following COPY) to a destination file (the very last file name). Here is the general form:

```
COPY source-file destination-file
```

For example, **COPY FILEA.DAT FILEB.DAT** will copy FILEA.DAT to FILEB.DAT. Using the COPY command to copy files will be explored more a bit later. It is included here because it can be used to create a new file, as you'll see next.

The COPY command allows you to use the special name CON, which stands for Console Keyboard, to enter text directly into a file. The source in this case is the keyboard. The format is:

> COPY CON *filename.txt*

The TYPE command can be used to display the contents of a file. It takes this general form:

> TYPE *filename.txt*

Figure 3.10 shows a screenshot of using the COPY CON command to create a file and the TYPE command to display the contents. Notice the case of the letters does not matter.

Figure 3.10: A Screenshot of Using the COPY CON and TYPE Commands

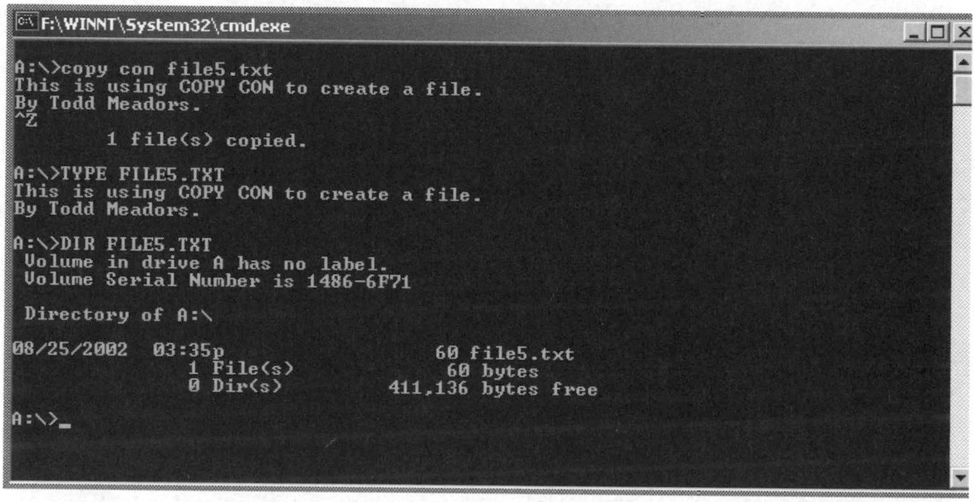

Now let's perform an exercise to help you understand the COPY CON and TYPE commands.

1. Use the appropriate steps to get to the DOS prompt on your floppy.
2. In order to create a file using the COPY CON command, type **COPY CON file2.txt**. The cursor will move to the beginning of the next line. There will be no prompt on that line.
3. Type the following, pressing **Enter** after each line of text is typed. If you make a mistake on a line and press Enter, you cannot change it using the COPY CON method. You would have to use EDIT to modify a previous line.

   Computers are powerful devices.
   DOS is fun and will come in handy on the job.
   The End

4. Press **F6**. A caret (^) and a Z (representing Shift + Z) appear on the screen. Press Enter. A message appears stating that one file copied. This is how you save the file. Note that F6 signifies an End-of-File (EOF).
5. To display the contents of the file using the TYPE command, type **TYPE file2.txt** and press Enter.
6. Create another file using the COPY CON method.
7. Use the TYPE command to display its contents.
8. Exit the system appropriately.

The COPY CON command is ideal for creating small files quickly. However, if EDIT is available, you should consider using it.

The EDIT Command

You can also create a file with the DOS EDIT command. This tool allows you to modify files as well. The COPY CON command will not allow you to change or delete text within a file, but the EDIT command will. To edit a file, type the EDIT command followed by a file name. If the file is present, it will be displayed on the screen. If it is not present, a new file will be created and a blue screen will appear. Figure 3.11 shows a screenshot of text created using the EDIT command.

Once the editor opens, you can begin typing your text. You need to press the Alt key to activate the menu commands. For example, you would press Alt + F + Save to save the file. Within EDIT, you can use either Save to save to the current file name that is opened, or Save As to a different file name. You can also search, cut and copy text, and perform other menu commands.

Figure 3.11: A Screenshot of Using the EDIT Command

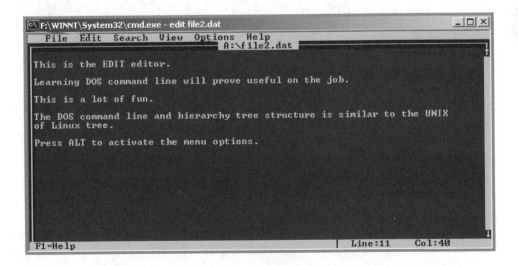

The COPY, REN, and MOVE Commands

You need to understand how to copy, rename, and move a file. To make a duplicate of a file, you would use the COPY command.

For example, if you wanted to copy a file named PAY4.DAT to one named PAY5.DAT, you would enter **COPY PAY4.DAT PAY5.DAT** at the command prompt. The source file name must be present. The destination file name is a name you determine. If it is present, DOS will ask you if you want to overwrite. You can use full or partial paths in conjunction with the source and destination. So if you wanted to copy a file from C:\SALES\MONTHLY\SALES1.DAT to A:\BACKUP\SALES1.DAT, you would enter:

COPY C:\SALES\MONTHLY\SALES1.DAT A:\BACKUP\SALES1.DAT.

Figure 3.12 shows a screenshot of using the COPY command to make a backup of a file named GAME1.DAT. The first DIR listing displays only GAME1.DAT. Then the file is copied to GAME2.DAT. Both files exist in the second DIR listing.

Figure 3.12: A Screenshot of the COPY Command

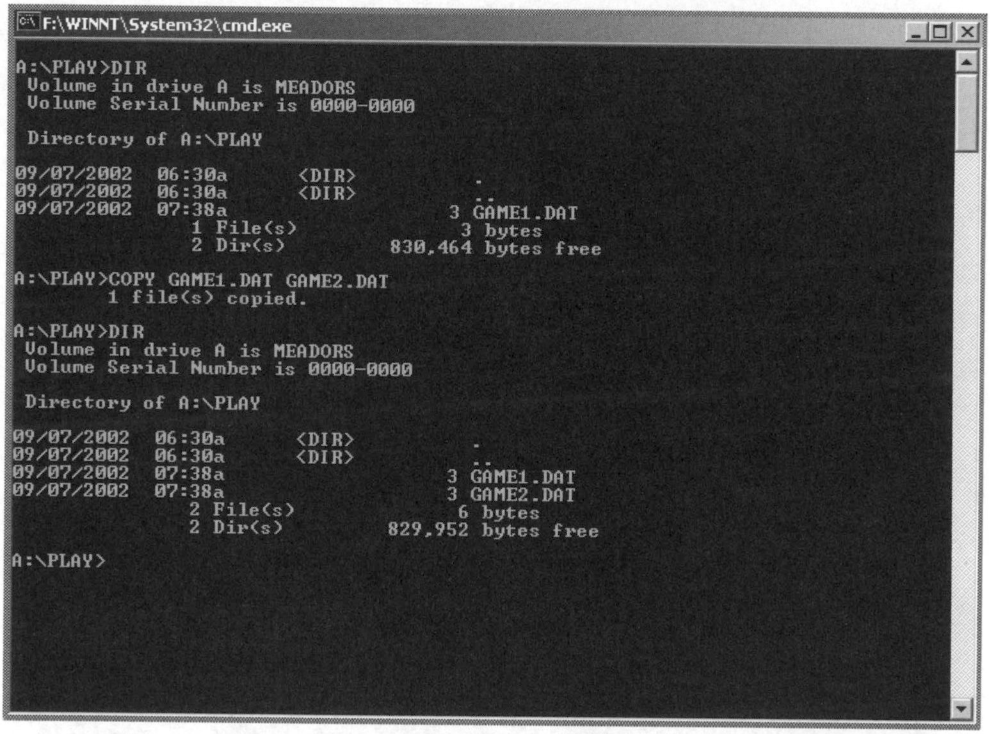

The REN command allows you to rename a file, but the destination directory must be the same as the source. In other words, you can only rename a file to another name within the same directory. The REN command cannot be used to rename a file to another directory. The syntax is:

> REN *source-filename destination-filename*

For example to rename a file named PAY7.DAT to PAY8.DAT, you would enter:

REN PAY7.DAT PAY8.DAT

The source file, PAY7.DAT, must present in this case. Figure 3.13 shows a screenshot of the REN command. The first DIR listing shows GAME1.DAT and GAME2.DAT. Next, the command **REN GAME1.DAT PLAYTIME.DAT** is used to rename the file. In the last DIR listing, the file is renamed. The old name no longer exists.

Figure 3.13: Using the REN Command

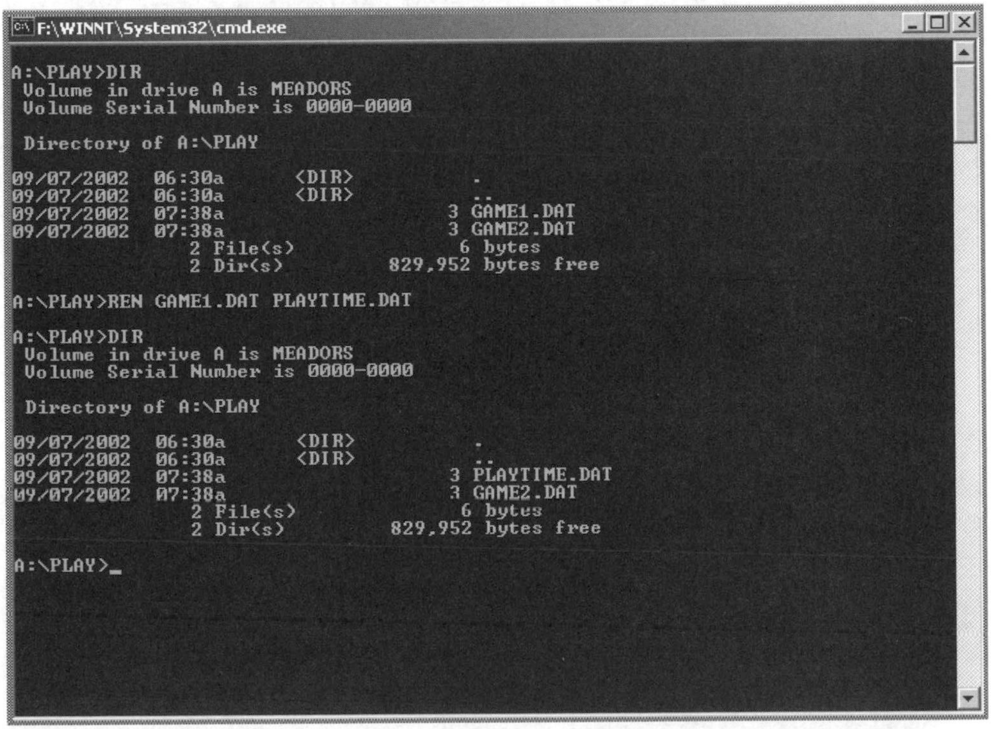

To move a file to another directory you would use the MOVE command. The syntax is:

> MOVE *source-filename destination-filename*

For example to move a file named SALES5.DAT to SALES6.DAT, you would enter:

MOVE SALES7.DAT SALES8.DAT

The source file, SALES7.DAT, must present in this case. Figure 3.14 shows a screenshot of using the MOVE command to move a file to another directory. The command DIR PLAYTIME.DAT displays the file. Next, **MOVE PLAYTIME.DAT A:\BEACH\PLAY TIME.DAT** moves the file to another directory. The command DIR PLAYTIME.DAT is run again to show you it not longer exits in the current directory. The command **DIR A:\ BEACH\PLAYTIME.DAT** shows you it does indeed exist in the A:\BEACH directory where it was moved.

Figure 3.14: Using the MOVE Command to Move a File to Another Directory

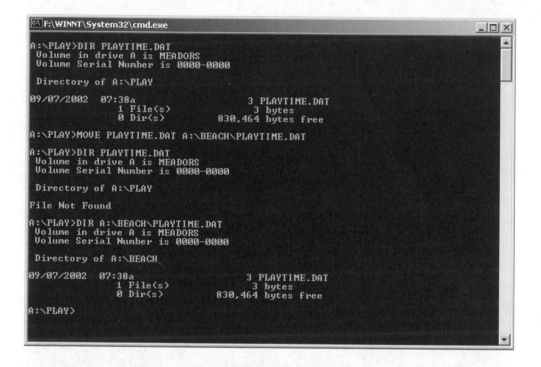

 The difference between the REN and MOVE commands is that with REN, the source and destination directory must be the current directory. With the MOVE command, the source and destination directory can be the same or different. When in doubt, use MOVE.

The DEL Command
In order to delete a file, you would use the DEL command. The syntax is:

DEL *filename*

One of the most useful options the DEL command allows is the /P option. With this option, you are prompted before the file is deleted. Without this option, DEL goes ahead and deletes the file. Think of the /P option as a safety switch. It gives you a chance to keep the file. Figure 3.15 shows a screenshot of the DEL command. You can see in the figure that **DEL /P FOOTBALL.DAT** prompts you to delete the file named FOOTBALL.DAT.

Figure 3.15: Using the DEL Command and the /P Option to Prompt Before Deleting

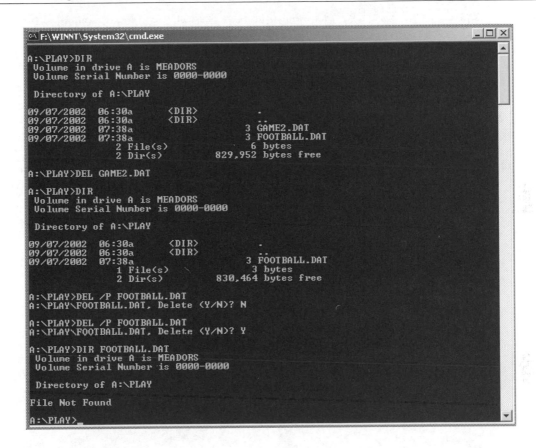

Now let's perform an exercise to help you understand EDIT, COPY, and MOVE.
1. Use the appropriate steps to get to the DOS prompt.
2. Go to the floppy drive.
3. Type **EDIT FILE4.TXT**.
4. Insert the following text:

Learning about DOS.
Having a lot of fun!

5. Press the Alt key and then the F key for File. Then press the S key for Save. This saves your file.
6. Press the Alt key and then the F key for File. Then press the X key for Exit. This exits DOS EDIT.
7. To copy the file to FILE4B.TXT, type **COPY FILE4.TXT FILE4B.TXT**.
8. To verify, type **DIR**.
9. To rename FILE4B.TXT, type **REN FILE4B.TXT FILE4Z.TXT**.
10. To verify, type **DIR**.
11. To move FILE4Z.TXT to another directory, first create a directory named DIR4 by typing **MKDIR A:\DIR4**.
12. Now move it. Type **MOVE FILE4Z.TXT A:\DIR4\FILE4Z.TXT**.
13. To verify, type **DIR A:\DIR4**
14. To delete FILE4.TXT, type **DEL FILE4.TXT**.
15. To verify, type **DIR**.
16. Exit the system appropriately.

3.4 Understanding Drive Related Commands

Drive letters are assigned to hardware devices when a computer boots. For example, the first floppy drive gets the drive letter A:. The colon is part of the device's drive letter. The first hard drive in a system gets the drive letter C:. The devices detected by the operating system can use drive letters A: through Z:.

The LABEL command allows you to change the label on a drive letter. It is like a software version of a sticker you would place on your floppy, with your name, to indicate the floppy is yours. Figure 3.16 shows a sample screenshot of the LABEL command.

Figure 3.16: Using the LABEL Command to Change the Label on the Floppy

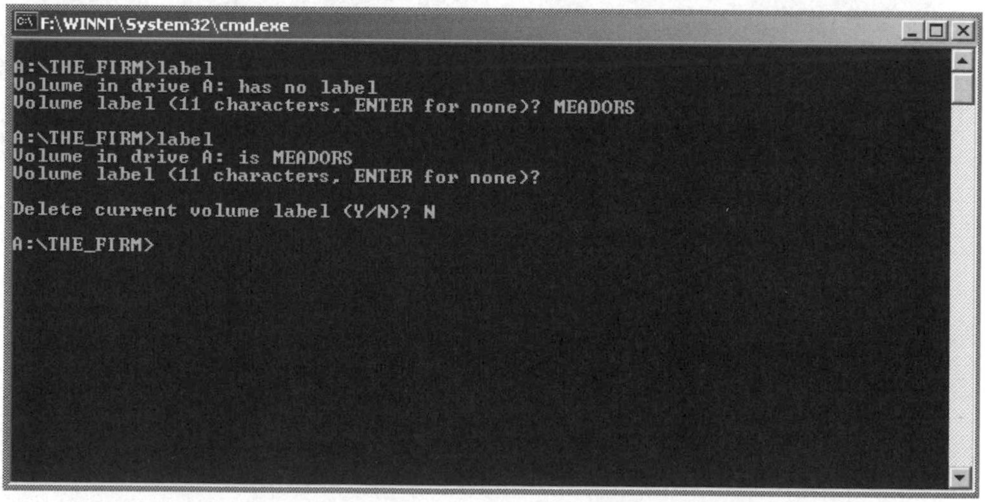

```
F:\WINNT\System32\cmd.exe                                    _□×

A:\THE_FIRM>label
Volume in drive A: has no label
Volume label (11 characters, ENTER for none)? MEADORS

A:\THE_FIRM>label
Volume in drive A: is MEADORS
Volume label (11 characters, ENTER for none)?

Delete current volume label (Y/N)? N

A:\THE_FIRM>
```

Another useful command is the CHKDSK command. It allows you to check the disk's status and reports statistics about the disk. Some statistics are the amount of space on the disk, the amount used, and any unused areas on the disk. You can use the /F option to fix errors. Figure 3.17 shows a sample screenshot of the CHKDSK command run on a floppy.

Figure 3.17: The CHKDSK Command

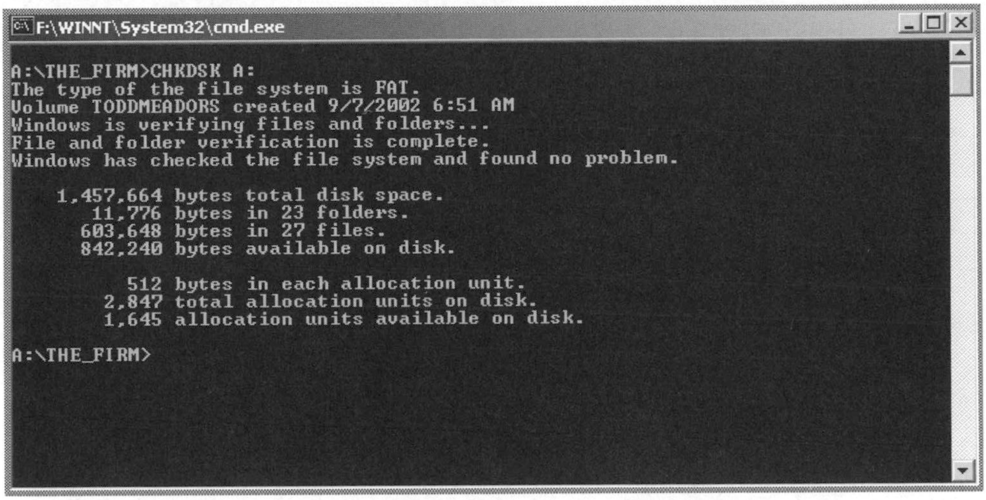

```
F:\WINNT\System32\cmd.exe

A:\THE_FIRM>CHKDSK A:
The type of the file system is FAT.
Volume TODDMEADORS created 9/7/2002 6:51 AM
Windows is verifying files and folders...
File and folder verification is complete.
Windows has checked the file system and found no problem.

   1,457,664 bytes total disk space.
      11,776 bytes in 23 folders.
     603,648 bytes in 27 files.
     842,240 bytes available on disk.

         512 bytes in each allocation unit.
       2,847 total allocation units on disk.
       1,645 allocation units available on disk.

A:\THE_FIRM>
```

You can change how your prompt appears by running the PROMPT command. Figure 3.18 shows the options available to change your prompt.

Figure 3.18: A Screenshot of the PROMPT Command Displaying Its Options

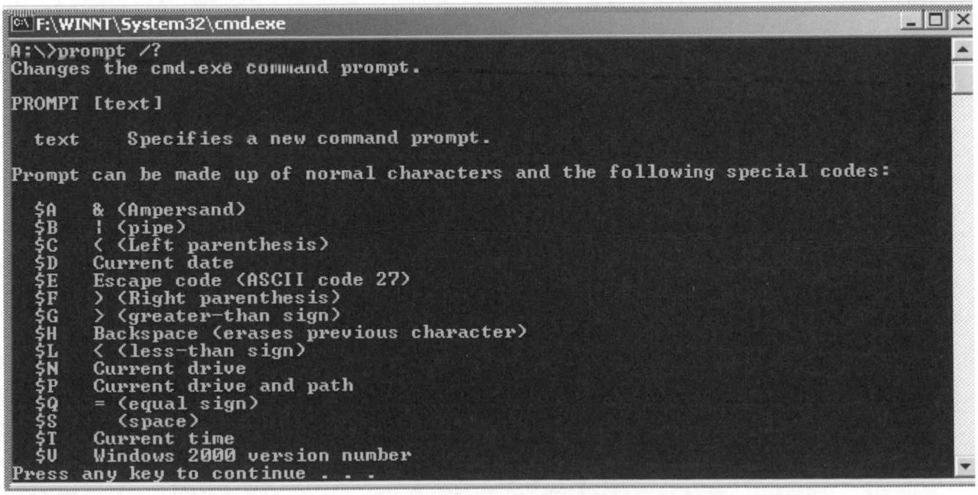

```
F:\WINNT\System32\cmd.exe

A:\>prompt /?
Changes the cmd.exe command prompt.

PROMPT [text]

  text    Specifies a new command prompt.

Prompt can be made up of normal characters and the following special codes:

  $A    & (Ampersand)
  $B    | (pipe)
  $C    ( (Left parenthesis)
  $D    Current date
  $E    Escape code (ASCII code 27)
  $F    ) (Right parenthesis)
  $G    > (greater-than sign)
  $H    Backspace (erases previous character)
  $L    < (less-than sign)
  $N    Current drive
  $P    Current drive and path
  $Q    = (equal sign)
  $S      (space)
  $T    Current time
  $U    Windows 2000 version number
Press any key to continue . . .
```

Let's do a brief exercise using the PROMPT command.
1. Use the appropriate steps to get to the DOS prompt.
2. Go to the floppy drive.
3. To change your prompt to include the text "Hi" followed by the date, type **PROMPT Hi $D** and press Enter.
4. To change your prompt to include text, the date, the time, and the greater than symbol, type **PROMPT Hi $D $T $G** and press Enter.
5. To change your prompt to "Enter command: >" type **PROMPT Enter command: $G** and press Enter.
6. To change your prompt to include the DOS version, the current drive and path, and the less than symbol, type **PROMPT $V $P $L** and press Enter.
7. To change your prompt back to its original state, type **PROMPT PG** and press Enter.
8. Exit the system appropriately.

3.5 Learning about Wildcard Characters

The DOS operating system allows for pattern matching of files. This means you can use certain symbols to match characters following a certain pattern. Suppose you want to see all of the files that begin with an S or all of the files that end in .DAT. How would you do this?

The operating system provides two pattern matching mechanisms, sometimes called "wildcarding," to accomplish this. These wildcard characters are:
• The asterisks symbol (*), which matches all characters.
• The question mark symbol (?), which matches a single character position.

You use the * symbol when you want to match all character positions. For example, if you wanted to display all files that begin with the letter S, then you would enter **DIR S***
at the prompt. If you wanted to display all the files that had DAT after the dot in the file name, then you would enter **DIR *.DAT**.

If you wanted to display all files that had a 5 in the fourth character position, and you didn't care about the previous or following characters, you have to use the question mark. To do this, you would type **DIR ???5*** at the command line prompt. You could not use the asterisks before the 5 in this case, because it would display all files with a 5 anywhere in the file name—not just in the fourth character position.

Figure 3.19 shows the use of wildcards. In the figure, the DIR T*.* command displays all files beginning with a T regardless of the remaining characters in the file name or extension. The next command, DIR *.TXT displays those files with a .TXT extension.

Figure 3.19: Working with Wildcards

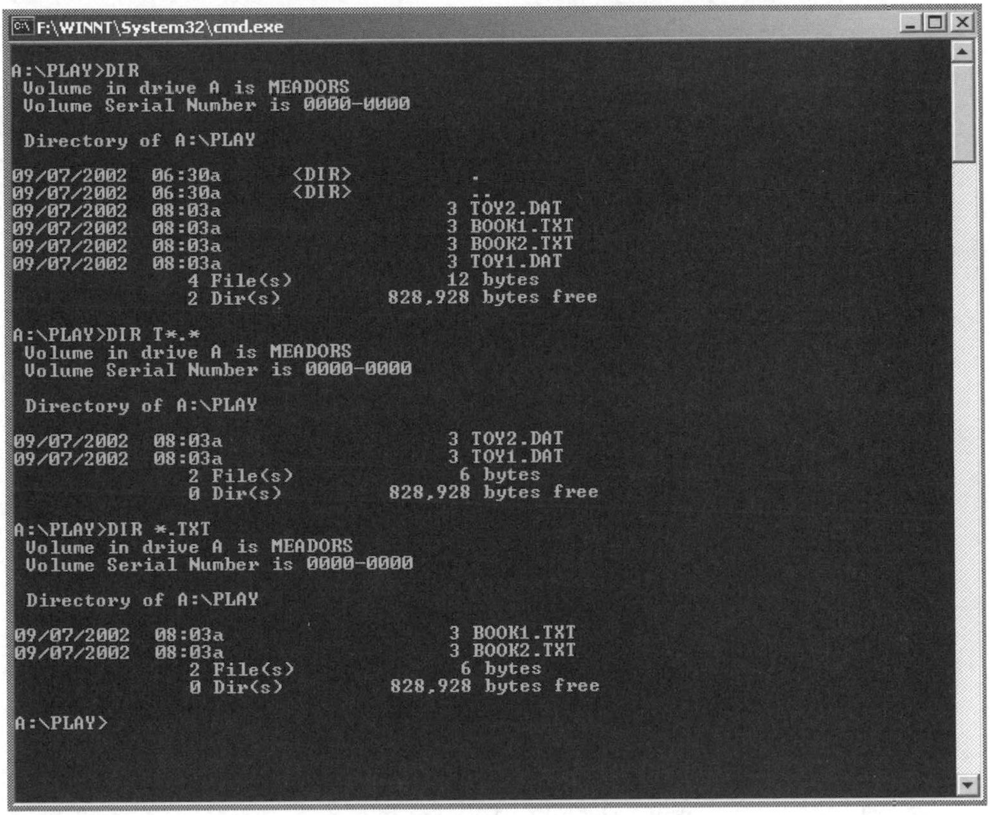

```
F:\WINNT\System32\cmd.exe

A:\PLAY>DIR
 Volume in drive A is MEADORS
 Volume Serial Number is 0000-0000

 Directory of A:\PLAY

09/07/2002  06:30a    <DIR>          .
09/07/2002  06:30a    <DIR>          ..
09/07/2002  08:03a                 3 TOY2.DAT
09/07/2002  08:03a                 3 BOOK1.TXT
09/07/2002  08:03a                 3 BOOK2.TXT
09/07/2002  08:03a                 3 TOY1.DAT
               4 File(s)           12 bytes
               2 Dir(s)       828,928 bytes free

A:\PLAY>DIR T*.*
 Volume in drive A is MEADORS
 Volume Serial Number is 0000-0000

 Directory of A:\PLAY

09/07/2002  08:03a                 3 TOY2.DAT
09/07/2002  08:03a                 3 TOY1.DAT
               2 File(s)            6 bytes
               0 Dir(s)       828,928 bytes free

A:\PLAY>DIR *.TXT
 Volume in drive A is MEADORS
 Volume Serial Number is 0000-0000

 Directory of A:\PLAY

09/07/2002  08:03a                 3 BOOK1.TXT
09/07/2002  08:03a                 3 BOOK2.TXT
               2 File(s)            6 bytes
               0 Dir(s)       828,928 bytes free

A:\PLAY>
```

3.6 Learning about Redirection and Filtering

When you run a command, the output generally goes to the screen. The input for most commands is usually the keyboard. With redirection, you change the normal input or output locations.

These are the redirection symbols:

- A single less-than symbol (<) to redirect input
- A single greater-than symbol (>) to redirect output
- Two greater-than symbols (>>) to redirect and append output

The syntax of redirecting input is:

command < filename.ext

For example, the SORT command will accept input from the command line. However, you can redirect input from a file by typing **SORT < UNSORT.DAT** at the command line prompt. This command will read input from the file named UNSORT.DAT and display the sorted results on the screen.

The syntax of redirecting output is:

> *command > filename.ext*

You would redirect the output of a command if you wanted to keep it in a file to print or view it later. For example, to redirect the output of the DIR command from the screen to a file named LISTING.TXT, type **DIR > LISTING.TXT**.

The output would be redirected to a file named LISTING.TXT. When you redirect standard output to a file using the > symbol, the contents of the file are deleted. If you wanted to add standard output from another command to the same file name, you would issue the redirect and append output symbols, >>.

The syntax of redirecting and appending output is:

> *command >> filename.ext*

For example, the command **DIR >> LISTING.TXT** means to append the directory listing to the file named LISTING.TXT. If you had issued **DIR > LISTING.TXT**, you would have written over the previously redirected listing of the DIR command.

The filter symbol is the pipe symbol (|). It appears as the broken vertical bar (it is located on the same key as the backslash symbol) on your keyboard but it prints as a solid vertical bar. The command on the left side of the pipe symbol sends its output as input to the command on the right side of the pipe symbol. Think of the pipe as a physical pipe beneath your kitchen sink. Water flows through the pipe and out the other end somewhere. For example, the DIR command displays files and directories, and the MORE command allows you to scroll through a page of data at a time. If the output of the DIR command scrolled several pages, how would you be able to see all of the files and directories? One answer is to use the pipe symbol.

The syntax of commands using the pipe symbol is:

> *command1 | command2*

The command **DIR | MORE** will display a list of file and directories one page at a time. You could also use the pause switch option /P on the DIR command.

Many users think the pipe symbol can only be used with the MORE command because they are frequently used together, but you can use the pipe symbol with other commands too. If you wanted to sort the directory listing, you could enter **DIR | SORT**. If this command displayed too many lines, then you could put several pipes together, such as **DIR | SORT | MORE**. This will allow you to scroll through the sorted listing. Several pipes can go on one command line. For example, you could have the following:

> *command1 | command2 | command3 | command4*

The way this works is that *command1's* output is pipe as input to *command2*. Then, that filtered output is piped to *command3* and then the filtered output of *command1*, *command2,* and *command3* is piped as input into *command4*.

Figure 3.20 shows a screenshot of redirecting output, redirecting and appending output, redirecting input, and using the pipe symbol. In the figure, the first ECHO command redirects output to the file named GREETING.TXT. The contents are displayed using the TYPE command. Next, the file is appended using the >> redirection symbols and the file's contents are displayed again. To redirect input, the command **SORT < UNSORT.DAT** is used. It simply sorts a file and displays the output to the screen. The command **TYPE UNSORT.DAT | SORT** is another version of sorting a file. It sends the output of the TYPE UNSORT.DAT command, which is the file's contents, to the SORT command for processing. The SORT command then sorts the data and displays the results on the screen. The command **TYPE UNSORT.DAT | SORT | MORE** displays the same information one screen at a time if there is more than one screen of data.

Figure 3.20: Using Redirection and Pipe Symbols

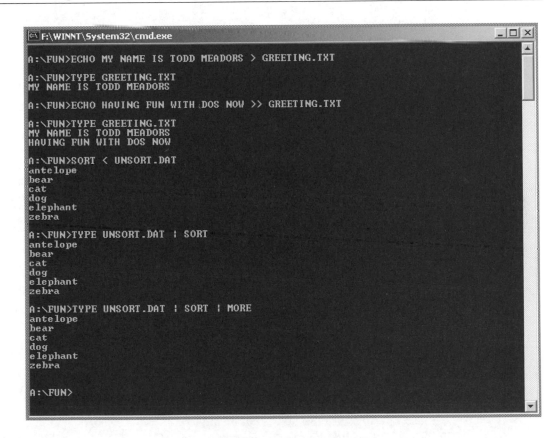

Let's do an exercise to help you understand the concepts of wildcarding, redirection, and filtering.

1. Use the appropriate steps to get to the DOS prompt.
2. Go to the floppy drive.
3. You will need to create some files to work with. Create the files named ZOO1.DAT, ZOO2.DAT, ZOO3.DAT, FUN1.TXT, and FUN2.TXT.
4. To display only the files beginning with an F, type **DIR F*.*** and press Enter.
5. To display only the files with an extension of .DAT type **DIR *.DAT** and press Enter.
6. To display only the files with a 2 in the fourth position, regardless of other characters, type **DIR ???2.*** and press Enter.
7. To redirect output, type **ECHO HI > ZAC12.DAT**.
8. To display the contents of the file, type **TYPE ZAC12.DAT**.
9. To redirect and append output, type **ECHO BYE >> ZAC12.DAT**.
10. To display the contents of the file, type **TYPE ZAC12.DAT**.
11. In order to redirect input, let's create a file named UNSORT.BAT using DOS EDIT. Type the following lines that contain a number ID (in column 1) followed by a Name (in column 3) and Phone Number (in column 25):

4 Todd Meadors	**555-555-5551**
2 Micki Meadors	**555-555-5552**
5 Zac Meadors	**555-555-5553**
3 Jessie Meadors	**555-555-5554**
1 Larry Meadors	**555-555-5555**

12. Save the file.
13. To sort the data, type **SORT < UNSORT.DAT**. By default, SORT sorts on the first column of data, although you can change which column it sorts on. So in UNSORT.DAT, the SORT command will sort on the ID. To sort on the first name column (column 3), you would enter **SORT /+3 <UNSORT.DAT**.
14. To sort the data and redirect the sorted output to another file named SORT.DAT, type **SORT < UNSORT.DAT > SORT.DAT**.
15. To view the contents of SORT.DAT, type **TYPE SORT.DAT**.
16. To reverse sort the data in reverse order and redirect the sorted output to another file, type **SORT /R < UNSORT.DAT > SORTR.DAT**.
17. To view the contents of SORTR.DAT, type **TYPE SORTR.DAT**.
18. To find Todd's phone information type **FIND "Todd" < UNSORT.DAT**. The ID, Name, and Phone Number for "Todd" are displayed on the screen.
19. To find Micki's phone number and redirect it to a file type **FIND "Micki" < UNSORT.DAT > Micki.dat**.
20. To view the contents of the file, type **TYPE MICKI.DAT**. The ID, Name, and Phone Number appear in the file.
21. You can use filtering with the SORT command to sort data. To sort the UNSORT.DAT file using the pipe filter symbol, type **TYPE UNSORT.DAT | SORT**.

22. To perform sort in reverse order using the pipe filter, type **TYPE UNSORT.DAT |
    SORT /R**.

23. To further understand filtering, you will need to create a file with about 30 lines of
    text. So create a file named LOTSA.DAT and place the numbers 1 to 30 on separate
    lines.

24. Type **TYPE LOTSA.DAT**. The data scrolls off the screen before you can view it all.

25. Type **TYPE LOTSA.DAT | MORE**. This time the data is displayed a page at a time.
    You can press the Spacebar or Enter key to continue viewing the contents of the file
    and return to your command prompt.

26. Exit appropriately.

3.7 Learning about Batch Files

Batch files are a specific type of file that contain other DOS commands. In effect, you are
creating your own program. A batch file must have an extension of .BAT. However, you
can leave off the .BAT extension when you execute it. Let's get started with an exercise to
help you understand batch files.

1. Use the appropriate steps to get to the DOS prompt.

2. Go to the floppy drive.

3. Create a file named BATCH1.BAT using DOS EDIT with the following lines:

CLS
DIR /P
DATE
VER

4. Save the file.

5. To run the batch file, type **BATCH1**. The batch file runs.

6. Exit appropriately.

The commands you place in a batch file are executed sequentially from the top to the
bottom, so you need to consider where you place your commands in order to get the
output you desire. For example, in BATCH1.BAT, the first command is CLS for clearing
the screen. If you place it last in the batch file, then it will be executed last and your
output will scroll off the screen. A few additional DOS commands that augment batch
files are listed in Table 3.5. These commands allow you to utilize conditions and loops in
your batch files. A **condition** allows you to test for criteria and perform commands based
upon it. A **loop** occurs in a program when commands are executed repeatedly. Table 3.5
shows a sample of these additional commands.

Table 3.5: Other Batch File Related Commands

Command	Description	Example
CALL	Calls one batch program within another.	CALL BATCH2
CHOICE	Waits for the user to choose one of a set of choices. The /C option allows you to specify allowable keys for the user to enter. Note: CHOICE is not available in Windows 2000 or in Windows XP.	CHOICE /C 123 Press a Number
FOR	Runs a specified command for each file in a set of files.	FOR %a IN (*.BAT) DO ECHO %a is a batch file.
GOTO	Directs the batch file to a label in a batch program.	GOTO LabelA
IF	Performs conditional processing in batch programs.	IF EXIST x.dat echo xx.dat exists >> yes.txt
PAUSE	Suspends processing in a batch file and displays the message "Press any key to continue…."	PAUSE
REM	Records a comment or remark.	REM This batch file copies files.
SHIFT	Allows you to shift parameters that are passed to a batch file at the command prompt. You can normally pass up to nine at one time on the command line following the batch file name.	SHIFT

Let's look at a few of these now. The CALL command is used to call a batch file program from another. The basic syntax follows:

CALL *filename*

For example, the following statement will call or run the SALES.BAT file from REPORTS.BAT batch file:

REM This is REPORTS.BAT
CALL SALES.BAT

The REM statement is used for placing comments or remarks in a file. Remarks should include the author's name, how the batch file is to be executed, and a general description as to what the batch file accomplishes. For batch file troubleshooting, you can place REM in front of a command when you are testing your batch file. See the partial batch file below using the REM command:

REM Author: Joseph Sunday
REM To run: MYBATCH file1 file2
REM This batch file will copy file1 to file2

The FOR command allows you to run a specified command for every file that is in a set of files. The syntax follows:

> FOR *%variable* IN *(set)* DO command

The *%variable* is a variable name such as %a, %b, or %c. It will contain the contents of the files in the *set*. For example, the following statement will display the message that a given file is a batch file for all batch files (*.BAT). If the file is not a batch file, no message will be displayed.

FOR %a IN (*.BAT) DO ECHO %a is a batch file.

The GOTO command is used to cause the batch file to branch to another portion of the batch file. So, if you wanted the batch file to branch to another location to perform a different set of commands, you would use GOTO. The basic syntax is:

> GOTO *label*

This is where *label* is a user-defined label. For example, in the following partial batch file the GOTO command is used to branch to a label named LabelA, thus creating a loop. During execution of this batch file, the label, LabelA, is the first command statement but the command interpreter ignores it. Next, commands following the label are executed. When they are finished, the GOTO command moves processing back to LabelA. Again, the commands following the label are executed.

> :LabelA
> *commands*...
> GOTO LabelA

Note that the name of the label must be preceded with a colon.

The IF command can be used for conditional processing. There are several variations of the IF command depending upon the operating system. For more information about the IF command, type IF /? or HELP IF. The following syntax will test whether a file exists (using EXIST) and if it does, then the command specified, as *command* will execute.

> IF EXIST *filename command*

For example, the following command will test if the file named x.dat exists and if so, its contents is displayed.

IF EXIST x.dat TYPE x.dat

Note that Windows XP supports the use of an ELSE clause to perform a command if the condition is false. However DOS and the DOS command mode in Windows 98 do not support the ELSE clause. Thus, the following would work in Windows XP to display the contents of the file if it exists and display "File not found" if it does not exist:

IF EXIST x.dat (type x.dat) ELSE echo File not found

The use of parentheses is required when using the ELSE clause.

The PAUSE command allows you to suspend processing in a batch program by displaying a message "Press any key to continue…". To use it, simply enter the command as follows:

PAUSE

One very useful concept with batch files is the ability to execute a batch file and add variables after its name on the command line. These variables are called **positional parameters**. You can pass up to nine positional parameters on the command line after the batch file name. The positional parameters are named %1 though %9. You can add more than nine but you must use the SHIFT command to shift or move the data one position to the left. For example, to give the batch file named PAYDATA.BAT three variables, "001" for Employee ID, $25.00 for Rate, and 35 for Hours, you would enter:

PAYDATA 001 $25.00 35

So, "001" is referenced in the batch file as %1, $25.00 is %2, and 35 is %3. To display these three positional parameters in a batch file, you would enter the following:

ECHO %1 %2 %3

Let's look at a practical example using positional parameters. In this next batch file PARAM1.BAT, the COPY command will copy the file name specified as the first parameter %1 to the file name specified as the second parameter %2. The TYPE command will be used to display the contents of the file name specified as %1.

COPY %1 %2
TYPE %1

Figure 3.21 shows a screenshot of PARAM1.BAT and the output of its execution.

Figure 3.21: A Batch Program File Using Positional Parameters

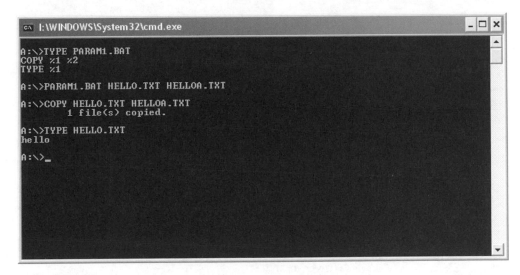

The CHOICE command allows you to create a menu in your batch file, but it is not available in Windows 2000 or Windows XP. The basic syntax is:

CHOICE /C *choices text*

The /C option lets you specify letters that are allowed. The default is Y for Yes or N for No.

The CHOICE command works by testing the key that was pressed. It uses the IF command to accomplish this. You use an option on the IF command called ERROR-LEVEL for testing the key that is pressed as in this example:

IF ERRORLEVEL *number command*

The *number* represents the key the user presses. Once the user presses a key, the *command* is executed.

Let's look at an example. In the following batch file a menu is created. The first command, @ECHO OFF turns off echo, which means that commands are not displayed on the screen as they are executed. The next few ECHO commands create the menu list that will be displayed on the screen. Next comes the CHOICE command. The /C options allows you to specify the letters ABC as valid keys the user can press. The first key assigned returns an ERRORLEVEL value of 1 when entered by the user. The second returns an ERRORLEVEL of 2 and so on. For example, if the user pressed A for a directory listing, because A is the first item number, then the ERRORLEVEL of 1 is set. Hence, the batch file causes execution to GOTO LabelA. You need to place the IF ERRORLEVEL commands in decreasing order. As you can see in the batch program, ERRORLEVEL 3 is tested first. Once the commands after the label name are executed, the GOTO TOP command causes processing to branch to the beginning of the batch file where :TOP is the second command in the batch file. This is the loop process in action. Ultimately pressing C causes ERRORLEVEL 3 to be set and the batch file ends.

```
@ECHO OFF
:TOP
CLS
ECHO                --------- MENU ---------
ECHO.
ECHO                A. DIR LISTING
ECHO                B. MEM LISTING
ECHO                C. EXIT
ECHO.
CHOICE /C:ABC SELECT A KEY
IF ERRORLEVEL 3 GOTO END
IF ERRORLEVEL 2 GOTO LabelB
IF ERRORLEVEL 1 GOTO LabelA
:LabelB
                MEM /C/P
                PAUSE
                GOTO TOP
:LabelA
                DIR A:\ /W/P
                PAUSE
                GOTO TOP
:END
```

You can terminate a batch file program by pressing Ctrl + C.

Figure 3.22 shows a screenshot of the execution of the batch file and running the first option, A, for a directory listing.

Figure 3.22: Screenshot of the Menu Batch File

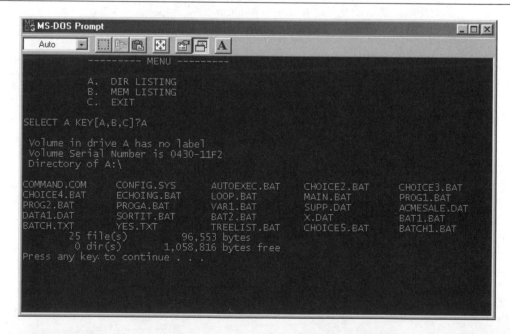

3.8 Learning about Additional Commands

The XCOPY command is more powerful and faster than the COPY command. The XCOPY command allows the copying of directories, subdirectories, and files. The syntax of the XCOPY command is:

> XCOPY *source-file destination-file*

If the destination is omitted, the XCOPY command copies the source files to the current directory. If you have a disk containing files in subdirectories to be copied to a different capacity disk, the XCOPY is a great command to use.

Two useful options the XCOPY allows are /S and /E. The /S option copies subdirectories except empty ones. The /E copies only empty subdirectories. Used together, you get all files and subdirectories in a directory. Figure 3.23 shows the XCOPY command with both the /S and /E options.

Figure 3.23: Using the XCOPY with the /S and /E Options

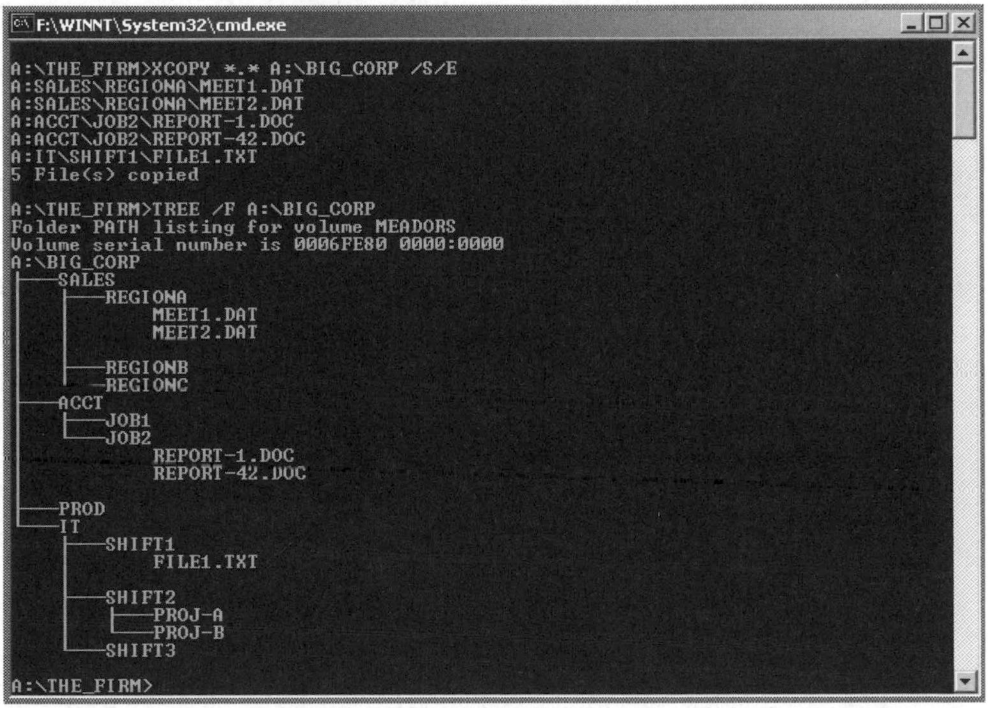

In Figure 3.23, you can see that the command **XCOPY *.* A:\BIG_CORP /S /E** copies everything from the current directory, which is THE_FIRM, to the directory named BIG_CORP. The command **TREE /F A:\BIG_CORP** displays the tree showing you the command worked.

The **ATTRIB** command sets, removes, or shows the attribute of a file or a directory. Attributes change how a file or directory displays on the screen or what can be done with the file or directory. Possible attributes include read-only, archive, system, and hidden. The **read-only attribute** protects files so they cannot be accidentally changed or deleted. For example, the AUTOEXEC.BAT and CONFIG.SYS files start and configure a computer. The **archive attribute** marks files that have changed since they were last backed up by a backup program. The XCOPY and MSBACKUP commands use the archive attribute as well as any other backup software program. The **system attribute** designates a file as a system file. Files with this attribute do not show in directory listings. The **hidden attribute** allows file hiding and even directory hiding. If someone has hidden files, and you need to see all files without having to change the attributes, the DIR /AH or ATTRIB commands display the files no matter what attributes they possess.

The syntax of the ATTRIB command is:

> ATTRIB *options filename*

Set each attribute using the +options where the +R option adds the read-only attribute, the +S option adds the system attribute, the +H adds the hidden attribute, and the +A adds the archive attribute. Remove each attribute using the -R, -S, -H, or -A option with the ATTRIB command. One command can set more than one attribute on files or directories. For example, to make the file PAY.TXT hidden and read-only, type **ATTRIB +R +H PAY.TXT**.

An excellent use of the archive attribute is to copy multiple files onto a disk. Normally, there is no warning when copying too many files to a disk that cannot hold them, until it is too late and the disk runs out of room. The BACKUP and RESTORE commands are traditionally used to get around this problem. However, the ATTRIB command (with the +A option) and the XCOPY command (with the /M option) together can copy files across multiple disks. For example, copying all the DOS files requires multiple disks. When copying all these files to a disk, the operating system copies as many files as possible, then produces an error saying the destination disk is full. To prevent this problem, use the ATTRIB command and assign all the files the archive attribute. Enter **ATTRIB +A C:\TESTDIR*.*,** then insert a disk into the A: drive and enter **XCOPY C:\TESTDIR*.* A: /M**. When the XCOPY command stops to display the error message saying the destination disk is full, insert a new floppy into the A: drive and repeat the command. The /M option tells XCOPY not to copy the same files again. The /M option only copies files that have the archive attribute set. Once the files are copied, they no longer have the archive attribute.

Let's look at a few screenshots to reinforce the material on ATTRIB. In Figure 3.24 the command **ATTRIB +R MANAGER.DAT** is used to set the read-only attribute. You cannot modify a file with this attribute; the file is protected. You can see the DEL MANAGER.DAT is used to attempt to delete it; however, an error message denying you access is displayed, indicating you cannot delete it. Only after the command **ATTRIB –R MANAGER.DAT** is used to turn off the read-only attribute can the file be deleted.

Figure 3.24: Using the ATTRIB Command with the +R and –R Options

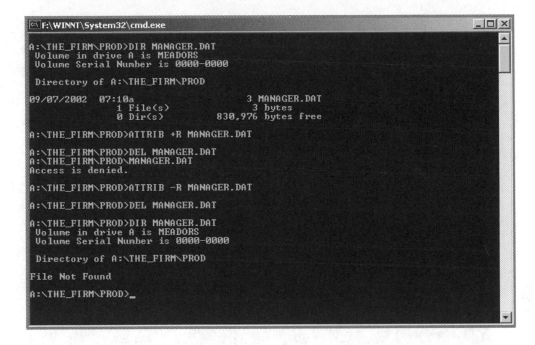

In Figure 3.25, the **ATTRIB +H VP.DAT** command is used to hide the file VP.DAT. With this attribute set you cannot even see the file with a DIR command, much less delete it as attempted in the figure. When the **DEL VP.DAT** command is attempted with the hidden attribute set, a "Could Not Find" message appears.

If a file is hidden, how do you unhide it? If you type ATTRIB command with no options, you will see all attributes for all files, hidden or not. This is done in Figure 3.25.

To unhide the VP.DAT file, type **ATTRIB –H VP.DAT**.

Figure 3.25: Using the ATTRIB Command with the +H and –H Options

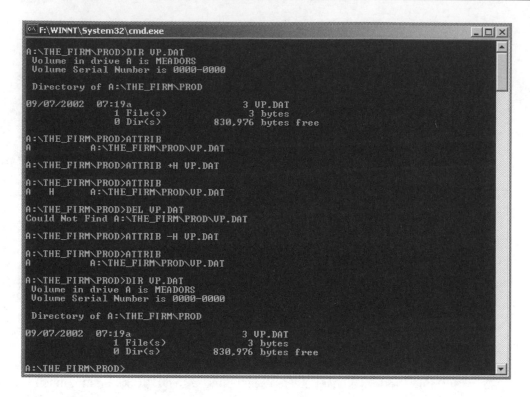

Chapter Summary

- All computers need an operating system to function. The Disk Operating System (DOS) runs on most PCs and was commercialized by Microsoft. There are two methods of accessing the DOS commands. You can boot up the DOS operating system or you can go into the DOS command prompt in the GUI of the Windows operating systems.
- The DOS file system is hierarchical (tree-like) in nature. The file system uses a directory to store files or other directories. A file cannot store another file or directory. The root directory (\), is the top level directory.
- DOS commands are either internal to COMMAND.COM or external, residing individually on the computer's disk. The two basic categories of commands are file and directory commands.
- In order to navigate the DOS tree, you use the CD command. A full path begins from root, \, and a partial path does not.
- You use the MD or MKDIR command to create directories. The RD or RMDIR command removes directories.
- The COPY command makes a copy of a file. The MOVE command allows you to change a file's location to another directory. The REN command renames a file. The

TYPE command is used to display the contents of a file. The DEL command deletes a file.

- The command prompt is where all commands are entered. When you get C:\> that is your command prompt. The C: drive is your hard drive. The A: drive is your floppy drive. Your prompt can be changed with the PROMPT command.
- You can manage a subset of the files by using wildcard characters. The * symbol matches all characters. The ? symbol matches a single character position.
- Redirection is the idea of changing a command's normal location for accepting input or displaying output. The > symbol redirects output. The >> symbols redirect and append output. The < symbol redirects input. Filtering uses the pipe symbol (|) to process data between commands.
- The AUTOEXEC.BAT is a boot file that will be executed during the system's boot sequence. The CONFIG.SYS file contains configuration information that enhances your PC's capabilities. Batch files have an extension of .BAT.

Review Questions

1. You want to make a duplicate of a file named data3.dat. The new file is to be named data4.dat located in the same directory. What command will do this for you?
   a) COPY data3.dat data4.dat
   b) COPY data4.dat data3.dat
   c) COPY data4.dat data4.dat
   d) MOVE data3.dat data4.dat

2. Which of the following is considered a full path?
   a) ..\.
   b) fun2\file2.txt
   c) \
   d) dir4

3. What symbol is used to redirect and append output to a file?
   a) >
   b) >>
   c) |
   d) ?

4. Which of the following will wildcard a single character position?
   a) ?
   b) *
   c) |
   d) >

5. Zac wants to display all the files that have a 0 for the fourth position and a 7 for the fifth position in their file name regardless of the characters in other positions. What command will accomplish this?
   a) DIR *05?
   b) DIR ?05*
   c) DIR 4 and 5
   d) DIR ???07*

6. Which of the following commands is used to display the contents of a file named FunA.dat?
   a) DIR FunA.DAT
   b) TYPE FunA.DAT
   c) MEM /C /P > FunA.DAT
   d) CD FunA.DAT

7. Which of the following will wildcard multiple character positions?
   a) ?
   b) *
   c) |
   d) <

8. Which of the following will delete all files with an extension of .DAT?
   a) DIR *.DAT
   b) DEL *.BAT
   c) DEL ?.DAT
   d) DEL *.DAT

9. Which of the following will create a directory?
   a) DIR
   b) FDISK
   c) MOVE
   d) MKDIR

10. Which of the following allows you to partition a hard disk drive?
    a) FDISK
    b) MKDIR
    c) SYS
    d) FORMAT

11. Which of the following allows you to transfer system files to the floppy?
    a) FDISK
    b) SYS C:
    c) SYS A:
    d) FORMAT A:

12. Which of the following is the correct load order for the DOS system files?
    a) MSDOS.SYS, IO.COM, CONFIG.SYS, COMMAND.COM, AUTOEXEC.BAT
    b) IO.SYS, MSDOS.SYS, CONFIG.SYS, COMMAND.COM, AUTOEXEC.BAT
    c) IO.SYS, MSDOS.SYS, COMMAND.COM, CONFIG.SYS, AUTOEXEC.BAT
    d) AUTOEXEC.BAT, IO.SYS, MSDOS.SYS, CONFIG.SYS, COMMAND.COM

13. Which of the following commands will run a batch program within a batch program?
    a) IF
    b) CHOICE
    c) CALL
    d) REM

14. Which of the following commands is a comment in a batch program?
    a) IF
    b) CHOICE
    c) CALL
    d) REM

15. What symbol is used to redirect but not append output to a file?
    a) >
    b) >>
    c) |
    d) ?

16. What command is used to sort a file named data1.dat in column 10 in reverse order and redirect and append the output to a file named data2.dat?
    a) TYPE data1.dat | SORT \+10 \R >> data2.dat
    b) TYPE data1.dat | SORT /+10 /R > data2.dat
    c) TYPE data2.dat | SORT /+10 /R >> data1.dat
    d) TYPE data1.dat | SORT /+10 /R >> data2.dat

17. What command is used to copy all the files that begin with an A, regardless of the middle characters with an extension of .TXT from A:\Prod to C:\Backup?
    a) COPY A?.TXT BACKUP
    b) COPY A:\Prod\A*.TXT C:\Backup\A*.TXT
    c) COPY C:\BACKUP\A*.TXT A:\A*.TXT
    d) COPY A:/A*.TXT C\Backup\A*.TXT

18. What key saves a file when using COPY CON?
    a) F5
    b) F6
    c) F8
    d) F7

19. What key bypasses the AUTOEXEC.BAT and CONFIG.SYS files after the POST during boot-up?
    a) F5
    b) F7
    c) F8
    d) F9

20. What key performs step-by-step confirmation of the lines in the AUTOEXEC.BAT and CONFIG.SYS files after the POST during boot-up?
    a) F5
    b) F7
    c) F8
    d) F9

Lab Projects

Note that these projects assume the PC is powered on. Although you can perform these projects in other versions of Microsoft Windows or DOS with minor changes, the assumed operating system is Windows 2000 unless otherwise specified.

Lab Project 1

The goal of this lab is to create a tree structure on the floppy.

1. Go to the DOS command prompt.

2. In order to change drive locations to your floppy, type **A:** and press Enter.

3. Type **MD Project1** and press Enter.

4. Type **CD Project1** and press Enter.

5. Type **MD dir2** and press Enter.

6. Type **MD dir3** and press Enter.

7. Type **CD dir2** and press Enter. Notice your prompt changes.

8. Type **CD** and press Enter.

9. Record the directory location.

   _____

10. Type **MD dir2a** and press Enter.

11. Type **CD dir2a** and press Enter.

12. Type **CD** and press Enter.

13. Type **COPY CON FILE2A.TXT** and press Enter. The prompt moves to the beginning of the next line.

14. Insert the following:

    This is DOS.
    I'm having fun!

15. Press **F6** and press Enter.

16. Record what happens when you press F6.

    _____

17. Type **CD ..** and press Enter.

18. Record the new directory location.

    _____

19. Type **COPY CON FILE2.DAT** and press Enter. The prompt moves to the beginning of the next line.

20. Insert the following:

This is FILE2.DAT.
Bye
The End!

21. Press **F6** and press Enter.

22. Type **COPY FILE2.DAT FILE22.DAT** and press Enter.

23. Record another way you could have created file22.txt.

_____

24. Type **CD ..** and press Enter. This should take you to A:\ Project1.

25. Type **CD** and press Enter.

26. Record the output.

_____

27. Type **MD dir3** and press Enter.

28. Type **ECHO Hello! > file32.txt** and press Enter. This is another method to create a file.

29. Type **COPY file32.txt file33.txt** and press Enter.

30. Exit appropriately.

Lab Project 2

In this lab project, you will review a list of commands and then draw the tree based upon the commands.

1. Review the list of commands below. Pay careful attention to the order of the commands listed. Create them on the floppy.

> **A:**
> **MKDIR Project2**
> **CD Project2**
> **MKDIR payroll**
> **MKDIR general**
> **MKDIR fun**
> **CD payroll**
> **MKDIR paychecks**
> **MKDIR payday**
> **CD payday**
> **ECHO Howdy, Todd Meadors > pay1.dat**
> **ECHO Hi, Zac, Jessie and Micki >> pay1.dat**
> **COPY pay1.dat pay2.dat**
> **CD ..\\..\\general**
> **ECHO Hi ya'll > gen1.dat**
> **COPY gen1.dat gen2.dat**
> **COPY gen2.dat \\Project2\\gen3.dat**
> **CD \\Project2\\payroll**
> **COPY payday\\pay1.dat paychecks\\payC.dat**
> **CD A:**

2. Draw the tree structure. You have been given a starting point, A:\\. The rest is up to you.

> A:\\
> | (Draw the tree here!)

Lab Project 3

The goal of this lab project is to have you gain additional practice creating tree structures in DOS. Note: Files have an extension and directories do not have an extension.

1. Go to the command prompt, create a directory named Project3, and then change locations to it.

2. Create the tree shown in Figure 3.26.

Figure 3.26: Sample Tree Structure to Be Created

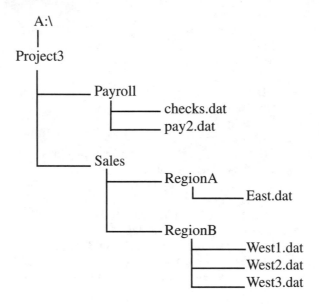

Lab Project 4

The goal of this lab project is for you to be able to distinguish between navigating to a parent, a child, and a sibling directory. A sibling directory is defined as a directory that is on the same hierarchical level as another. This lab project is dependent upon completion of Lab Project 3.

1. Go to the DOS command prompt.

2. Insert the floppy with the tree structure from Lab Project 3.

3. Change directory locations to A:\. Your prompt will reflect the change.

4. Type **CD Project3** and press Enter.

5. Type **CD Sales** and press Enter. Your prompt will reflect the change.

6. In order to change to a child directory, you must use the child directory's name.

   Type **CD RegionA** and press Enter. Your prompt will reflect the change.

7. In order to change to a sibling directory, type **CD ..\RegionB** and press Enter. The format is CD ..*sibling-name* where *sibling-name* is the name of a sibling directory. In Figure 3.26, RegionA and RegionB are sibling directories.

8. Answer this question: There is another sibling relationship in the Figure 3.26. What are the sibling names?

   _____

9. In order to change back to RegionA, type **CD ..\RegionA** and press Enter.

10. Record what this command accomplishes.

    _____

11. In order to change to the parent directory, type **CD ..** and press Enter.

12. Record what this command accomplishes.

    _____

13. Exit.

Lab Project 5

The goal of this lab project is to utilize some of the features of the DOS editor.

1. Go to the DOS command prompt.

2. Type **MD Project5** and press Enter.

3. Type **CD Project5** and press Enter.

4. Type in **EDIT file27.txt** and press Enter.

   Computers are exciting to learn.
   DOS is fun but I need to brush up on my keyboarding skills.
   See ya!
   Todd Meadors

5. Press Alt + F to access the File menu.

6. Press S to Save the file.

7. Move your cursor to the first letter of the first line.

8. To search for a word, press Alt + S to access the Search menu.

9. Press F for Find. The Find dialog box appears.

10. To search for the text **exciting**, type **exciting** and press Enter. The text **exciting** highlights.

11. Press Alt + E to access the Edit menu.

12. Press t for Cut.

13. Type the word **fun**. The text **exciting** was just replaced with the text **fun**.

14. Press Alt + F to access the File menu.

15. Press S to Save the file.

16. Explore some of the other EDIT commands.

17. List the ones you find most useful.

   _____

   _____

18. Press Alt + F and then X to Exit.

19. Exit.

Lab Project 6

The goal of this lab project is to help you correctly use the XCOPY, RD, and DEL commands.

1. Go to the command prompt.

2. Insert a disk into the A: drive and create two files, one called **FRIEND.TXT** and another called **FOE.TXT**, and place the files in the A: drive's root directory. Use an editor to create the two files. Type two lines of text in each file.

3. From the C:\> prompt, type **XCOPY A:\F*.* A:\TEST1**. A message might appear asking if TEST1 is a file or a directory. If so, press **D** for directory. A message appears stating the number of files copied. If the message does not appear, redo steps 2 and 3.

4. From the C:\> prompt, type **DEL A:\TEST1*.*** and press Enter.

5. From the C:\> prompt, type **RD A:\TEST1**.

6. Verify that the directory is gone by typing **DIR A:\TEST1**.

7. Type the following commands:

> **MD \TEST2**
> **CD \TEST2**
> **MD SUB1**
> **MD SUB2**
> **MD SUB3**

8. Using an editor, create three files called **FILE1**, **FILE2**, **FILE3** and place the files in the SUB1 subdirectory. Verify that the files are created and in the current directory before proceeding.

9. Using the XCOPY command, copy all of the files located in SUB1 subdirectory to the SUB2 and SUB3 subdirectories on the floppy disk.

10. Exit.

Lab Project 7

The goal of this lab project is to help you correctly use the ATTRIB command.

1. Go to the DOS command prompt.

2. Create a directory name Project7.

3. Change directory locations to Project7.

4. Create two files, Secret1.DAT and Secret2.DAT.

5. To make the Secret1.DAT read-only, use the ATTRIB command with the +R switch. Type **ATTRIB +R A:\PROJECT7\Secret1.DAT**.

6. To prove that the +R switch works, attempt to delete the files. Type **DEL A:\PROJECT7\Secret1.DAT**.

7. Record the message.

_____

8. Try to modify the file with the DOS EDIT command.

9. Record the message.

_____

10. To make Secret1.DAT readable, type **ATTRIB –R A:\PROJECT7\Secret1.DAT**.

11. Attempt to modify the file by using the DOS EDIT command.

_____

12. Record whether or not you could edit the file.

13. To hide the file named Secret2.dat, type **ATTRIB +H A:\PROJECT7\Secret2.DAT**.

14. To verify that the file is hidden, type **DIR A:\PROJECT7\Secret2.DAT**. The file should not appear.

15. View the file named Secret2.dat by typing **ATTRIB**.

16. Record the output and identify each column.

_____

17. View the file named Secret2.dat by typing **DIR /AH**.

18. Record the output and identify each column.

_____

19. Unhide Secret2.dat.

20. Record the command used.

_____

21. Use the DIR command to view the file.

22. Record whether or not you could view it with the DIR command (with no options).

_____

23. Exit.

Lab Project 8

The goal of this lab project is for you to utilize the SORT command to sort data in ascending (lower to higher) and descending (higher to lower) order.

1. Get to the DOS command prompt.

2. Type **MD Project8** and press Enter.

3. Type **CD Project8** and press Enter.

4. Type **EDIT INPUT.DAT** and press Enter. The DOS editor opens.

5. Enter the following text making sure you press the Enter key at the end of each line. Each row is considered a Part Record and contains four fields. The first field is the Part Number and begins in column 1. The second field is the Part Name and begins in column 14, the third field is the Part Cost and begins in column 35, and the last field is the Part Quantity and begins in column 43. You are required to place the fields in the correct columns.

102A	Wrench	12.00	76
105T	Drill	99.00	17
103F	Saw	55.00	38
101A	Hammer	35.00	64

6. Save the file and exit the editor.

7. Your prompt will return.

8. Sort the data in ascending order. Type **SORT < INPUT.DAT** and press Enter.

9. Record the output.

_____

10. Sort the data in descending order. Type **SORT /R < INPUT.DAT** and press Enter.

11. Record the output.

_____

12. Sort the data in descending order and redirect the output to another file. Type **SORT /R < INPUT.DAT > REVERSE.DAT** and press Enter.

13. Record the output.

_____

14. To sort the file by the Part Name, type **SORT /+14 < UNSORT.DAT**.

15. Record the output.

_____

16. Sort by the Part Cost name column in reverse order.

17. Record the command you used.

_____

18. Sort by the Part Quantity column reverse order and redirect the output to a file.

19. Record the command you used.

20. Exit.

Lab Project 9

The goal of this lab project is to create an AUTOEXEC.BAT file using COPY CON.

1. Get to a DOS prompt.

2. Create a bootable floppy.

3. Note that when Enter is pressed after each line, NO messages appear!

4. Go to A:\.

5. Create an AUTOEXEC.BAT file using either the COPY CON method or EDIT method. Type **COPY CON AUTOEXEC.BAT** or **EDIT AUTOEXEC.BAT.**

6. Insert the following text:

    REM The AUTOEXEC.BAT file.
    ECHO This is the AUTOEXEC.BAT file executing...
    VER
    DATE
    PAUSE
    TIME
    PROMPT My DOS Prompt: $G

7. Reboot the computer and watch the screen as the AUTOEXEC.BAT file automatically executes.

8. Close your DOS command prompt.

Lab Project 10

The goal of this lab project is to understand DOSKEY.

1. Get to the DOS prompt.

2. To install the DOSKEY program from the prompt, type **DOSKEY**.

3. Insert the floppy disk into the floppy drive.

4. From the command prompt, type

   A:
   MD \SPECIAL

5. The DOSKEY program allows use of the arrow keys and function keys (the F1 through F12 keys on the keyboard). Press the **up arrow** once. The last typed command appears at the prompt. If the last command does not appear, then DOSKEY is not loaded properly. Perform Steps 1 through 6 again.

6. Press the **up arrow** until the MD\SPECIAL command appears at the command prompt. *Do not press Enter* when the command appears.

7. Press the **left arrow** key 3 times until the blinking cursor is under the *I* in SPECIAL. *DO NOT PRESS ENTER*. The arrow keys, when used with DOSKEY, allow movement through the command line.

8. While the cursor is blinking under the letter I, type the letter **T**. *DO NOT PRESS ENTER*. The command should now read MD\SPECTAL. DOSKEY is automatically in the type over mode.

9. Press **Ins** (Insert), but *DO NOT PRESS ENTER*. The cursor changes to a blinking box.

10. While the cursor is a blinking box over the letter A, type the letter **R**. The command changes to SPECTRAL, because pressing the Ins key causes DOSKEY to go into the insert mode.

11. Press Enter to create a directory named SPECTRAL.

12. Press the **up arrow** key once. The last command typed appears on the screen.

13. Press **Esc** (Escape) once. Esc erases the current command from the command prompt.

14. Press **F7** once. A list of all typed commands appears on the screen.

15. Press **F9** once. The Line Number prompt appears. While in DOSKEY, F9 allows you to enter a specific command based on a line number (that showed when you pressed F7).

16. Type the number corresponding to the **A:** command, then press **Enter**. The A: command appears at the command prompt.

17. Press **Esc** to clear the command.

18. Press and *hold down* **Alt**. While holding Alt down, press **F7** once, then release both keys. DOSKEY uses Alt + F7 to clear the list of commands DOSKEY tracks. Sometimes, pressing the F7 key is cumbersome to use because the command list is so long. Clearing the command list allows DOSKEY to start over tracking commands.

19. Press **F7** to verify the command list is clear.

20. Exit.

Lab Project 11

The goal of this lab project is for you to be able to utilize the redirection and pipe symbols.

1. Get to a DOS command prompt.

2. Type **MD Project11** and press Enter.

3. Type **CD Project11**and press Enter.

4. Create 10 files and five directories within Project11. Use appropriate names.

5. Record the name of the command you used to create one of the files.

_____

6. Record just the name of the command you used to create one of the directories.

7. In order to redirect output and create a new file, type **ECHO Directory Listing > DIRL-IST.DAT** and press Enter.

8. In order to redirect and append output, type **DIR >> DIRLIST.DAT** and press Enter.

9. In order to redirect and append output again, type **ECHO All Done! >> DIRLIST.DAT** and press Enter.

10. Verify the contents of the file named DIRLIST.DAT.

11. Record the command used.

_____

12. In order to overwrite the contents of the file named DIRLIST.DAT, type **ECHO > DIRL-IST.DAT** and press Enter.

13. Issue another command to redirect output.

14. Record the command used.

_____

15. Issue another command to redirect and append output.

16. Record the command used.

_____

17. To use the pipe symbol, type **SORT /? I MORE**. The listing is displayed one screen at a time.

18. Issue another command to filter using the pipe symbol.

19. Record the command used.

_____

20. Close the DOS command prompt.

Lab Project 12

The goal of this lab project is for you to be able to utilize the wildcard symbols.

1. Go to the DOS command prompt.

2. Change directory locations to A:\. Your prompt will reflect the change.

3. Type **MD Project12** and press Enter.

4. Type **CD Project12** and press Enter.

5. Create the following 12 files in Project12. These represent payroll files with the three-character month followed by the two-digit day of the month.

JAN07.DAT	**DEC04.DAT**	**JUL04.DAT**
JAN02.DAT	**DEC21.TXT**	**JUN06.DAT**
JAN03.DAT	**JUL04.TXT**	**JUL13.TXT**
JAN04.TXT	**FEB07.DAT**	**DEC07.DAT**

6. In order to display all files ending in .DAT, type **DIR *.DAT** and press Enter.

7. Record the output.

   _____

8. Issue a command to display all files ending in .TXT.

9. Record the command.

   _____

10. Record the output.

    _____

11. In order to display all files that begin with a J and end with .DAT, type **DIR J*.DAT** and press Enter.

12. Record the output.

    _____

13. Issue a command to display all files that begin with a D and end in .DAT.

14. Record the command.

    _____

15. Record the output.

    _____

16. In order to display all files for January, type **DIR JAN*** and press Enter.

17. Record the output.

    _____

18. Issue a command to display all files for July.

19. Record the command.

_____

20. Record the output.

_____

21. In order to display files for the fourth day of the month (a 0 in the fourth position and a 4 in the fifth position), type **DIR ???04*** and press Enter.

22. Record the output.

_____

23. Issue a command to display just the files for the seventh day of each month.

24. Record the command.

_____

25. Record the output.

_____

26. Exit.

Lab Project 13

The goal of this lab project is to help you understand batch files.

1. Get to a DOS command prompt.

2. Open the DOS editor and create a file named Project13.dat with the following data: Employee ID in column 1, Employee Name in column 6, and Pay Rate in column 20.

0006	Zelda, Jack	44.20
0001	Smith, Zac	40.25
0003	Brown, Mike	29.50
0009	Smith, Zoe	35.00
0008	Addams, Al	24.00
0002	Cook, Jim	20.01
0004	Cook, Jinny	22.01
0005	Adams, Art	17.50
0007	Meade, Timmy	16.11

3. Save the file and close the editor.

4. Open the DOS editor and create a file named PROJ13.bat.

5. Enter the following commands. Replace the remarks with appropriate comments.

```
REM "YOUR NAME"
REM "PROGRAM NAME"
REM "PROGRAM DESCRIPTION"
@ECHO OFF
ECHO COPYING THE FILE TO MAKE A BACKUP:
COPY %1 %2
ECHO SORTING THE DATA:
TYPE %2 | SORT | MORE
ECHO REVERSE SORT:
TYPE %2 | SORT /R | MORE
ECHO SORTING THE DATA BY PAY RATE:
TYPE %2 | SORT /+20 | MORE
ECHO SORTING THE DATA BY LAST NAME:
TYPE %2 | SORT /+6 | MORE
ECHO SORTING BY PAY RATE AND REDIRECTING TO A NEW FILE:
TYPE %2 | SORT /+20 > PAY.DAT
ECHO CONTENTS OF PAY.DAT:
TYPE PAY.DAT | MORE
```

6. Save the file and close the editor.

7. Run the batch program by typing **PROJ13 PROJ13.DAT PROJ13C.DAT**. The parameter %1 is PROJ13.DAT and %2 is PROJ13C.DAT. Watch the screen as the batch file executes.

8. Create another batch program file named PROJ13A.BAT with the following:

   ECHO RUNNING PROJ13A.BAT
   SORT PAY.DAT

9. Modify PROJ13.BAT and append these two statements:

   CALL PROJ13A.BAT
   ECHO RETURNED TO PROJ13.BAT

10. Rerun the batch program by typing **PROJ13 PROJ13.DAT PROJ13C.DAT**. This time PROJ13A.BAT is called from PROJ13.BAT.

11. Exit.

Lab Project 14

The goal of this lab project is to combine your understanding of the DOS tree structure and batch files. You will write a batch file that will be used to create the tree structure from Project 3.

1. Create a directory named Proj14 and change to it.

2. Create a batch file named TREEMAKE.BAT to create the tree structure in Figure 3.26 of Project 3.

3. On a separate piece of paper record your batch file.

Lab Project 15 Challenge

The goal of this lab challenge project is to have you create a bootable floppy disk and add an AUTOEXEC.BAT and CONFIG.SYS file.

1. Format a floppy and make it bootable.

2. In the AUTOEXEC.BAT file, do the following:

      a. Turn echoing of command off.
      b. Place an appropriate PATH command that includes the references to the external commands depending upon your operating system.
      c. Display the version of the operating system.
      d. Enable DOSKEY.
      e. Set the PROMPT to include the current drive path and the greater than symbol.

3. In the CONFIG.SYS file do the following:

      a. Enable XMS.
      b. Enable EMS.
      c. Place DOS into HMA.
      d. Enable UMBs.
      e. Set concurrent files to 28.

4. Boot the floppy.

5. Immediately after the POST, press F5 to bypass AUTOEXEC.BAT and CONFIG.SYS.

6. Record the amount of all types of memory.

    _____

    _____

    _____

7. Boot the floppy again.

8. This time press F8 to perform the step-by-step confirmation. This gives you the chance to decide whether or not you want to execute a particular command within AUTOEXEC.BAT and CONFIG.SYS. This time do not load XMS and EMS.

9. Record the amount of all memory.

    _____

    _____

    _____

10. The amounts in step 6 and step 9 should differ. Record why.

    _____

    _____

    _____

11. Exit appropriately.

Lab Project 16 Challenge

The goal of this lab challenge project is to create a batch file menu. You will need to use either DOS or the Windows 98 command prompt mode.

1. Create a menu on the floppy that does the following:
   Allows the user to enter:

   - The number 1 for a directory listing of files on the floppy with an extension of
   - .BAT and redirect the output to a file named Proj15.dat.
   - The number 2 for changing the date.
   - The number 3 for changing the time.
   - The number 4 for formatting the floppy.
   - The number 5 for transferring system files.
   - The number 6 for running another batch file that contains this command:
       FDISK
   - The number 7 for exiting the menu.

2. Include appropriate remarks.

3. Run and test your menu batch program.

Internet Discovery

Internet Discovery Lab 1

The goal of this Internet Discovery Lab is to help you use the Internet to discover batch file samples and concepts.

1. Connect to the Internet.

2. Open your web browser

3. Access the **www.google.com** web site.

4. Search for DOS Batch files.

5. Visit at least three sites that discuss batch files.

6. Record the web site addresses.

_____

_____

_____

7. Look for a batch file under 10 lines in length. Record the commands in it and explain each line.

8. Exit.

Internet Discovery Lab 2 Challenge

The goal of this Internet Discovery Lab is to use the Internet to download a free copy of DOS.

1. Connect to the Internet.

2. Open your web browser

3. Access the **www.freedos.org** web site.

4. Download a copy of the free DOS.

5. Install and run it.

Soft Skills: Help Desk Support

1. A user passes you in the hall and asks you how to make a floppy bootable and keep the current files that are already on it. How do you handle this situation?
2. You are working the midnight help desk when a user calls and asks you for assistance. They user wants to partition a floppy with the FDISK command. What assistance do you offer the user?
3. You are the Team Leader for the second shift help desk for Arbornomics, Inc. A user calls and asks you and your team members for assistance. There is a mix of DOS, Windows 98, Windows NT, Windows 2000, and XP operating systems on the PCs. The user wants to create a batch file that contains the TREE command. The user wants to be able to run this on any operating system. What do you tell the user?

Critical Thinking

1. Prove why the **CD ..** command issued at the root directory of a drive changes you to root.
2. Compare the DOS file system to a company's organization chart.
3. Explain the difference between the COPY, MOVE, and REN commands.
4. Why does each directory have a dot (.) directory except the root directory?

 # Study Skills

Relying on Your Textbook and Notes

- Do you read your textbook? You should. If you read your textbook, your test scores will improve. Although reading technical information such as a textbook can be difficult, it is important that you spend time each day reading your textbook. Allocate at least one to two hours a day!

- As you read your textbook, use a brightly colored (such as yellow) marker to highlight definitions, facts, explanations of statements, and any other information that you feel is important. This will provide a great "summary" of facts because the important ones will be highlighted.

- Make sure you answer the questions at the back of the book—at least twice. This will improve your score on a test. The first time, try to do them on your own. Then, if you have trouble, research the chapter. Additionally, go through the Soft Skills: Help Desk Support and the Critical Thinking sections in each chapter to further your understanding of material.

- Do **not** wait until the night before a test to do your major studying. If you've read your textbook chapters all along, you should not have to cram the night before. Get a good night's rest the night before the test and eat a good breakfast the morning of the test. Review the Chapter Summary right before the test, to help you get the most important facts.

- Take good lecture notes. Your lecture notes should include any class discussion that is relevant to the chapter. If the instructor says something more than once, it is probably important so write it down.

- Rewrite the chapter, page by page, in your own words. Although this takes a little time, it proves invaluable for test preparation. By rewriting the chapter in your own words, you are going over it at least twice—once for reading and again for rewriting. So, when it comes time for a test, you've already gone over the material. So, at this point, you can simply review your notes. Plus, some instructors allow you to use your notes (but not your textbook) for a test!

Study Skills: Self-Study Question(s)

1. Identify at least one Study Skills item from this section that you performed prior to a test.
2. Did you read your textbook today?

4

Chapter 4
Introduction to
Microsoft
Windows 98

OBJECTIVES

The goal of this chapter is twofold:
- To introduce you to Windows 98.
- To help you prepare and pass the following sections of the A+ Operating System Technologies Exam:

A+ Operating System Technologies Exam Objectives
covered in this chapter (and corresponding page numbers)

1.1 Identify the major desktop components and interfaces, and their functions. Differentiate the characteristics of Windows 9x/Me, Windows NT 4.0 Workstation, Windows 2000 Professional, and Windows XP.

1.2 Identify the names, locations, purposes, and contents of major system files.

1.5 Identify the major operating system utilities, their purpose, location, and available switches.

Domain 2 Installation, Configuration and Upgrading

2.1 Identify the procedure for installing Windows 9.x/Me, Windows NT 4.0 Workstation, Windows 2000 Professional, and Windows XP and bringing the operating system to a basic operational level.

2.2 Identify steps to perform an operating system upgrade from Windows 9.x/Me, Windows NT 4.0 Workstation, Windows 2000 Professional, and Windows XP. Given an upgrade scenario, choose the appropriate next steps.

2.3 Identify the basic system boot sequences and boot methods, including the steps to create an emergency boot disk with utilities installed fro Windows 9.x/Me, Windows NT 4.0 Workstation, Windows 2000 Professional, and Windows XP.

2.4 Identify procedures for installing/adding a device, including loading, adding, and configuration device drivers, and required software.

2.5 Identify procedures necessary to optimize the operating system and major operating system subsystems.

Domain 3 Diagnosing and Troubleshooting

3.1 Recognize and interpret the meaning of common error codes and startup messages from boot sequence, and identify steps to correct the problems.

3.3 Recognize common operational and usability problems and determine how to resolve them.

In this chapter, you will complete the following sections:
- 4.1 Understanding Microsoft Windows 98
- 4.2 Managing Files and Folders
- 4.3 Understanding the Windows 98 Registry
- 4.4 Understanding the Pre-Installation Steps of Windows 98
- 4.5 Understanding How to Install and Upgrade Windows 98
- 4.6 Troubleshooting the Windows 98 Installation
- 4.7 Understanding the Dual-Boot Feature of Windows 98 Systems
- 4.8 Configuring Windows 98
- 4.9 Understanding the Boot Process
- 4.10 Troubleshooting Application Problems

4.1 Understanding Microsoft Windows 98

Microsoft's Windows 98 is a popular operating system normally used by home users and small businesses. It is a 32-bit operating system that supports plug and play, DVD drives, TV tuner adapters, multitasking, 16-bit applications, and 32-bit applications. Windows 98 supports older (16-bit) applications better and takes less hard drive space and memory than NT Workstation and 2000 Professional, but it does <u>not</u> have the built-in security capabilities that NT and 2000 offer.

With Windows 98, Microsoft introduced **Internet Explorer** or **IE**. IE is a web browser for connecting to the Internet. It is integrated into Windows 98. Windows 98 also supports **Internet conferencing** and **Internet Connection Sharing** or **ICS**. With Internet conferencing, you can use your computer and the Internet to communicate with others around the world. With ICS, multiple computers can connect to the Internet via a computer using Windows 98. The Windows 98 computer is connected to both the Internet and the other computers.

With Internet Connection Sharing, you are able to share an Internet connection using your Windows 98 computer. This is a very useful feature of Windows 98 because it allows you to have one connection to an Internet Service Provider (ISP). The only downside is if the Windows 98 computer that is connected to the Internet crashes, you lose access to the Internet for the other computers.

The file systems that Windows 98 supports for hard drive partitions are FAT16 and FAT32. A FAT16 partition can be converted to a FAT32 partition using a command called CVT1.EXE. Once a partition is converted to FAT32, it cannot be reconverted to FAT16.

When a computer's drive has been formatted with an NTFS partition, a FAT16 partition must be created before installing Windows 98.

The Windows 98 Desktop

After booting Windows 98 the desktop appears. The **desktop** is the area where all work in Windows 98 begins. Refer to Figure 4.1 for a screenshot of the Windows 98 desktop. It is the interface between the user and the computer files, applications, and hardware. The desktop is part of the operating system's **GUI (Graphical User Interface)** environment. The desktop consists of icons and shortcuts. An **icon** is a graphical representation of an application, file, folder, or utility. A **shortcut** is a special type of icon created by the user to quickly access a file or application.

Figure 4.1: The Windows Desktop

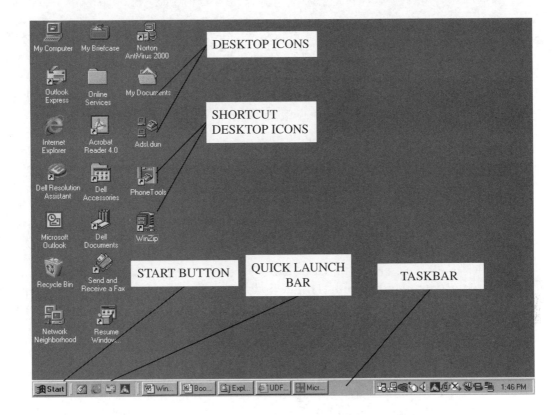

When you double-click on an icon or shortcut, you open an application, file, or window that allows you to interface with various devices installed or connected to the computer. A **window** is a part of the screen that belongs to a specific application or utility. Windows are a normal part of the working environment through the desktop.

A **shortcut** icon looks like any other desktop icon except that it has a bent arrow in the lower left corner. A shortcut represents a path (location on the drive) to a file, folder, or program. It is a link to where the file or application resides on a disk. When you double-click on a shortcut icon, the shortcut icon tells Windows where to find the specific file that the icon represents. If the shortcut represents an application, the application opens. If the shortcut represents a document, the application used to create the document appears along with the document. A shortcut offers faster access to an application or file than going through the Start button or through the My Computer icon. Users frequently place shortcuts on the desktop and you must know how to create one. Users create shortcuts to their favorite applications or to documents used frequently.

Your Windows 98 desktop is like your own desk where you study or work. The icons represent your tools such as your pens, staplers, paper clips, etc. Think of a shortcut as a yellow sticky note with a commonly called phone number on it. You store the phone number on the sticky note so you can access the number quickly.

Sometimes the desktop is cluttered with things you put on it. For example, you may clutter you desktop when you download files from the Internet and place them on your desktop.

Let's go through a brief exercise to help you understand how to maintain an uncluttered Windows 98 desktop. In this exercise, you will work with the **Auto Arrange** option to force your icons on your desktop to be automatically arranged towards the left hand side of the desktop. When you attempt to relocate an icon, it will automatically be arranged back to its original position—on the left hand side of the desktop. It is assumed that Windows 98 is installed and running on a computer with the Windows 98 desktop displayed on your screen.

1. Right-click on a blank desktop space. Once a menu appears, point to the **Arrange Icons** option. If the **Auto Arrange** option does have a check mark beside it, click on an **empty portion of the desktop** to cancel because Auto Arrange is already in use. If the **Auto Arrange** option does not have a check mark beside it, click on the **Auto Arrange** option to add the check mark and enable **Auto Arrange** capability.
2. Note the position of the icons on the desktop.
3. Click on the **My Documents** icon and while continuing to hold the mouse button down, drag the icon to the center of the screen. You will see that the My Documents icon moves to its original position.
4. Click on the **Recycle Bin** icon and while continuing to hold the mouse button down, drag the icon to the top right portion of the screen. This icon moves to its original position too.
5. Right-click on a **blank desktop space.** A menu appears.
6. Using step 1 as a guide, set Auto Arrange back to its original setting.

One of the Windows 98 features you can set for the desktop is Auto Arrange. When the **Auto Arrange** option is enabled, the desktop icons cannot be moved. When it is not enabled, the desktop icons can be moved but this setting may clutter your desktop.

One way to modify the desktop appearance is by changing the wallpaper scheme. A **wallpaper scheme** is a background picture, pattern, or color. Other changes to the desktop are altering the **color scheme** (which is used in displaying folders), and enabling a

screen saver (which is the picture, color, or pattern that displays when the computer is inactive).

Let's go through an exercise to help you understand how to modify or view your desktop appearance. It is assumed that Windows 98 is installed and running on a computer with the Windows 98 desktop displayed on your screen.

1. Right-click on an **empty desktop area.** A menu appears.
2. Click on the **Properties** option. The Display Properties window appears. (Note: this screen can also be accessed by clicking on the **Start** button, point to the **Settings** option, click on the **Control Panel** option, and double-click on the **Display** control panel icon.)
3. Click on the **Background** tab. This tab controls how the background color and pattern appears.
4. To change your wallpaper to Black Thatch, click on the **Black Thatch** option.
5. Click on the **Apply** button.
6. To view the **Screen Saver** tab settings, click on the **Screen Saver** tab. This tab controls what screen saver, if any, loads and is used to control monitor power settings. A screen saver password can also be applied.
7. Click on Screen Saver down arrow.
8. To view the **Appearance** tab settings, click on the **Appearance** tab at the top of the window. This tab is good to use when people with vision problems need adjustments.
9. Click **Cancel** to close the Display Properties window.

As mentioned before, icons are an important part of the Windows 98 desktop. The desktop consists of various icons such as My Documents, My Computer, Recycle Bin, Internet Explorer, the Start button, the Quick Launch bar, and the taskbar. These can be seen in Figure 4.1. The **My Documents** icon is used to quickly access the My Documents folder on the hard drive. The My Documents folder is the default location for files the user saves. The **My Computer** icon is used to access the hardware, software, and files located on the computer.

The **Recycle Bin** is used to hold files and folders that the user deletes. When a file or folder is deleted, it is not really gone. Instead, it goes into the Recycle Bin. The deleted file or folder can be removed forever from the Recycle Bin, just as a piece of trash can be removed from a real trash can. The deleted files and folders in the Recycle Bin take up hard drive space.

The **Internet Explorer** icon is used to start the Internet Explorer application, which is used when communicating across the Internet. Internet Explorer is Microsoft's Internet communications package.

Other common desktop items include the Start button, Quick Launch bar, and the taskbar. The **Start button** is located in the lower left corner of the desktop and is used to launch applications and utilities, find files and other computers, get help, and add/remove hardware and software.

The **Quick Launch bar** is a set of icons to the right of the Start button that allows you to launch applications with one click on a Quick Launch icon. The Quick Launch bar is a great addition to the GUI operating system for those people who have many windows open at one time. An important icon on the Quick Launch bar is the **Show Desktop** icon. This icon looks like a desk with a pencil touching a piece of paper. A click on the Show Desktop icon reduces all windows and displays the desktop. Another Quick Launch icon is the **Internet Explorer** icon. Click once on this icon and the Internet Explorer application opens. If a user does not prefer the Quick Launch bar, right-click on an empty taskbar space, point to the Toolbars option, and click on Quick Launch.

An important desktop item is the taskbar. The **taskbar** is the bar that runs across the bottom of the screen. The taskbar holds buttons that represent applications or files currently loaded into memory. The taskbar also holds icons that allow access to system utilities. These utilities can include a clock icon for the date and time, an icon of a speaker for volume control, and an icon for a virus utility. Look back to Figure 4.1 to identify the taskbar.

The Start Button

Now let's look at an important tool used by Windows 98—the Start button. The Start button contains additional buttons. They will be discussed next. The **Shut Down** Start button option is used to shut down the computer, restart the computer, restart the computer in MS-DOS mode, and possibly put the computer in standby mode. The **Restart the computer in MS-DOS mode** option restarts the computer and boots to a command prompt. From there, you can shut off the computer and type commands at the prompt, or type WIN to start the GUI Windows 98 environment. The **Standby** option is available on computers that support power-saving features. Standby is helpful on laptop computers to save on battery life.

The **Log Off Start button** option stops the current environment and brings up a dialog box for a user name and password. This is so others can log into a network or onto the computer with a user name and password and display their own customized desktop. Press the Esc key to bypass this dialog box. The **Run** button (sometimes called the Run command) option starts an application or brings up a command prompt window. The **Help** option is for Windows 98 general usage and troubleshooting assistance. The **Find** button option helps to locate files and remote network devices.

The Settings Option

The **Settings** option is used to access various sub-options that allow computer configuration. This is one of the most commonly used Start menu options. Refer to Figure 4.2 for a sample screenshot. The **Control Panel** option is used to access various utilities that customize the Windows 98 environment, such as the display, mouse, and CD-ROM. You get to Control Panel by going through the Settings option. The **Printers** selection is used to install, configure, or monitor a printer. The **Taskbar** and **Start Menu** item is used to customize the taskbar or Start button menu. **Folder Options** brings up a dialog box that allows you to change how files and folders appear on the screen. The **Active Desktop** option allows enabling, disabling, or customizing the Active Desktop environment. The

Windows Update submenu item is used to connect to the Microsoft web site to access the most recent drivers and operating system updates.

The **Documents** selection contains the 15 most recently used files (provided the application supports this option). The **Favorites Start** menu item is a list of preferred web sites. The **Programs** choice allows access to applications installed in the computer.

Figure 4.2: The Windows 98 Start Button

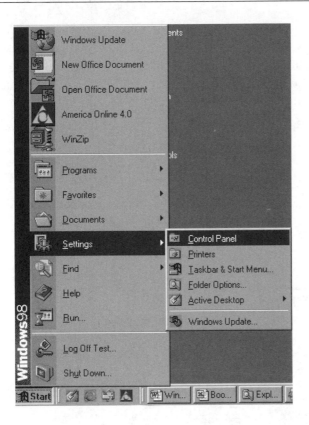

Frequently you will interact with the Windows 98 operating system through a dialog box. A **dialog box** is used by the operating system and with Windows applications and allows you to configure application or operating system preferences. The most common features found in a dialog box are a text box, tabs, a Help button, a Close button, a check box, a radio button, a drop-down menu, an OK button, a Cancel button, and an Apply button.

A **text box** is an area where you can type a specific parameter. When you click inside a text box, a vertical line appears, which is the insertion point. Any typed text is placed to the right of the insertion point. Notice in Figure 4.3 how the Top, Bottom, Inside, Outside, etc., options are text boxes. Text boxes sometimes have up or down arrows that can be used to select a preset option or you can simply type your own parameter.

Tabs normally appear across the top of a dialog box. Each tab normally holds a group of related options. Click once on the tab to bring that particular major section to the window forefront. The tabs in Figure 4.3 are Margins, Paper Size, Paper Source, and Layout.

The **Close button**, which is an X located to the right of the Help button, is used to close the dialog box. When the Close button is used, changes that have been made in the dialog box are not applied.

When checked, a **check box** option is enabled or turned on. As shown in Figure 4.3, clicking inside the check box enables the Mirror margins option. When an option is enabled, a check mark appears in the check box. A **radio button** is a round circle, which behaves in the same way. A radio button is enabled when a solid dot appears in the radio button. Click once on a blank radio button and a solid dot appears in the radio button center. Click once on a radio button with a dot in it, and the dot disappears and the option is disabled.

Figure 4.3: The Windows 98 Dialog Box Components

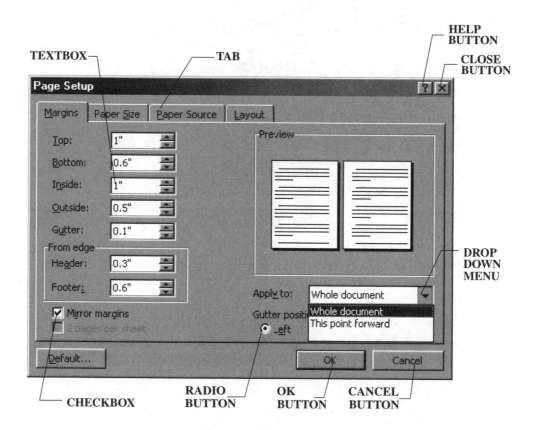

Drop-down menus are presented when you click on a down arrow. The example in Figure 4.3 shows a drop-down menu that appears when you click on the down arrow in the section marked Apply to. Clicking on the Preferred option in the drop-down menu causes that option to appear in the drop-down window.

The **OK button** and the **Cancel button** are standard in a dialog box. When you click on the OK button, all options selected or changed within the dialog box are applied. When you click on the Cancel button, all changed options in the dialog box are not applied; the options are left in their original state.

Another related button that can be found in a dialog box, but is not shown in Figure 4.3, is the Apply button. The **Apply button** is used to make changes immediately (before clicking on the OK button). One example is when changes are made to the desktop's background. New operating system specialists often make the mistake of clicking on the Close button (the button with an X). When a dialog box is closed with the Close button, no changes in the dialog box window are saved or applied.

4.2 Managing Files and Folders

Users are always creating, deleting, and moving files and folders. It is important that you are able to do these tasks quickly and without error. The important thing to remember is to think about what file and folder you want to work with, where the files and folders are located now, and where you want the files or folders to be eventually.

A drive letter followed by a colon represents every drive in a computer. For example, the floppy drive is represented by A: and the first hard drive partition is represented by C:. Disks or drives hold files. A **file** is an electronic container that holds computer code or data. Another way of looking at it is thinking of a file as a box of bits. A file is kept on some type of media, such as a floppy disk, hard drive, tape, or CD. Each file is given a name called a **file name**. An example of a file name is 98CHAP.DOC.

Files are kept in folders. A **folder** holds files and can also contain other folders. In older operating systems, a folder was called a directory. Every file and folder is given a name. It is easier to understand file and folder names if we look at how older operating systems named files and folders and then look at how Windows 98 names differ.

With Windows 95/98/NT/2000, the application normally adds an extension to the end of the file name. In most windows, Windows 98 does not automatically show the extensions. To view the extensions in Windows Explorer, click on the View menu option. Click on the View tab and on the Hide file extensions for known file types check box, which will remove the check from the box. Click on the OK button.

When Windows recognizes an extension, it associates that extension with a particular application. File name extensions can tell you a lot about a file, such as which application created the file or what purpose the file has. Table 4.1 lists the most common file extensions, their purposes, and which application typically creates the extension.

Table 4.1: Common File Extensions

Extension	Purpose or Application
AI	Adobe Illustrator
BAT	Used in DOS environments for batch files. Combines executable commands into one file and uses one command to start the batch file.
BMP	Bitmap file
CAB	Cabinet file—a compressed file that holds operating system or application files.
COM	Command file or executable file that opens an application or tool.
DLL	Dynamic Link Library file contains executable code that can be code already running.
DOC	Microsoft Word
DRV	Device driver—a piece of software that enables an operating system to recognize a hardware device.
EPS	Encapsulated postscript file
EXE	Executable file that opens an application
GIF	Graphics Interchange File
HLP	Windows-based help file
INF	Information or setup file
INI	Initialization file—Used in Windows 3.x environment to control applications and the operating environment. Used in 95, 98, NT, and 2000 to be backward compatible with Windows 3.x.
JPG or JPEG	Joint Photographic Experts Group file format graphics file
MPG or MPEG	Movie clip file
PCS	Microsoft Paintbrush
PDF	Adobe Acrobat portable document format
PPT	Microsoft PowerPoint
RTF	Rich Text Format
TIF or TIFF	Tag Image File Format

TXT	Text file
VXD (also VxD)	Virtual device driver
WPS	Microsoft Works file
WRI	Microsoft WordPad
XLS	Microsoft Excel
ZIP	Compressed file

Windows 98 Files

File names in Windows 98 can be up to 255 characters in length. These extended file names are commonly called **long file names**. Folders and file names can contain all characters, numbers, letters, and spaces *except* the following:

/ (forward slash) " (quotation marks) \ (backslash)
| (vertical bar) ? (question mark) : (colon)
***** **(asterisk)**

As you can see, the list is much shorter (which means more characters are allowed) in Windows 98 when compared to a similar list for DOS file name characters in Chapter 3.

An example of a long file name is WINDOWS 98 CHAPTER.DOC. Any time a document has been saved with one of these long file names and is taken to an older computer with an operating system that does not support long file names, the file name is shortened to a maximum of eight characters. Windows does this by using the first six characters of the file name, deleting any spaces, and using two special characters—a tilde (~) and a number. For example, WINDOWS 98 CHAPTER.DOC would be shortened to WINDOW~1.DOC. If there were two files named WINDOWS 98 CHAPTER.DOC and WINDOWS 98 INDEX.DOC, the two files would be saved as WINDOW~1.DOC and WINDOW~2.DOC, respectively.

When saving a file in a Windows application, the application automatically saves the file to a specific folder. This is known as the **default folder**. With Windows 98, this folder is the My Documents folder.

A file's **path** is like a road map to the file and includes the drive letter plus all folders and subfolders, as well as the file name and extension. For example, if the CHAP1.DOC file is in the MY DOCUMENTS folder on the first hard drive partition, the full path is C:\MY DOCUMENTS\CHAP1.DOC. The first part is the drive letter where the document is stored, C:. The C: represents the first hard drive partition. The name of the document is always at the very end of the path. In the example given, CHAP1.DOC is the name of the file. Everything in between the drive letter and the file name is the name of one or more folders where the CHAP1.DOC file is located. The folder in this example is the MY DOCUMENTS folder.

If the CHAP1.DOC file is located in a subfolder called COMPUTER BOOK, which is located in the folder called MY DOCUMENTS, then the full path is C:\MY DOCUMENTS\COMPUTER BOOK\CHAP1.DOC. Notice how the backslashes in the path are always used to separate the folder names as well as separate the drive letter from the first folder name.

Windows Explorer is the most common application used to create, copy, or move files or folders; however, the My Computer window can also be used in a similar fashion. When you are copying a file or folder, use the Copy and Paste functions. When you are moving a file or folder, use the Cut and Paste functions.

Let's go through an exercise to help you understand how to manipulate files and folders. It is assumed that Windows 98 is installed and running on a computer with the Windows 98 desktop displayed on your screen.

1. Insert your floppy in the floppy drive.
2. Right-click **Start** and then click **Explore**. Windows Explorer opens with left and right windowpanes.
3. Click the icon associated with the floppy.
4. Click **File**, click **New,** and then click **Folder**. The name of the folder appears as **New Folder**. Type **My Folder** and press Enter.
5. In the left windowpane, expand the floppy by clicking the plus sign to the right of it. If you see a minus sign to the right, it already is expanded.
6. Click **My Folder**.
7. To create a folder within **My Folder**, click **File**, click **New,** and then click **Folder**. The name of the folder appears as **New Folder**. Enter your own name for the name of the folder and press Enter.
8. To create a file within **My Folder**, click **File**, click **New,** and then click **New Text Document**. The name of the text document appears as **New Text Document**. Press Enter to keep this name.
9. To rename the new text document, right-click the document and click **Rename**. The document name is highlighted. Type **My Document** and press Enter. Note if the extension (.txt) appears, then the folder is set to show extensions. In this case, leave it otherwise you will receive a message that changing the file extension may cause the file to become unusable.
10. To delete the text document, right-click the document (note its name has changed) and click **Delete**. A confirmation message will appear.
11. Click **Yes** to delete. The file is deleted.
12. Close Windows Explorer.

A text document <u>is</u> a file.

When you delete a file or folder from a floppy, the file or folder is permanently deleted. When you delete a file or folder from a hard drive, the file or folder is automatically sent to the Recycle Bin. The contents of the Recycle Bin take up hard drive space and many users do *not* realize that they are not really deleting the file, but simply moving it to the Recycle Bin. To delete a file permanently from the hard drive, hold down the Shift key while pressing the Delete key on the keyboard. Otherwise, you will have to remember to empty the Recycle Bin periodically. An exercise at the end of the chapter illustrates how to copy, move, and delete files and folders. Figure 4.4 shows how the A+ COMPLETE BOOK.DOC long file name looks in graphical form using the Windows Explorer application.

Figure 4.4: Windows Explorer Document

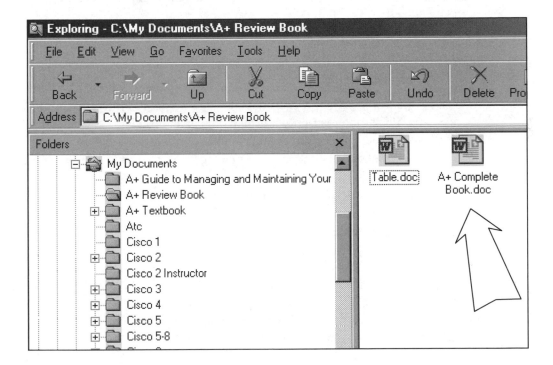

File Attributes

My Computer and **Windows Explorer** can be used for setting attributes for a file or folder. The file and folder attributes are read-only, hidden, archive, and system. The read-only attribute marks a file or folder so that it cannot be changed. The hidden attribute marks a file or folder so that it is not visible through My Computer or Windows Explorer unless someone changes the default view for the window. Some applications use the archive attribute to control which files or folders are backed up. The system attribute is placed on certain files used to boot Windows 98.

Let's perform an exercise to help you understand a file's attributes. It is assumed that Windows 98 is installed and running on a computer with the Windows 98 desktop displayed on your screen.

1. Insert your floppy in the floppy drive.
2. Right-click **Start** and then click **Explore**. Windows Explorer opens with left and right windowpanes.
3. Click the icon associated with the floppy.
4. To create a file, click **File**, click **New,** and then click **New Text Document**. The name of the text document appears as **New Text Document**. Press Enter to keep this name.
5. To set the file's attributes, right-click the new text document and click **Properties**. The Properties window for the file opens.
6. In the **Attributes** section of the page, check **Read-only** and click **OK**. The Properties window closes.
7. Next, you will attempt to modify the file. Remember it is set to read-only. Double-click the document. The document opens.
8. On the document screen, type your name and press Enter. This is considered text.
9. Attempt to save the document. Click **File** then click **Save**. The **Save As** window opens. Normally, this window does not appear—the file is simply saved. However, because the file has the read-only attribute, you cannot save the text to this file. Windows 98 wants you to enter a new file name.
10. To see the error message, click **Save**. A message is displayed indicating you cannot save this file.
11. Click **OK** and then click **Cancel**. The document screen remains.
12. Close the document. Do not save changes if prompted.
13. Close Windows Explorer.

Determining the Windows 98 Version

The operating system version is very important to you as an operating system specialist. With Windows 95, 98, NT, and 2000, upgrades or patches to the operating system are provided by way of service packs. A **service pack** fixes problems with the operating system. You must determine what version of operating system is on the computer so you can research whether or not a service pack is needed. There are two ways to determine what version of 98 is loaded on a computer. They are listed below.

• Click on the Start button. Point to the Programs option. Point to the MS-DOS Prompt option. At the command prompt, type **ver.** The 98 version appears on the screen.
• Right-click on the My Computer desktop icon. Click on the Properties option. A window appears. On the General tab under the System section, the version of Windows 98 appears.

It is important to know how to determine the version of your operating system. The reason is that many applications require a specific operating system version in order to run properly.

4.3 Understanding the Windows 98 Registry

All Windows 98 software and hardware configuration settings are stored in a set of files collectively called the registry. The **registry** contains such things as folder and file property settings, application preferences, and user profiles. A **user profile** contains a user's specific configuration settings, such as what applications the user has access to, desktop settings, user preferences, and the user's network configuration. The registry loads into memory during the boot process but once in memory, the registry is updated continuously by changes made to software and hardware. Many times when you make a change to the registry, you must reboot in order to make the changes take effect. The new registry settings will then be "seen" by the operating system kernel and stored in memory.

Be careful when modifying the registry. You could make a change that could render your system uscless. Because most of the things you do on the system ultimately change the registry, when in doubt, simply use the graphical method to change a setting.

The registry is made up of two files—**SYSTEM.DAT** and **USER.DAT**. The registry files are stored in the folder where the Windows 98 files are located, which is normally the C:\WINDOWS folder. The SYSTEM.DAT holds computer-specific hardware settings, plug and play configurations, and application settings. USER.DAT holds user-specific files (the user profile settings) such as log-on names, desktop settings, and Start button settings. Both files have the hidden attribute set by default. If the system is configured for multiple users (user profiles), a USER.DAT file is built for each user and is normally kept in the C:\WINDOWS\PROFILES*USER*\USER.DAT (where *USER* is the user name).

In the Windows 3.x environment, initialization files (having an extension of .INI) were used instead of the registry. They were called initialization files because they were read by the system during the initial boot stages of the operating system. Each .INI file contained application-specific data, hardware configuration information, the computer environment configuration, and so forth. Windows 98 still has the SYSTEM.INI, PROTOCOL.INI, and WIN.INI files located in the default Windows 98 folder (normally C:\WINDOWS) as well as the C:\COMMAND.COM, C:\AUTOEXEC.BAT, and C:\CONFIG.SYS files so that older 16-bit applications can operate under Windows 98. To see the contents of these files in Windows 98, click on the Start button, click on the Run option, and type **SYSEDIT** in the dialog box. The System Configuration Editor opens with the SYSTEM.INI, WIN.INI, PROTOCOL.INI, AUTOEXEC.BAT, and CONFIG.SYS files open in separate windows.

The registry is divided into five subtrees. **Subtrees** are like folders and are also sometimes called branches or hives. Think of the hive as a collection of folders. The five standard subtrees are as listed in Table 4.2. Each of these subtrees has **keys** that contain values related to hardware and software settings.

Table 4.2: TWindows 98 Registry Subtrees

Registry Subtree	Function
HKEY_LOCAL_MACHINE	Holds global hardware configuration. Included in the branch is a list of hardware components installed in the computer, the software drivers that handle each component, and the settings for each device. This information is not user-specific.
HKEY_USERS	Keeps track of individual users and their preferences.
HKEY_CURRENT_USER	Holds a specific user's configuration such as software settings, how the desktop appears, and what folders the user has created.
HKEY_CURRENT_CONFIG	Holds information about the hardware profile that is used when the computer first boots.
HKEY_CLASSES_ROOT	Holds file associations and file links. The information held here is what allows the correct application to start when you double-click on a filename in Explorer or My Computer (provided the file extension is registered).

Editing the Windows 98 Registry

Most changes to Windows 98 are made through various control panels, but sometimes the only way to make a change is to edit the registry directly. Windows 98 comes with a registry editor called REGEDIT. Some technical problems can only be corrected by editing the registry. The registry editor is not a tool for average users, but it is a tool for use by operating system specialists; thus it is not readily available through the Start menu or through System Tools. In Figure 4.5, you can see that the subtree HKEY_LOCAL_MACHINE\Hardware\Description\System\CentralProcessor\0 has an Identifier value of x86 Family 6 Model 8 Stepping 3. An exercise will not be given here due to the potential danger of accessing the registry. Lab Project 6 at the end of the chapter is available but ask your instructor for permission before performing this lab project.

Figure 4.5: The Windows 98 REGEDIT Window

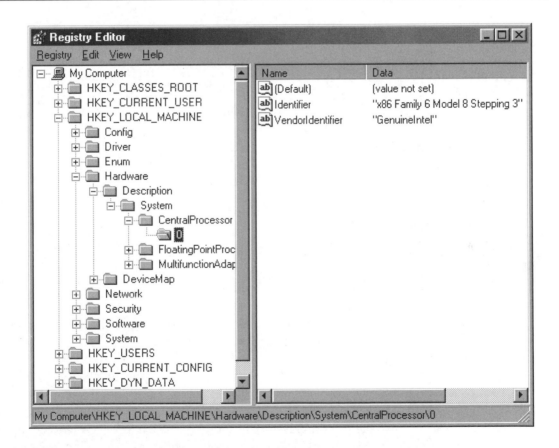

Backing Up and Restoring the Windows 98 Registry

The registry should be backed up whenever the computer is fully functional and when any software or hardware changes are made. The registry should be backed up and stored on a different working computer <u>before</u> disaster hits. The time to learn how to restore the registry is <u>not</u> when the computer is down.

The registry can be backed up and restored these different ways:
- SCANREG Registry Checker utility
- REGEDIT utility
- Backup utility

Using the SCANREG Registry Checker Utility

The most common method used is the SCANREG Registry Checker utility. The Windows 98 Registry Checker utility automatically starts every time the computer starts. The Registry Checker automatically scans the registry for any problems. If no errors are found, the Registry Checker backs up the registry once a day. By default, five backups are kept in a hidden folder, which is normally located in C:\WINDOWS\SYSBCKUP. The backups are labeled RB000.CAB, RB001.CAB, RB002.CAB, RB003.CAB, and RB004.CAB where RB000.CAB is the oldest registry backup. These backup files contain a compressed version of USER.DAT, SYSTEM.DAT, WIN.INI, and SYSTEM.INI. When a new backup is created and five backups already exist, the oldest backup is replaced.

The Registry Checker utility has two versions—SCANREGW.EXE and SCANREG.EXE. The SCANREGW version is used when the Windows 98 GUI interface is active. The SCANREG.EXE is the 16-bit real mode DOS version. The version that runs during startup is SCANREGW. To help you differentiate the two, just remember the one with a "W," as in SCANREGW, in its name is the one used for Windows. The one without the "W" is for the DOS command line—not Windows.

The command SCANREGW can be executed from the Run dialog box if you want to make your own registry backup. This would be important if you are going to install new software, install hardware, remove software, or before editing the registry.

The Registry Checker runs and verifies that the registry does not have any errors, displays a message that the registry has already been backed up that current day, and prompts with the question, "Would you like to back it up again?" Click on the *Yes* button to make a backup of the registry. A prompt on the screen appears when the backup has completed successfully.

You can customize the Registry Checker utility by editing the SCANREG.INI file. The SCANREG.INI file is normally located in the C:\WINDOWS folder and you can edit it by opening the SCANREG.INI file in any text editor program such as Notepad or WordPad. Table 4.3 lists the SCANREG.INI file entries and possible values.

Table 4.3: TWindows 98 SCANREG.INI Entries

Entry	Explanation and Values
Backup=	0 = disables registry backup 1 = registry is backed up the first time the computer is started on a specific day
BackupDirectory=	Default is blank, which means that backups are saved to the WINDOWS\SYSBCKUP hidden folder.
Files=	By default, this value does not exist, but is used to add the registry CAB backup file.
MaxBackupCopies=	Default is 5 and valid values are 0 to 99.
Optimize=	Default is 1, which means that the Registry Checker utility automatically optimizes the registry if it contains 500 KB or unused space. Value of 0 means the Registry Checker utility does not automatically optimize unused space in the Registry.

The SCANREG.EXE program has several options that can be used to customize the way it works. Table 4.4 lists the switches that can be used with the SCANREG command.

Table 4.4: Windows 98 SCANREG Switches

Switch	Explanation
/"comment=*text*"	*text* is a descriptive comment that is added to the Registry backup.
/autoscan	The Registry is automatically scanned and backed up without displaying any prompts.
/backup	Backs up the Registry.
/fix	Repairs the Registry and optimizes it by removing unused space.
/opt	Optimizes the Registry by removing unused space.
/restore	Brings up the Registry Checker window and allows a specific Registry backup file to be selected for repairing a system that will not boot or operate correctly.
/scanonly	Checks the Registry for errors and displays a message if problems are found.

You need to be familiar with the information in Table 4.3 and Table 4.4 because you are likely to see this on the A+ Operating System exam.

Using the REGEDIT Command

The second method of backing up the registry is to use the REGEDIT command. The REGEDIT program allows you to export the registry to a file that has an extension of .REG. The backed up file can be edited with a text editor, if necessary. The file can also be imported back into the computer. Because accessing the registry is potentially danger-ous, an exercise will not be given here. As stated earlier in this chapter, Lab Project 6 at the end of the chapter deals with the registry. However, you should ask your instructor before performing the lab project.

Using the Backup Utility

The Backup utility is accessed through Start button, Programs, Accessories, System Tools, Backup. With the Backup utility, you can either back up the entire computer, or just back up the registry.

4.4 Understanding the Pre-Installation Steps of Windows 98

Windows 98 can be installed from a central location or locally. Because this book focuses on the A+ Certification exam, only local installation is covered. The pre-installation of any operating system is more important than the installation process. If you grab a disk or CD and load a new operating system without going through a logical process, you are asking for trouble. The overall steps to take before installing Windows 98 are as follows:

1. Decide whether the installation will be an upgrade or a clean install.
2. Decide whether the computer will have more than one operating system installed.
3. Scan the computer for viruses.
4. Determine if the hardware is compatible.
5. Obtain necessary software drivers.
6. Back up any data files necessary.
7. Disable any unnecessary TSRs before upgrading to Windows 98.

The first decision to make when planning to install Windows 98 is whether you are upgrading from another operating system or performing a clean install. A **clean install** is when you install an operating system on a computer that does not have one, or the exist-ing operating system has been removed (the hard drive is formatted). If the decision is to do an upgrade, then determine what operating system is already installed. Windows 98 supports upgrading from DOS 5 and higher, Windows 3.x, and Windows 95. When Win-dows 98 is installed as an upgrade from Windows 3.x or Windows 95, the user's applica-tions and data are preserved if the operating system is installed in the same folder (directory) as the original operating system. If Windows 98 is installed in a different folder, then all applications must be reloaded.

A related issue, if upgrading to 98, is whether or not to convert the hard drive partition to FAT32. Once a partition is converted to FAT32, the partition <u>cannot</u> be changed. If you are unsure whether or not to convert the partition, leave it unchanged and later use the Drive Converter wizard to upgrade. You would want to convert the partition to FAT32 for these reasons:

- Use of cluster space is more efficient.
- FAT32 supports larger hard drive partition sizes.
- Supports larger hard drives (up to 2TB).
- With FAT32, the root folder can be relocated and a backup copy of the FAT can be used.
- FAT32 is more flexible than FAT16 because the root folder is an ordinary cluster that can be located anywhere on the drive instead of the outer hard drive track. This also means that the previous root folder entries limitation is no longer an issue.

Sometimes a clean install is the best choice, especially if the current operating system is DOS or Windows 3.x. Because a clean install involves formatting the hard drive, the user's data must be backed up and all applications reinstalled once the Windows 98 installation is complete. In addition, all user-defined settings are lost if Windows 98 is installed into a different folder.

The second decision that must be made is whether or not Windows 98 will be installed along with one or more other operating systems. This is often called a dual-boot or multi-boot scenario. **Dual-boot** means that the computer can boot from two (or more) different operating systems. Windows 98 can be dual-booted with DOS 5 and higher, Windows 3.x, NT, and 2000. If this is desired, a separate hard disk partition should be created and used for each operating system. Dual booting is discussed in further detail in Section 4.7.

The third pre-installation step is to scan the drive for viruses. The next section covers viruses in detail.

The fourth step when installing Windows 98 is to determine what computer hardware is installed and whether it is compatible with Windows 98. Table 4.5 lists the minimum and preferred hardware requirements for installing Windows 98.

Table 4.5: Windows 98 Minimum Requirements

Component	Minimum
CPU	Intel 486 (or compatible) 33 MHz
RAM	16 MB
Free hard drive space	120 MB
Input Device	Keyboard, mouse, or other pointing device
Multimedia Drive	CD-ROM
Floppy drive	3.5" 1.44 MB
Video	VGA

An upgrade from Windows 95 or Windows 3.x takes a minimum of 120MB free hard disk space. If you are doing a new installation to a FAT16 file system, 165MB of free hard disk space is the minimum. If you are doing a new installation to a FAT32 file system, 140MB is the minimum free hard disk space. If you are installing Windows 98 to a partition or drive other than the C:, the C: drive still needs a minimum of 25MB free hard disk space for Windows 98 system and log files.

The Windows 98 CD contains a document called HCL (Hardware Compatibility List). Use this list to see compatible hardware. The most current list is on Microsoft's web site. Once you have verified all hardware, you may have to get Windows 98 device drivers (step 5) from the hardware device manufacturer or its web site. There are also notes about hardware in the Windows 98 README and SETUP.TXT files contained on the 98 Setup disks.

The fifth step involves obtaining any necessary software drivers that are compatible with Windows 98. At a minimum, you may need a modem driver that is compatible with Windows 98. That way, if you have to load other drivers, such as a video driver, you will be able to download it from the Internet. A **modem** is a device that allows you to access the Internet.

The sixth step involves backing up your data. As with any upgrade, hardware change, or software change, data should be backed up. A user's data is very valuable to him or her. Whether you are doing a clean install or an upgrade, if the user has data on the computer, it must be backed up before starting the installation process. Also, before backing up data, remove any unwanted files and/or applications to free up hard drive space.

The final step is to disable old TSRs no longer needed. If Windows 98 is being upgraded, disable any unnecessary TSR (Terminate and Stay Resident) programs and device drivers loaded through CONFIG.SYS, AUTOEXEC.BAT, or the STARTUP folder. Windows 98's SETUP.TXT file on the CD-ROM contains information about TSRs that are incompatible with the upgrade.

Viruses

When installing a new operating system on a computer that already has an operating system loaded, it is a good time to run a virus scan. A **virus** is a computer program that is designed to do something to your computer that changes the way the computer operates. The virus could infect the computer so it does not boot, infect a particular application so it operates differently, or erase files. Some viruses are written to cause mischief rather than harm. An example of this could be a program that puts a picture on the screen. Some people think that they can eliminate viruses by high-level formatting their hard drive. This is a mistake. Do not take a chance; take a few moments and scan the hard drive for viruses! Common types of viruses include the boot sector virus, the file virus, the macro virus, and the Trojan horse virus.

A **boot sector virus** is a program placed in the computer's boot sector. Because the computer loads boot sector code into memory, the virus loads into RAM at the same time. Once in memory, the virus can take control of computer operations and spread to other drives, such as floppy drives, hard drives, and drives located on a network.

A **file virus** replaces or attaches itself to a file that has a COM or EXE extension. COM or EXE files are commonly known as executable files. Exccutable files are used to start applications. By attaching itself to this type of file, a virus can cause the program to not start or operate properly as well as load into RAM and affect other COM or EXE files.

A **macro virus** is written for a specific application such as Microsoft's Excel or Word. A macro virus is written in a specific language and attaches itself to a document that was created in the application. Once the document is opened and loaded into memory (along with the virus), the virus can attach itself to other documents.

A **Trojan horse virus** pretends to be a normal application. When the virus executes, the computer does something that the user does not expect such as put a message or a picture on the screen or put a new screen saver up. A Trojan horse virus does not replicate (copy itself somewhere else). The virus can be used to gather information such as user names and passwords that can be later used to hack into your computer.

Three other types of viruses that infect computers are the stealth virus, polymorphic virus, and a worm virus. A **stealth virus** is a program written to avoid being detected by anti-virus software. When an anti-virus program executes, the stealth virus provides the anti-virus program with a fake image that makes the anti-virus program believe that no virus is present. A **polymorphic virus** is a virus that constantly changes its own program to avoid detection by an anti-virus program. A **worm virus** makes a copy of itself from one drive to another and can use a network to replicate itself. The most common types of worm viruses today are in the form of an e-mail message. Once the e-mail is opened, the worm virus is sent to every other user that is in an address book. Common symptoms of a virus are as follows:

- Computer does not boot.
- Computer hard drive space is reduced.
- Applications will not load.
- An application takes longer to load than necessary or longer than normal.
- Hard drive activity increases, especially when nothing is being done on the
- computer.
- An anti-virus software message appears.
- The number of hard drive sectors marked as bad steadily increases.
- Unusual graphics or messages appear on the screen.
- Either files are missing or files are created (taking up excessive hard disk space).
- A message appears that the hard drive cannot be detected or recognized.
- Strange sounds come from the computer.

If a virus is detected or even suspected, run an anti-virus program. Follow the program's directions for installing and executing. The time to get an anti-virus program is *before* a virus infects the computer, because the damage may be irreversible, especially if backups are not performed. Back up data often! Always back up data files before upgrading to a new operating system. Backups are an important part of any computer support plan.

Maintaining the anti-virus program and keeping it up-to-date with the latest virus signatures is also very important. New viruses are constantly created, so the virus software must be kept current as well. Some anti-virus software can be set to load into memory when the computer boots and runs continuously. Make sure you disable this feature when installing an operating system patch (called a **Service Pack** or **SP** by Microsoft). The anti-virus software can prevent the upgrade or patch from installing. Other types of software that can prevent an operating system from being upgraded are power management and disk management software/tools. Disable these utilities and applications before attempting an operating system installation or upgrade.

4.5 Understanding How to Install and Upgrade Windows 98

Once all of the pre-installation checklist steps are complete, you are ready to install Windows 98. The installation process is easy if you performed the pre-installation steps. The program to install Windows 98 is called Setup or **SETUP.EXE**. The Setup program can be run from a Windows 95 GUI environment, from a Windows 3.1x GUI environment, or from a command prompt. If you are upgrading from Windows 95 and keeping the current configuration, run the Setup program from the Windows 95 interface by closing all programs including any anti-virus programs. Insert the Windows 98 CD into the CD-ROM drive. Normally, the Setup program automatically starts. If the Setup program does not automatically start, click on the **Start** button, click on the **Run** option, and type the drive letter for the CD-ROM followed by a colon and **\setup** in the text dialog box and press **Enter**. An example of the command is **D:\setup**. The Windows 98 Setup wizard starts.

During the installation process, you are asked to enter a Product Key. The Product Key is located either on the certificate found on the back of the Windows 98 book (that shipped with the computer) or on the back of the 98 CD. In addition, during the setup process, you are prompted for the type of setup. The installation types are listed below.

- The **Typical** option installs the most common components for Windows 98. This is the most frequently chosen option.
- The **Portable** option is used for laptop computers and installs Windows 98 components used on portable computers.
- The **Compact** option is the smallest possible number of Windows 98 files that can be loaded and Windows 98 still work.
- The **Custom** option allows you to select the components to install.

The SETUP.EXE file can be used with various switches that control the setup process. Table 4.6 lists some of the switches that can be used. If networking is installed, an Identification window appears prompting for the computer name, workgroup name, and a description of the computer. The computer name prompt is used to uniquely identify the computer across a network. The name can be up to 15 characters long and should not contain blank characters. The workgroup name can also be up to 15 characters and must be the same for all computers networked together. Contact the network administrator for what to type in this dialog box. The computer description box can contain up to 48 characters and contain spaces to describe the computer or the computer's location.

Table 4.6: Some Windows 98 SETUP.EXE Switches

Switch or Option	Purpose
/?	Displays switches available
/c	Bypasses SmartDrive (used for drive caching)
/d	Bypasses using the existing Windows configuration
/ie	Skips the Startup Disk screen
/is	Skips the disk check (ScanDisk)
/it	Bypasses checking for TSRs
/nr	Skips the Registry check

The A+ Operating System exam will test your knowledge of these switches. So learn them.

4.6 Troubleshooting the Windows 98 Installation

The key to troubleshooting the Windows 98 installation is to get the operating system installed as best you can and troubleshoot the problem with the tools provided with Windows 98. Windows 98 ships with a feature called **Safe Recovery** that allows the installation to continue after a failure. Installation problems can be caused by a number of factors. The following are the most common causes and their associated solutions during the installation process:

- **Incompatible drivers**—Obtain Windows 98 drivers from the hardware manufacturer.
- **Existing drivers are incompatible or halt the installation/upgrade process**—Obtain Windows 98 drivers from the hardware manufacturer. If upgrading from a previous operating system, edit the AUTOEXEC.BAT and CONFIG.SYS and put the REM command before each line so the drivers do not load.
- **Incompatible TSRs**—Remove TSRs or obtain updated ones from the software manufacturer. Make sure to remove power management and anti-virus programs.
- **Minimum hardware requirements have not been met**—Upgrade the hardware. The top items to check are the CPU (486DX 66MHz minimum), RAM (16MB minimum), and hard drive space (120MB minimum).
- **A virus is on the hard drive**—Run a virus-checking program and remove the virus.
- **Pre-installation steps have not been completed**—Go back through the list!
- **The Windows 98 boot disk or CD is corrupted** (not as likely as the other
- causes)—Try the disk in another machine and see if you can see the contents. For the CD, check to see if any scratches or dirt are on the surface. Clean the CD as necessary.
- **Incorrect Product Key**—Type in the correct CD key to complete the Windows 98 installation. The Product Key is located either on the certificate found on the back of the Windows 98 book (that shipped with the computer) or on the back of the 98 CD.

Several text files can be helpful in determining the installation problem. Table 4.7 lists the text files and a description of each. You can view the contents of a file to determine what occurred at various stages of the installation.

Table 4.7: Windows 98 Startup Log Files

Log File	Description
BOOTLOG.TXT	Hidden text file in the root directory that is created when Windows 98 boots for the first time. Contains a list of Windows 98 components and drivers that load during the boot process.
DETCRASH.LOG	Hidden text file in the root directory that is used when Setup fails during the installation process.
DETLOG.TXT	Hidden text file in the root directory that contains a chronological list of hardware devices found during the hardware detection phase.
NETLOG.TXT	Text file stored in the root directory that chronologically lists network components found during the network detection phase.
SETUPLOG.TXT	Hidden text file stored in the root directory that chronologically lists what happens during the installation process.

If Windows 98 gets an error message that states "Invalid System Disk" after the system reboots the first time, a virus may be present; an anti-virus program is running and has not been disabled either through AUTOEXEC.BAT, STARTUP folder, or CMOS; or disk management software is enabled.

4.7 Understanding the Dual-Boot Feature of Windows 98

Sometimes users would like to try a new operating system, but keep their old operating system loaded as well. If this is the case, two operating systems can reside in one system and it is called dual-booting. Windows 98 can be dual-booted with DOS, Windows 3.x, Windows 95, NT, and 2000 Professional, but care must be taken. NT supports and uses either NTFS or FAT16 and Windows 98 supports and uses either FAT16 or FAT32. NT cannot read FAT32 partitions and Windows 98 cannot read NTFS4 partitions. If Windows 98 is to access the files stored when using NT, both operating systems need to use FAT16 (but the FAT32 and NTFS features are lost). It is best if the two operating systems are placed on different hard drive partitions and Windows 98 loaded before NT is installed. If the computer currently dual-boots with NT from another operating system or environment such as DOS, Windows 3.x, or Windows 95, the Windows 98 Setup program can be executed from the other operating system.

The dual-booting feature is a misnomer. You cannot boot more than one operating system. What it really means is that you can install multiple operating systems (yes, more than two). However, you can only boot one operating system at a time. For the most part, in business, it is not practical to dual-boot because only one operating system can be booted. For home use, dual-booting is a good idea because you can learn different operating systems simply booting to the one you want.

4.8 Configuring Windows 98

One of the most common windows in Windows 98 is the Control Panel window. A **control panel** is a method for configuring various Windows 98 components. The Add New Hardware, Add/Remove Programs, and Printing control panels are used when installing or removing hardware and software. Each control panel icon represents a Windows utility that allows you to customize a particular part of the Windows 98 environment. The number of control panels displayed depends on the type of computer and the components contained within the computer. Figure 4.6 shows some of the more common Windows 98 control panels.

Figure 4.6: The Windows 98 Control Panel

You must know which control panel to use for changing a computer's configuration. Table 4.8 shows some common Windows 98 control panels and the function of each.

Table 4.8: Common Windows 98 Control Panel Functions

Control Panel	Function
Accessibility	Controls keyboard, sounds, display, and mouse behavior for people with vision, hearing, or mobility disabilities.
Add New Hardware	Used when installing new hardware devices.
Add/Remove Programs	Used to install or uninstall software applications.
Display	Used to install a monitor driver and configure monitor settings.
Keyboard	Used to install a keyboard driver and configure keyboard.

Multimedia	Used to install multimedia device drivers, configure properties for audio and video, and configure settings for playing audio CDs.
Network	Used to install NIC drivers, add network software, and configure network connections.
Passwords	Used to configure user preferences and passwords.
Power Management	Used to reduce electrical power use in devices such as a monitor or hard drive.
Printers	Used to add, remove, and modify printer settings.
Regional Settings	Used to set the time zone and set the format for numbers and currency.
System	Allows viewing system information and changing parameters related to system performance such as virtual memory. It also allows access to Device Manager where you can manage devices.
Users	Used to configure the computer for more than one person.

Operating system specialists must frequently add new hardware and software using the operating system. Windows 98 has specific tools for these functions. Using the correct procedure is essential for success on the job. The following sections handle many of the tasks an operating system specialist must perform:
• Adding Plug and Play Devices
• Adding Non-Plug and Play Devices
• Adding a Printer
• Removing Hardware Devices
• Installing and Removing Software

   Hardware devices are physical components that connect to the computer. Hardware devices can be either plug and play or non-plug and play. A device driver is a piece of software that allows hardware to work with a specific operating system. Some device drivers are automatically included with Windows 98. An operating system specialist must be aware of what hardware is installed into a system so that the latest Windows 98-compatible drivers can be downloaded and installed.

Adding Plug and Play Devices
Plug and play devices are hardware and software designed to be automatically recognized by the operating system. The key to a successful plug and play device installation includes the following:
• Possessing the most up-to-date device driver

• Following the directions provided by the device manufacturer

Always make sure that the computer is turned off when installing any new hardware component. Some plug and play devices can be inserted into the computer without restarting. These include a PC Card, a laptop into a docking station, and a Universal Serial Bus (USB) device. However, devices such as internal modems or network cards require the computer to be turned off during installation.

Install the device according to the device manufacturer's instructions. Once installed, power on the computer. The Windows 98 Found New Hardware wizard appears. Windows 98 attempts to find a driver for the new device. If Windows 98 detects the plug and play device, the device gives Windows 98 a device ID. Windows 98 uses this device ID to look for the appropriate .INF file. A .INF file contains information used to install and configure the device. If a driver cannot be found, a dialog box appears. The best policy with any operating system is to use the latest driver even if the operating system detects the device.

Adding Non-Plug and Play Devices

Legacy devices are also called **non-plug and play** devices. For devices that are not plug and play, use the Add New Hardware control panel. The Add New Hardware wizard allows hardware configuration and is used for hardware that is not automatically recognized by Windows 98. It is also used for plug and play devices that do not install properly with Windows 98's automatic detection. Windows 98 prompts with the question asking if you want Windows 98 to search for non-plug and play devices. Select the *Yes* option. If Windows 98 finds the device, follow the rest of the prompts to install a driver for the device. If Windows 98 does not find the device, you must select the type of device that is being installed. Then, you must select the device manufacturer and device model from a list. A generic device or "other" option is available for most device categories. Have the Windows 98-compatible device driver ready, click on the **Have Disk** button and specify the drive and path to the driver.

You should let Windows 98 automatically detect any new hardware you add to your system. As a result, you won't need to do locate the correct driver—unless, of course, Windows 98 cannot locate an appropriate driver for the hardware.

Adding a Printer

Printers can be connected to a computer through the printer's parallel or USB port through a local area network. Only local printers (printers directly connected to a computer port) will be covered in this chapter. Networked printers will be explained in the Chapter 9. Windows 98 can automatically detect printers. If Windows 98 detects the printer, the operating system installs the drivers, updates the registry, and allocates system resources to the printer. Most printer are automatically detected by Windows 98.

To install a printer, connect the printer to the appropriate computer port with the appropriate cable. Power on the computer, and the Windows 98 wizard normally detects it and leads you through the installation process. However, if it does not, have the Windows 98-compatible printer driver ready, click on the **Start** button, point to the **Settings** selection, and click on the **Printers** option. When the Printers window opens, click on the **Add Printer** icon. Click on the **Next** button and select the **Local Printer** selection. When prompted for the printer driver, insert the CD or disk that contains the printer driver and use the Browse button to locate the driver. Continue through the Add Printer wizard until the printer is installed.

Removing Hardware Devices

When a plug and play hardware device is removed, Windows 98 automatically detects this and removes the resources and registry entries assigned to the device. When removing most non-plug and play hardware devices (all but printers), use Device Manager. Right-click on the **My Computer** desktop icon to access Device Manager, then click on the **Properties** option, and then click on the **Device Manager** tab. Through Device Manager, a device can either be removed or disabled. When a non-plug and play device is removed, the driver does not load and any resources assigned to the device are now free. When a non-plug and play device is disabled, the resources assigned to the device are kept, but the device's driver does not load when the computer restarts.

To remove a device, click on the plus sign located beside the appropriate device category, then click on the name of the device being removed, and then click on the **Remove** button. To disable a device, click on the plus sign located beside the appropriate device category, then click on the device name, and then click on the **Properties** button. On the **General** tab in the Device Usage section is a check box that allows you to disable the device for a specific hardware profile. Click once in this check box to disable the device.

If you are removing a printer from the system, use the **Printers** control panel. Access this control panel by clicking on the **Start** button. Point to the **Settings** option and click on the **Printers** option. Right-click on the printer you want to delete and choose the **Delete** option.

Installing and Removing Software

No computer is fully functional without software and Windows 98 supports 16-bit and 32-bit applications. Most software today is 32-bit and comes on CD and includes an **Autorun** feature. If the CD has the Autorun feature, an installation wizard steps you through installing the software when the CD is inserted into the drive. If there is not an Autorun feature on the CD or if the software is on a disk, then the Add/Remove Programs control panel is normally used to install or remove the software.

Remember to always to consult the application documentation for installation procedure.

To access the Add/Remove Programs control panel, click on the **Start** button and point to the **Settings** option. Click on the **Control Panel** option and then double-click on the **Add/Remove Programs** control panel icon. To install an application, click on the **Install/Uninstall** tab, click on the **Install** button, insert the application disk or CD, and click on the **Next** button. Windows 98 searches the floppy drive and CD for a Setup program. If one is found, continue the installation process. If one is not found, type the drive letter and path for the application's Setup program. Use the **Browse** button if necessary. Click on the **Finish** button to complete the process.

Close all active applications before starting the installation process. This will eliminate some complications. The computer must frequently be rebooted after an application or Windows component has been installed.

To remove a software application, use the same Add/Remove Programs control panel; however, instead of clicking on the **Install** button, select the application to be removed and click on the **Add/Remove** button. Do not forget to check the application's documentation for specific removal procedures.

The Add/Remove Programs control panel can also be used to add operating system components, add programs across your network, and add or remove Windows components. The Windows Setup tab is used to add or remove operating system components. The Startup Disk tab is used to create a Windows 98 disk that can be used when Windows 98 does not boot properly.

4.9 Understanding the Boot Process

Every operating system needs specific files that allow the computer to boot. These files are known as system files or startup files. The common **system files** and their specific location on the hard drive are listed in Table 4.9. The locations listed in Table 4.9 assume that Windows 98 is loaded in the default folder (WINDOWS). If Windows 98 is loaded in a different folder, substitute the location WINDOWS for the name of the folder in which Windows 98 was initially loaded.

Table 4.9: Common Windows 98 Startup Files

Startup File Name	File Location and Purpose
AUTOEXEC.BAT	Root directory—used to load TSRs not designed to run under Windows 98
BOOT.INI	Root directory—used when multiple operating systems are present
CONFIG.SYS	Root directory—used to load 16-bit drivers not designed to run under Windows 98
DRVSPACE.BIN	WINDOWS\COMMAND—supports compressed drives

GDI.EXE	WINDOWS\SYSTEM—provides support for the graphical environment; one of the Windows 98 core files
GDI32.DLL	WINDOWS\SYSTEM—provides support for the graphical environment
HIMEM.SYS	WINDOWS—driver for extended memory
IFSHELP.SYS	WINDOWS—driver for 32-bit Installable File System Manager
IO.SYS	Root directory—boot file for real mode that loads drivers and TSRs listed in CONFIG.SYS and AUTOEXEC.BAT
KERNEL32.DLL	WINDOWS\SYSTEM—loads the main Windows components
KRNL386.EXE	WINDOWS\SYSTEM—loads Windows device drivers
MSDOS.SYS	Root directory—contains boot parameters and provides backwards compatibility for applications
SYSTEM.DAT	WINDOWS—part of the Registry
SYSTEM.INI	WINDOWS—only exists to be backward compatible with older applications
USER.EXE	WINDOWS\SYSTEM—provides the user interface
USER32.DLL	WINDOWS\SYSTEM—provides user interface code
WIN.COM	WINDOWS—the file that starts Windows
WIN.INI	WINDOWS—only exists to be backward compatible with older applications

The boot process is actually quite involved, but the major steps are as follows:

1. The computer powers on.
2. POST executes.
3. If the computer has a plug and play BIOS, plug and play adapters and devices are configured. If the computer does not have a plug and play BIOS, all ISA bus devices are enabled.
4. BIOS searches for an active partition on the hard drive.
5. BIOS reads the Master Boot Record, then locates and loads the information into sector 0 of the system partition. The contents of sector 0 define the type of file system and the location of the boot files, then start loading the file system.
6. Real mode starts.
7. MSDOS.SYS boot configuration loads.
8. DRVSPACE.BIN loads if it is needed for compressed drives.
9. Prompts for a hardware profile if multiple profiles exist.

10. Bitmap image stored in LOGO.SYS loads.
11. SYSTEM.DAT loads.
12. DOS drivers and TSRs needed for backward compatibility load as specified by the CONFIG.SYS and AUTOEXEC.BAT files.
13. Initializes static VxDs in real mode specified by VMM32.VXD and the registry.
14. Starts protected-mode and loads WIN.COM.
15. Loads protected-mode VxDs as specified by VMM32.VXD, the registry, and the settings in the SYSTEM.INI file.
16. Loads KRNL386.EXE, GDI files, user libraries, Explorer shell, and network support (if needed).
17. Executes any programs located in the Startup folder and those referred to in the registry.

The only reason the AUTOEXEC.BAT, CONFIG.SYS, WIN.INI, and SYSTEM.INI files load is to make Windows 98 backward compatible with older operating systems, 16-bit TSRs, and 16-bit drivers.

The IO.SYS File

Windows 98's IO.SYS file is responsible for loading key files that were previously loaded through the CONFIG.SYS file. Table 4.10 lists the files that are automatically loaded through IO.SYS and the default setting. If the computer needs a setting change, modify the setting by adding the line in the CONFIG.SYS file.

Table 4.10: Windows 98 Default IO.SYS Settings

Setting	Default Parameter and Purpose
buffers=	30 — sets the number of file buffers to create for 16-bit programs
dos=high	No default setting—loads part of the operating system into the HMA
fcbs=	4 — sets the file control blocs that can be open
files=	60 — the number of concurrent open files
himem.sys	No default setting—enables access to XMS
ifshlp.sys	No default setting—used to load device drivers
lastdrive=	Z: — specifies the last drive letter available to be assigned by the operating system
server.exe	No default setting—included for DOS compatibility
shell=	No default setting—sets the default command interpreter

stacks=	9,256 — specifies the number and size of data stacks to handle hardware interrupts. For example, the default setting of 9,256 indicates that 9 stacks are available and they are 256 bytes in size.

The MSDOS.SYS File

Windows 98 has a hidden, read-only system file called MSDOS.SYS that is located in the root directory. MSDOS.SYS enables backward compatibility with older DOS applications. The file holds multiple operating system boot options as well as the paths to important Windows 98 files such as the registry. There are lines of Xs in the MSDOS.SYS file. Do not remove the Xs. Table 4.11 lists some of the more common options found in the MSDOS.SYS file.

Table 4.11: Windows 98 MSDOS.SYS Settings

Section	Setting	Default Setting and Purpose
[Options]	AutoScan=	1 — enables a prompt to run ScanDisk; a value of 0 disables it; a value of 2 runs Scan-Disk without prompting
[Options]	BootDelay=	2 — seconds delayed after the "Starting Windows" message appears; a value of 0 disables the delay
[Options]	BootGUI=	1 — enables the GUI interface; a value of 0 boots to a prompt
[Options]	BootKeys=	1 — enables startup function keys; a value of 0 prevents startup keys from functioning and overrides the BootDelay= setting
[Options]	BootMenu=	0 — enables/disables Startup Menu automatically appearing; a value of 1 displays the Startup menu
[Options]	BootMenuDefaults=	3 or 4 — the number of the menu item to automatically start; a value of 3 is the default for a computer without networking installed; a value of 4 is the default for a computer with networking installed
[Options]	BootMulti=	0 — enables dual-booting; a value of 1 enables the F4 key (for DOS) or the F8 key (for the Windows 98 Startup menu)

[Options]	BootWin=	1 — enables Windows 98 as the default operating system; a value of 0 makes another operating system the default; this setting is only useful with DOS-dual booted with Windows 98
[Options]	LoadTop=	1 — enables loading COMMAND.COM or DRVSPACE.BIN at the top of conventional memory; a value of 0 disables this feature
[Paths]	HostWinBootDrv=	C — the drive letter for the boot drive root directory
[Paths]	WinBootDir=	Varies — defines the location of the startup files
[Paths]	WinDir=	Varies — specifies the folder where many of the Windows 98 files are located

Troubleshooting the Boot Process

Windows 98 has a wealth of tools to help you when troubleshooting the boot process. One of the more common startup problem solving tools is the **Startup menu**. This menu is used to determine how to boot Windows 98. Access the Windows Startup menu by pressing the Ctrl key when you boot the computer. If this does not work, press the **F8** key during the boot process. Each item in the Startup menu is used in different situations. The Startup menu options are listed in table 4.12.

Table 4.12: Windows 98 Startup Menu Options

Mode	Purpose
Normal	Used to boot Windows normally.
Logged	Used to track the boot process and log each event in a file located in the root directory called BOOTLOG.TXT; used to determine where the boot failure occurs.
Safe	Prevents the CONFIG.SYS and AUTOEXEC.BAT files from loading, prevents the [Boot] and [386enh] sections of the SYSTEM.INI file from loading, prevents the Load= and Run= parameters of the WIN.INI file from loading, prevents the items found in the Startup folder from loading, prevents portions of the registry from loading, prevents all device drivers except for the keyboard and mouse from loading, and loads a standard VGA driver.
Step-by-Step confirmation	Allows performing the boot process one step at a time to see where the problem occurs.

Command Prompt Only	Used to troubleshoot DOS applications (only CONFIG.SYS, AUTOEXEC.BAT, COMMAND.COM, and the Registry are loaded; the GUI does not load).
Safe Mode with Command Prompt Only	Used if the computer does not boot to Safe Mode; does not load the HIMEM.SYS or IFSHLP.SYS files, and does not execute WIN.COM to start the GUI. Various switches can be used at the command prompt with the WIN.COM file to troubleshoot the problem.
Previous Version of MS-DOS	Used to perform a DOS function that does not operate correctly under Windows 98.

Other tools that can be used for startup problems include the System Configuration utility, Microsoft System Information tool, Automatic Skip Driver, System File Checker, Startup and Shutdown Troubleshooter, and Registry Checker. Determining which tool to use and how to access the tool is the challenge.

The **System Configuration** utility is used when you suspect a problem with old device drivers or TSRs especially on a computer that has been upgraded to Windows 98. The System Configuration utility allows you to enable or disable entries in the AUTOEXEC.BAT, CONFIG.SYS, SYSTEM.INI, and WIN.INI files. The order in which the file entries load can also be controlled. These files are available in Windows 98 to be compatible with older operating systems and older applications. Access the System Configuration utility by clicking on the **Start** button | **Programs** | **Accessories** | **System Tools** | _**System Information**. Click on the **Tools** menu option and select the **System Configuration Utility** from the drop-down menu. The easiest way to determine if one of the four files is causing the problem is to click on the **Diagnostic startup** radio button and click on the **OK** button. When the Windows 98 Startup menu appears, press the number corresponding to the **Step-by-Step Confirmation** setting. The command that starts the System Configuration utility is MSCONFIG.EXE located in the WINDOWS\SYSTEM subfolder. Figure 4.7 shows the System Configuration Utility.

Figure 4.7: Windows 98 System Configuration Utility

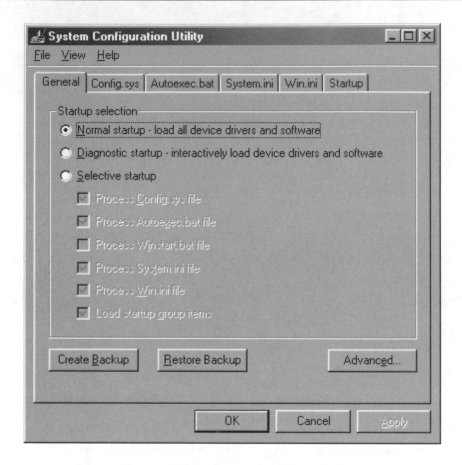

The **MSINFO (Microsoft System Information)** tool is used to display information about system resources and can be used to detect conflicts between devices. Follow the steps below to access MSINFO from a fully operational Windows 98 computer:

1. To access the MSINFO tool, click the **Start** button, point to **_Programs**, point to **Accessories**, point to **System Tools**, and click **System Information**. The System Information window opens.
2. Click on the plus sign beside the **Hardware Resources** option in the left window. The option expands.
3. Click on the **Conflicts/Sharing** setting to see resource conflicts between devices and resources.
4. To view the software environment, click **Software Environment** to expand it.
5. Click **16-bit modules loaded** to view the 16-bit programs that are currently loaded into memory.

6. Click **32-bit modules loaded** to view the 16-bit programs that are currently loaded into memory.
7. Close the System Information window.

Another popular option is the **Forced Hardware** view that lists devices that have been assigned resources manually. The Problem Devices selection allows you to view any devices that have problems and the History option shows seven days of driver history. The command to execute MSINFO tool is MSINFO32.EXE. Figure 4.8 shows the Microsoft System Information tool.

Figure 4.8:　Windows 98 System Information Utility

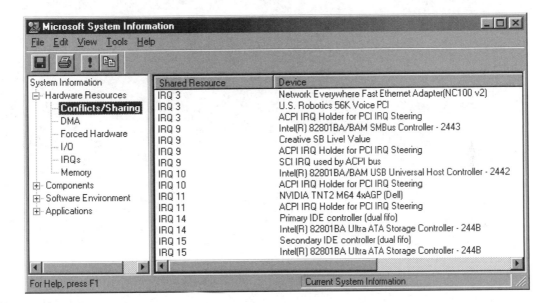

The MSINFO and the Windows Report tools update a file called HWINFO.DAT every time these tools execute. The HWINFO.DAT file is created during the installation process and holds a record of the computer's current hardware configuration, device drivers, and system resources. The MSINFO tool can be used to open the HWINFO.DAT file at any time.

The **ASD.EXE (Automatic Skip Driver Agent)** program is found in the same folder as where Windows 98 is installed. The Automatic Skip Driver tool determines which drivers fail to load during startup. After two failed attempts, ASD marks the device as defective and turns off the device driver. Clicking the Tools menu in System Information accesses ASD. If no problems are found, a dialog box appears on the screen.

Sometimes applications overwrite system files or files become corrupt. The **System File Checker (SFC.EXE)** is located in the WINDOWS\SYSTEM subfolder and protects your system files by checking them and repairing them if necessary. A prompt appears before the original files are restored.

Many error messages can appear in Windows 98. Some of them are listed in Table 4.13.

Table 4.13: Windows 98 Troubleshooting

Error Message	Solution
A device referenced in SYSTEM.INI, WIN.INI, or registry could not be found	1)Edit SYSTEM.INI and WIN.INI and look for all references to the device. Put a semicolon before the line. For Windows 9x, see if there is a 9x driver for the device. If so, leave the semicolon before the SYSTEM.INI line and reload the 9x driver if necessary. After rebooting the computer, if an application displays a message that it cannot find this file, the file needs to be recopied (usually to the WINDOWS SYSTEM folder) and the semicolon removed from the SYSTEM.INI line. 2) To see which file caused the error, use MSCONFIG to choose Selective Startup and step through the startup process. 3) If a Registry error occurs, back up the Registry, use a Registry Editor to locate the driver in the HKLM \System\Current\ControlSet\Services\Vxd key. If you know the driver is no longer needed, highlight the Open subkey and delete it. If any application needs the missing file, a message appears that the file is missing. Reload the file the application disk or CD.
Bad or missing COMMAND.COM	The COMMAND.COM file is missing. Replace the boot disk or CD.
Error in CONFIG.SYS line *xx*	Edit CONFIG.SYS and look at the line referenced in message. Check if the referenced file is in the listed folder. Check for typing errors on the line. Replace file in appropriate folder if necessary.
Error loading kernel. You must reinstall Windows.	Extract the KERNEL32.DLL file from Windows disk or CD and copy it to the WINDOWS\SYSTEM folder.
HIMEM.SYS not loaded or missing or corrupt HIMEM.SYS	Check the WINDOWS folder for the HIMEM.SYS file. Reload file from original disk, boot disk, or CD if the file is missing or corrupt.

Incorrect DOS version	The command or utility is from a different system version than the command interpreter loading during startup. Replace the command or COMMAND.COM and system files with the appropriate version or use the SETVER command to fool the application into using the command interpreter installed.
No operating system found	Boot from the appropriate boot disk and replace the operating system files by typing SYS C:.

Another error resulting from different causes is the Windows Protection Error. A **Windows Protection Error** is usually caused by a virtual device driver being loaded or unloaded. Sometimes the specific device driver (VxD) is mentioned in the error message. The following list cites the most common causes of a Windows Protection Error:

• Real mode driver conflicting with a protected mode driver
• Corrupt registry
• A virus has infected WIN.COM or COMMAND.COM or one of the files is corrupt
• Driver referenced in the registry has initialized and a conflicting SYSTEM.INI driver loads
• I/O or DMA address conflict (use Device Manager or MSINFO)
• Motherboard has been replaced. (Sometimes Windows 98 must be reinstalled to recognize the new motherboard capabilities.)

Windows 98 comes with many troubleshooting wizards. One that relates to the boot process is the Startup and Shutdown troubleshooter. Access this troubleshooting wizard follow these steps on a fully operational Windows 98 computer:

1. Click the **Start** button and then click the **Help** button. The Help window opens.
2. Click on the **Index** tab.
3. Type **startup** in the text box.
4. Double-click on the **troubleshooting** option in the left windowpane.
5. In the right window, click on the **Click here** option to start the wizard.
6. Go through the various screens by clicking a radio button for a problem and then clicking **Next**.
7. When finished, click **Close**.

Sometimes, when the computer will not boot, a startup disk (boot disk) is needed. Another time to use a startup disk is when the computer has a virus. Create a Windows 98 startup disk by following these steps on a fully operational Windows 98 computer:

1. Click the **Start** button, point to Settings, and then click Control Panel.
2. Double-click **Add/Remove Programs**. The Add/Remove Programs window opens.
3. Click **Startup Disk**.
4. Insert a floppy that you don't mind formatting. CAUTION: All data will be lost because the disk will be formatted!

5. Click **Create Disk**. The disk is created.

6. When finished, close the window.

4.10 Troubleshooting Application Problems

Most computer problems occur when installing an operating system or a new hardware component, but applications can cause problems too. Application problems can occur when a new application is installed, a new operating system is installed, an operating system service pack or upgrade is installed, and during normal computer operation. The way in which an application problem is tackled depends on whether the application is 16-bit or 32-bit. This is because Windows 98 uses two types of multitasking—preemptive and cooperative. **Preemptive multitasking** is the operating system determining which application gets the microprocessor's attention and for how long. Preemptive multitasking is used with 32-bit applications. **Cooperative multitasking** relies on the application to relinquish control of the CPU and is the type of multitasking that 16-bit applications use.

An important concept when dealing with applications is the virtual machine. A **virtual machine** is a way for the operating system to appear as a separate computer to each application. Each 32-bit application runs in its own virtual machine. Each 16-bit DOS application runs in its own virtual machine. However, every 16-bit Windows application (an application designed for Windows 3.x) runs in one virtual machine. When a 16-bit Windows application crashes, all 16-bit applications that are loaded into memory crash as well.

Another important concept to remember is that 16-bit applications were not designed to interact with the registry. DOS applications are designed to interact and update the AUTOEXEC.BAT and CONFIG.SYS files. 16-bit Windows applications are designed to interact and update the AUTOEXEC.BAT, CONFIG.SYS, and various .INI files such as SYSTEM.INI and WIN.INI. Windows 98 still contains and supports these files in order to be backward compatible with older applications.

General Protection Fault (GPF)

When an application error occurs, a **GPF (General Protection Fault)** error message appears. When a Windows 16-bit application GPF occurs, all other 16-bit Windows applications are halted until you exit the application that caused the error. In most cases, the other 16-bit applications must be closed as well. When a 32-bit application GPF occurs, no other application is affected. Sometimes when the GPF occurs, you are allowed to terminate the application from the GPF error window, but sometimes you must use a different method to quit the application. In that case, press the Ctrl + Alt + Del keys and the Close Program window appears. Click on the offending application and then click on the **End Task** button.

Dr. Watson

Another useful utility to help with application problems is Dr. Watson. **Dr. Watson** has the ability to take a snapshot of the computer system when a fault occurs. For software applications, Dr. Watson can provide information about the problem cause. The Dr. Watson utility does not load by default in Windows 98. To start Dr. Watson on a fully operational Windows 98 computer, follow these steps:

1. To access the MSINFO tool, click the **Start** button, point to **Programs**, point to **Accessories**, point to **System Tools**, and click **System Information**. The System Information window opens.

2. From the System Information window, click on the **Tools** menu option and choose the **Dr. Watson** menu item. A Dr. Watson icon appears in the taskbar system tray in the bottom right corner.

3. To view Dr. Watson, double-click the Dr. Watson icon on the taskbar. It builds information about the system and opens.

4. When finished, close the window.

On the Diagnosis tab is information relating to the system snapshot. Figure 4.9 illustrates the Dr. Watson window. The Dr. Watson window also has blank space at the bottom so you can insert notes.

Figure 4.9: The Dr. Watson Window

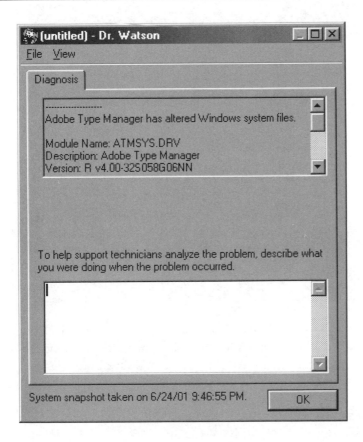

Dr. Watson also includes an advanced view that provides detailed information about the computer system. Click on the **View** menu option and select **Advanced view**. The Advanced view has many tabs across the top. To view more tabs, click on the right or left arrows located to the right of the tabs. Several important tabs include the Kernel drivers, User Drivers, MS-DOS drivers, and 16-bit Modules. These tabs separate the various drivers that can be used on Windows 98. Figure 4.10 illustrates the Dr. Watson Advanced view.

Figure 4.10: Dr. Watson Advanced View Window

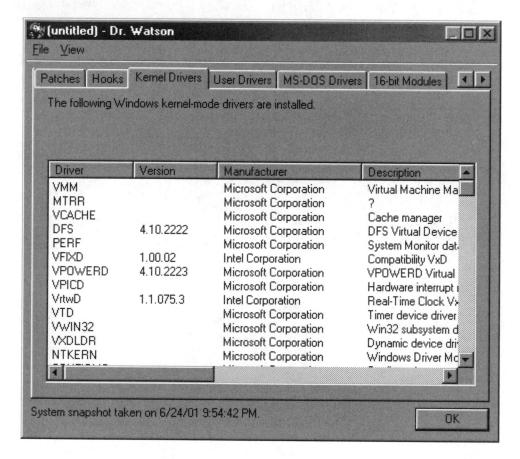

When Dr. Watson is running and an application error occurs, Dr. Watson logs the information into a file called WATSON*xx*.WLG file. The *xx* in the file name is a number that is automatically incremented by Dr. Watson. To view a saved file, click on the *File* menu option and select the Open Log File menu item. Any system snapshot can also be saved using the **Save** or **Save As File** menu options. To have Dr. Watson start every time the computer boots, create a shortcut to the WINDOWS\DRWATSON.EXE file and place the shortcut in the WINDOWS\START MENU\PROGRAMS\STARTUP folder.

System Monitor

Many tools have already been covered that help with resource management, but two new ones are System Monitor and Resource Meter. **System Monitor** is a utility that tracks performance of individual system components, such as disks, file system, operating system kernel, and memory manager. Under each category are numerous individual selections. System Monitor helps identify where performance bottlenecks are located. To access System Monitor, click on the **Start** button, point to **Accessories** and **System Tools**, and click on the **System Monitor** option.

System monitor has three views—ine chart, bar chart, and numeric chart. Figure 4.11 shows System Monitor using a bar chart. An exercise at the end of this chapter demonstrates System Monitor usage.

Figure 4.11: System Monitor Bar Chart

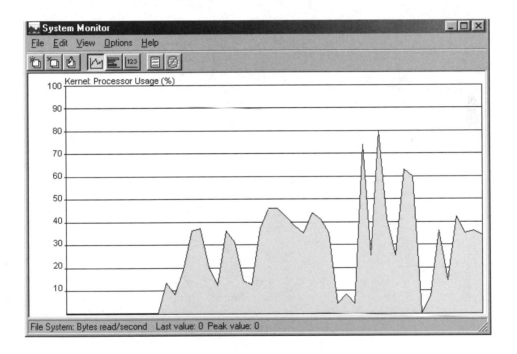

Resource Meter is a very simple graphical display of how Windows 98 is using memory. To access Resource Meter, click on the **Start** button, point to **Programs, Accessories**, and **System Tools** options, and click on the **Resource Meter** selection. A window with a Resource Meter description appears on the screen and then the Resource Meter becomes an icon on the taskbar system tray. Double-click on the icon that looks like a set of stairs with a colored bar across the bottom to display Resource Meter. Keep in mind that Resource Meter consumes system resources just like any utility. After viewing Resource Meter, right-click on the Resource Meter icon in the taskbar system tray and click on the **Exit** option.

The three Windows 98 core components—Kernel (system), User, and GDI—are shown with individual lines. The **System line** represents the Windows 98 kernel, which is the part of the operating system that supports input/output, task scheduling, and virtual memory management. The **User line** represents how much memory is being used to manage user input such as mouse usage, window sizing, etc. The **GDI (Graphics Device Interface) core component line** represents the amount of memory being used for screen and printer graphics. Figure 4.12 shows the Resource Meter window.

Figure 4.12: Resource Meter

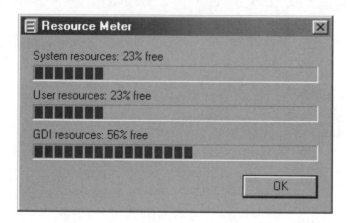

Sometimes you must adjust the paging file size for optimum computer performance. The **System** control panel is used to set the virtual memory size. Once in the System control panel, click on the **Performance** tab. Click on the **Virtual Memory** button and two options are available—Let Windows manage my virtual memory settings (the default setting) and Let me specify my own virtual memory settings. Click on the **Let me specify my own virtual memory settings** radio button, select the hard drive partition and minimum/maximum amount of hard drive space. Click on the **OK** button when finished. The minimum/maximum settings should contain the same value for maximum computer performance.

Chapter Summary

- Windows 98 is a 32-bit operating system used by home users and businesses. It supports Plug and Play, USB, and both 16-bit and 32-bit applications.
- Windows 98 provides a hierarchical file system allowing you to create folders and store files within them.
- The Registry is a database, which stores the hardware and software information installed on your computer.
- As part of the pre-installation steps of Windows 98, you should decide whether or not to perform a clean installation or an upgrade. You should consider installing virus protection software, which can detect and remove common software viruses.

- The SETUP.EXE program allows you to install Windows 98. It has numerous switches that can be uses for various purposes.
- Troubleshooting your Windows 98 installation includes entering the wrong Product Key and having incompatible hardware or hardware drivers.
- You can dual-boot Windows 98 and Windows NT or Windows 2000. However, it is advisable to install each on separate partitions. It is recommended to install Windows 98 before you install the other operating systems.
- Configuring Windows 98 involves using Control Panel to change the look and feel of your Windows 98 system. Examples include: changing mouse, keyboard, and sound settings. You can also add and remove devices.
- You need to understand the boot process of Windows 98 so you know how to trouble-shoot when problems occur.

Review Questions

1. What Microsoft browser software allows you to connect to the Internet?
   a) Internet Connection Sharing
   b) Internet Explorer
   c) Windows Explorer
   d) Control Panel

2. What Microsoft feature allows you to connect to the Internet through another Windows 98 computer?
   a) Internet Connection Sharing
   b) Internet Explorer
   c) Windows Explorer
   d) Control Panel

3. The _____ is the area where all of your work in Windows 98 begins. This area contains icons and shortcuts.
   a) Desktop
   b) Window
   c) Control Panel
   d) Long File Names

4. What Windows feature prevents you from moving icons on your desktop?
   a) Auto Unclutter
   b) Auto Arrange
   c) Arrange Order
   d) Automatically Detect

5. Where are files held when they are deleted from the hard disk drive?
   a) Trash Can
   b) They are not held anywhere because they are permanently deleted.
   c) Recycle Bin
   d) Quick Launch Bar

6. Where is the name of the horizontal bar the runs across the bottom of your screen?
   a) Recycle Bin
   b) Taskbar
   c) Quick Launch Bar
   d) Task Manager

7. A file with an extension of _____ is a compressed cabinet file.
   a) COM
   b) TXT
   c) CAB
   d) DLL

8. A file with an extension of _____ is a command file or executable.
   a) COM
   b) TXT
   c) CAB
   d) DLL

9. Which of the following are invalid Windows 98 long file names? Choose all that apply.
   a) Payroll.txt
   b) F/Le:100.tx*t
   c) Sales.dat
   d) X.DOC

10. Which of the following commands will display the Window 98 version?
    a) ver
    b) version
    c) update
    d) vers

11. Which of the following commands allows you to install Windows 98?
    a) SETUP.COM
    b) SETUP.BAT
    c) SETUP.XLS
    d) SETUP.EXE

12. Which Windows 98 registry files contain computer-specific hardware settings?
    a) SETUP.DAT
    b) SYSTEM.DAT
    c) SYSTEM.BAT
    d) USER.DAT

13. Which Windows 98 registry files contain user profile settings?
    a) PROFILE.INI
    b) SYSTEM.DAT
    c) SYSTEM.BAT
    d) USER.DAT

14. Which registry subtree contains global hardware configuration?
    a) HKEY_USERS
    b) HKEY_CURRENT_USER
    c) HKEY_CLASSES_ROOT
    d) HKEY_LOCAL_MACHINE

15. Which registry subtree contains user preferences?
    a) HKEY_USERS
    b) HKEY_CURRENT_USER
    c) HKEY_CLASSES_ROOT
    d) HKEY_LOCAL_MACHINE

16. Which command allows you to backup the registry from the 16-bit real mode (DOS mode)?
    a) REGEDIT
    b) SCANREGW
    c) SCANREG
    d) SCANREG.INI

17. Which term is used to define the capability to install more than one operating system on your computer?
    a) Dual-boot
    b) Partitioning
    c) Formatting
    d) TSR

18. Which hidden startup log file is stored in the root directory and chronologically lists what occurred in the installation process?
    a) DETLOG.TXT
    b) SETUPLOG.TXT
    c) NETLOG.TXT
    d) DETCRASH.LOG

19. Which Control Panel function allows you to install a new software application?
    a) Add New Hardware
    b) Add/Remove Programs
    c) Regional Settings
    d) Printers

20. Which Windows 98 Startup file is located in the WINDOWS\SYSTEM directory and provides support for the GUI environment?
    a) HIMEM.SYS
    b) GUI.SYS
    c) GDI.SYS
    d) WIN.COM

Lab Projects

Unless otherwise indicated, all projects assume the computer is powered on, you are logged in, and your desktop appears on the screen.

Lab Project 1

The goal of this lab is to help you determine system and disk information.

1. To determine CPU type, operating system and version, and RAM capacity, right- click the **My Computer** icon and click **Properties**. The System Properties screen appears with your system information.

2. Record the operating system type and level, the processor type, and RAM.

   _____

3. Click **OK** when complete.

4. In order to determine disk capacity and file system type, double-click **My Computer**. The My Computer screen appears with icons. Note the drive icons.

5. Right-click on the C: drive and then click **Properties**. Make sure the **General** tab is selected. The Properties page for the C: drive appears.

6. Record the used space for drive C:.

   _____

7. Record the free space for drive C:.

   _____

8. Record the file system type for drive C:.

   _____

9. Draw the pie chart representing how the space is used.

10. Insert a floppy in the floppy drive.

11. Repeat steps 6 through 9 for the floppy. Write your answers on a separate piece of paper.

12. Log out.

Lab Project 2

The goal of this lab is to help you understand Task Manager.

1. First you will start a task. Press Start in the lower right-hand corner. Next, click **Run** and in the Open dialog box, type **calc** and press Enter. The calculator program appears on the screen.

2. To see the **calc** task (along with other jobs), press Ctrl + Alt + Del.

3. Record three other tasks.

_____

4. You can scroll up and down this list and click **End Task** to delete a task. Be careful because ending a task will terminate it and you do not want to terminate a system task.

5. Highlight **calc** and press End Task.

6. Now, see what happens when you remove a system task. Terminate the **explorer** task. Follow the steps that appear on the screen. Hint: The shutdown screen will appear.

7. Record what happens.

_____

8. Log out.

Lab Project 3

The goal of this project is to associate a file extension with a file type.

1. Click on the **Start** button and point to the **Programs** selection.

2. Point to the **Accessories** option.

3. Click on the **Notepad** menu selection.

4. Type in the following:

   However far modern science and techniques have fallen short of their inherent possibilities, they have taught mankind at least one lesson: Nothing is impossible.
   —Lewis Mumford

5. Click on the **File** menu option.

6. Click on the **Save** option from the drop-down menu.

7. Insert a formatted disk into the floppy (A:) drive.

8. Click on the **down arrow** in the **Save in** textbox.

9. Click on the **3_ Floppy (A:)** option.

10. In the File name text box, type **Junk**.

11. Click on the **Save** button.

12. Click the Notepad application by clicking on the **Close** button (which is a button in the upper right corner with an X).

13. Right-click on the **Start** button.

14. Click on the **Explore** option.

15. Click on the **View** menu option.

16. Click on the **Folder Options** drop-down menu option.

17. Click on the **View** tab in the Folder Options dialog box.

18. If the **Hide file extensions for known file types** check box contains a check mark, click inside the check box to remove the check mark. If the check box is empty, ignore this step.

19. Click on the **OK** button.

20. In the left Explorer window, use the vertical scroll bar to locate the A: drive. Click on the **3_ Floppy (A:)** drive option.

21. Locate the **Junk.txt** file in the right window and double-click on the icon.

22. Record what happened. Did Notepad open with the Junk file open?

_____

23. Close the Notepad application by clicking on the **Close** button (which is a button in the upper right corner with an X).

24. In the right Explorer window, right-click on the **Junk** file name.

25. Click on the **Rename** option. The name of the file, Junk.txt, is highlighted.

26. Type in **junk.abc** and press the **Enter** key. Junk.txt is renamed to junk.abc. A Rename warning box appears stating that if you change a file name extension, the file may become unusable. It also asks, "Are you sure you want to change it?" Click on the **Yes** button.

27. What does the junk.abc file icon look like now?

_____

28. Double-click on the **junk.abc** file icon.

29. What happened when you double-clicked on the junk.abc file icon?

_____

30. In the Choose the program you want to use section, scroll down until you reach the Notepad icon. Click on the **Notepad** icon and then click on the **OK** button.

31. What happened when you clicked on the OK button?

_____

32. In the Notepad application, click on the **File** menu option. Then click on **New** from the drop-down menu.

33. Type in the following:

   Technology is dominated by two types of people: Those who understand what they do not manage, and those who manage what they do not understand.
   —Source Unknown

34. Click on the **File** menu option.

35. Click on the **Save** option from the drop-down menu.

36. Click on the **down arrow** in the **Save in** text box.

37. Click on the **3_ Floppy (A:)** option.

38. In the File name text box, type **Junk2**.

39. Click on the **Save** button.

40. Close the Notepad application by clicking on the **Close** button (which is a button in the upper right corner with an X).

41. Using Explorer, rename the **Junk2.txt** file to **Junk2.abc**. Notice the file icon after the change.

42. How is the JUNK2.ABC icon different from before?

_____

43. Double-click on the **Junk2.abc** icon.

44. What happened when you double-clicked on the Junk2.abc icon?

_____

Lab Project 4

The goal of this project is to help you understand how to create a tree structure from a screenshot.

1. Using Windows Explorer, create the following tree structure in Figure 4.13 on your floppy. To get to Explorer, right-click Start and then click Explore.

Figure 4.13: A Sample Tree Hierarchy

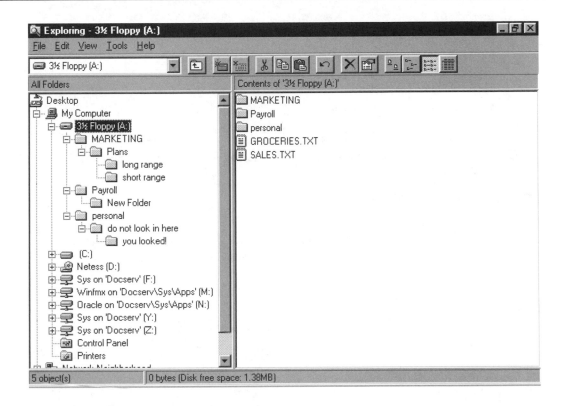

2. Using Notepad, create the following files as Text Documents in the **long range** folder. Place your name in each file.
   Monday.txt
   Tuesday.txt
   Wednesday.txt
   Thursday.txt
   Friday.txt

3. Record the steps you took to create them.

4. Rename the **New Folder** to **Weekly**.

5. Record the steps you took.

   _____

6. Create a **Medium Range** folder in **Plans**.

7. Record the steps you took.

   _____

8. In the **personal** folder, create a file called **resume.txt**.

9. Hide this folder.

10. Record the steps you took.

    _____

11. View all files.

12. Record the steps you took.

    _____

13. Make a complete copy of the files in **Plans** and put them in a folder called **BACKUP of PLANS**.

14. Record the steps you took.

    _____

15. In Explorer, right-click on the floppy to make a disk copy of the floppy. Insert another floppy disk and make a disk copy of this floppy to another floppy.

16. Record the steps you took.

    _____

17. Create another folder in the root directory of the floppy that has your name in it.

18. Create another folder in the root directory of the floppy named Windows 98 Project 4.

19. Show your instructor.

20. Close all windows.

Lab Project 5

The goal of this project is to reinforce your understanding of how to create a tree structure. You will also work with attributes to help you understand how they work.

1. Using Windows Explorer, create the following tree structure in Figure 4.14 on your floppy.

Figure 4.14: A Sample Tree Hierarchy

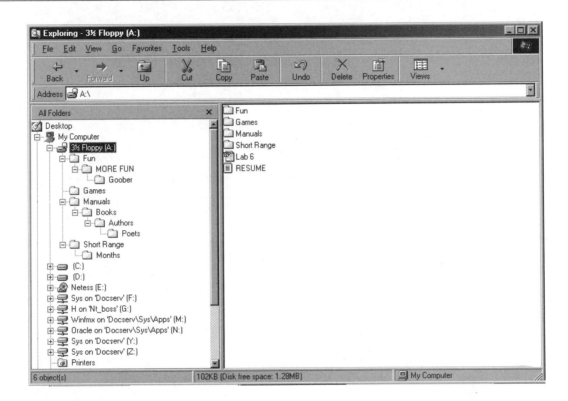

1. Using Notepad, place these files in the **Poets** folder:
    Poe.txt
    Walker.txt
    Smith.txt

2. In Explorer, click **View**, then **Folder Options,** and then **View**. The **Folder Options** window opens. In the **Advanced Settings** box, activate the setting so hidden and system files are not displayed.

3. Right-click on **Poe.txt** and click **Properties**. The Properties page appears. Click **Hidden** and then click **OK**. The files are hidden.

4. In Explorer, click **View**, then **Folder Options,** and then **View**. The **Folder Options** window opens. In the **Advanced Settings** box, activate the setting to show all files. This will display hidden files.

5. Record whether or not you can now see **Poe.txt**.

_____

6. Using the previous step as a guide, make the file **Walker.txt** read-only.

7. Record the steps you took.

_____

8. Open the file named **Walker.txt** in Notepad and attempt to add data.

9. Record what occurred.

_____

10. Close Notepad.

11. Make **Walker.txt** readable.

12. Open the file named **Walker.txt** in Notepad and attempt to add data.

13. Record what occurred.

_____

14. Close Notepad.

15. Create another folder in the root directory of the floppy that has your name in it.

16. Create another folder in the root directory of the floppy named **Windows 98 Project 5**.

17. Show your instructor.

18. Close all windows.

Lab Project 6 (Caution!)

The goal of this project is to familiarize you with the REGEDIT registry editing utility.

CAUTION: Editing the registry can cause your computer to run erratically, or not run at all! When performing any registry editing, follow ALL directions carefully, including spelling, syntax use, etc. Failure to do so may cause your computer to fail. *Ask your instructor for permission before proceeding!*

1. From the Start menu, choose **Run**, type **REGEDIT**, and click **OK**. The REGEDIT utility opens.

2. In the left window, expand **HKEY_LOCAL_MACHINE**, **Hardware**, and **Description**, **System**, **CentralProcessor**, and then select the **0** folder.

3. In the right window, the **Identifier** and **VendorIdentifier** information display.

4. Record the CPU identifier.

   _____

5. In the left window, expand **HKEY_LOCAL_MACHINE**, **Software**, and **Microsoft**.

6. List three Microsoft applications loaded under the Microsoft folder.

7. In the left window, expand **HKEY_USERS**, **.DEFAULT**, **Control Panel**, **Appearance**, and then select the **Schemes** folder.

8. Name three schemes available to be chosen from the Display control panel as listed in the right window.

9. When finished viewing the information, close the **REGEDIT** utility.

Lab Project 7

The goal of this project is to use the Windows 98 Registry Checker utility to back up the registry.

1. Click on the **Start** button.

2. Point to the **Programs** option.

3. Point to the **Accessories** option.

4. Point to the **System Tools** option.

5. Click on the **System Information** option. The Microsoft System Information window opens.

6. Click on the **Tools** menu option and select **Registry Checker**. The Registry Checker tool opens. Click on the **Yes** button. A window appears with a message that the backup was successful. Click on the **OK** button.

7. Notice the system data and time by double-clicking on the time display on the taskbar. Write the system date and time in the space below.

_____

8. Using Windows Explorer, locate the WINDOWS\SYSBCKUP folder. In the left Explorer window, click on the **WINDOWS\SYSBCKUP** folder.

9. In the right window, locate the RB000.CAB, RB001.CAB, RB002.CAB, RB003.CAB, and RB004.CAB files.

10. Right-click on each of these file icons and determine when the backup file was created.

11. Record the backup file name that was used to manually back up the registry.

_____

12. Close the **System Information** window.

Lab Project 8

The goal of this project is to install an updated driver under the Windows 98 operating system. You will need the latest Windows 98 driver for the hardware device.

Note: The installation process outlined below may differ due to computer differences. If the process is different, follow the directions on the screen. Contact your instructor or a lab assistant if you are unsure of the step.

1. Turn the computer on and verify that the operating system loads. If necessary, log in to Windows 98 using the userid and password provided by the instructor or lab assistant.

2. Click on the **Start** button, point to **Settings**, and click on the **Control Panel** menu option.

3. When the Control Panel window opens, scroll down to the **System** icon and double-click on it.

4. Click on the **Device Manager** button.

5. Click on the **+ (plus sign)** beside the class of device you want to upgrade.

6. Right-click on the specific device you want to upgrade.

7. Click on the **Properties** option.

8. Click on the **Driver** tab.

9. Click on the **Update Driver** button. The Upgrade Device Driver wizard appears.

10. Click on the **Next** button.

11. Click on the **Display a list of the drivers in a specific location, so you can select the driver you want** radio button. Click on the **Next** button.

12. Click on the **Have Disk** button.

13. Insert the media that contains your updated driver into the floppy drive, CD-ROM, or DVD drive.

14. In the **Copy manufacturer's files from** box, type in the drive letter for the device that contains the updated driver followed by a colon and click on the **OK** button. An example of a driver device is A:. The Browse button can be used to locate a device driver as well.

15. Click on the **Next** button and follow the instructions on the screen to complete the upgrade.

Lab Project 9

The goal of this project is to use the Windows 98 System Information tool to troubleshoot device conflicts.

1. Turn on the computer and verify that Windows 98 loads.

2. If necessary, log on to Windows 98 using the userid and password provided by the instructor or lab assistant.

3. Click on the **Start** button and point to the **Accessories** and **System Tools** selections.

4. Click on the **System Information** option. If the System Information item is not highlighted, click on it.

5. Record the percentage of free system resources.

_____

6. Click on the + (**plus sign**) next to the **Hardware Resources** item.

7. Click on the **Conflicts/Sharing** item.

8. Are any devices sharing an IRQ (for Interrupt Request)? If so, list two that are sharing an IRQ and the IRQ they are sharing.

_____

9. Click on the **IRQs** item located under the Hardware Resources folder in the left window.

10. List any IRQ that is available for use by a new device.

_____

11. Click on the **Memory** item located under the Hardware Resources folder in the left window.

12. List one memory address range used by a device.

_____

13. In the left window, click on the + (**plus sign**) next to the **Components** item.

14. In the left window, click on the **Problem Devices** item. Any devices with hardware conflicts are listed in the right window. Keep in mind that you use the Device Manager program to correct any problems with resource allocation.

15. Record whether any devices have conflicts. If a device conflict exists, record the devices in conflict.

_____

16. Close the System Information window by clicking on the **Close** button (the one with the X) in the upper right corner of the window.

Lab Project 10

The goal of this project is to create a boot disk that can be used when the computer does not boot properly. You will need a blank floppy disk.

1. Click on the **Start** button.

2. Point to the **Settings** option.

3. Click on the **Control Panel** option.

4. Double-click on the **Add/Remove Programs** control panel.

5. Click on the **Startup disk** tab.

6. Click on the **Create disk** button. The Insert Disk window opens. Insert a floppy disk into the A: drive. Click on the **OK** button.

7. Record when you think the startup disk would be useful.

_____

8. When the boot disk is created, use Explorer to examine the contents of the disk, the items in the AUTOEXEC.BAT file, and the items in the CONFIG.SYS file.

9. Does the CONFIG.SYS include commands for CD-ROM support? How do you know?

_____

Lab Project 11

The goal of this project is to use Windows 98 tools effectively when problems occur. Dr. Watson, System File Checker, and Registry Checker should be installed.

1. Click on the **Start** button, point to the **Accessories** and **System Tools** option, and click on the **System Information** selection.

2. Click on the **Tools** menu item and select the **Dr. Watson**. A new icon appears on the right end of the taskbar next to the clock icon.

3. Double-click on the **Dr. Watson** taskbar icon. Dr. Watson records a snapshot of the system. The Diagnosis tab displays any problems found during the analysis.

4. Click on the **View** menu item and select **Advanced view**. Ten tabs appear across the top.

5. Click on the **System** tab. The System tab shows an overall view of the computer including operating system version, amount of RAM installed, amount of resources available, amount of free space for the swap file, etc.

6. How much RAM does the computer have?

   _____

7. Click on the **Startup** tab. The Startup tab shows the applications that run every time the computer boots.

8. List three applications that load when the computer boots as found on the Startup tab.

9. Click on the **MS-DOS Drivers** tab. The MS-DOS Drivers screen displays a list of 16-bit drivers that load when the computer boots.

10. List three MS-DOS drivers.

   _____

11. Close the Dr. Watson tool by clicking on the **OK** button. Notice that the Dr. Watson utility is still active because the icon is still located on the taskbar. Right-click on the **Dr. Watson** icon located on the taskbar. Select the **Exit Dr. Watson** item. Dr. Watson closes.

12. Click on the **Microsoft System Information** button located on the taskbar. The Microsoft System Information window reappears on the screen.

13. Click on the **Tools** menu item and select the **System File Checker** option. The System File Checker window opens.

14. Ensure the **Scan for altered files** radio button is selected and click on the **Start** button. The system takes about a minute to scan the system files. When the scan is finished, a message displays. Click on the **OK** button. Click on the **Close** button to exit from the System File Checker window.

15. Click on the **Tools** menu item and select the **System File Checker** option again.

16. Click on the **Settings** button. The Settings window allows you to customize how the System File Checker tool works by allowing you to add or remove files to be checked,

prompt before restoring system files, determine the default location for the good system files are stored, etc.

17. Click on the **View Log** button on the Settings tab. You may be prompted that the file is too big for Notepad and asked if it is okay to open the document in WordPad. If so, click on the **Yes** button. The default name of the log file is SFCLOG.TXT and it is stored in the WINDOWS folder by default. Close the **SFCLOG.TXT** file window.

18. Click on the **Search Criteria** tab. The Search Criteria tab lists files and folders that are checked through the System File Checker tool.

19. Locate the folder that contains the Windows 98 operating system (normally C:\WINDOWS). Are subfolders located in the WINDOWS folder checked by the System File Checker tool?

_____

20. Click on the **Advanced** tab. The Advanced tab is where you can define the verification data file's default location.

21. What is the default location for the verification data file?

_____

22. Click on the **Cancel** button to leave the System File Checker Settings window. Click on the **Close** button to exit the System File Checker utility.

23. Click on the **Close** button to exit the Microsoft System Information window.

Lab Project 12 Challenge

The goal of this project challenge is to install Windows 98 on a computer that does not have an operating system or one on which the old operating system will be replaced (removed). You will need a Windows 98 CD, a blank floppy disk, and a computer with a formatted hard drive. You can refer to Chapter 3 for a discussion on partitioning and formatting your hard disk. The installation process outlined below may differ due to computer differences. If the process is different, follow the directions on the screen. Contact your instructor or a lab assistant if you are unsure of the step.

1. Insert the Windows 98 setup disk into the floppy drive and boot the computer. The Startup menu appears.

2. Insert the Windows 98 CD into the CD-ROM and type **1** and press the **Enter** key.

3. At the prompt, type **x:setup**, where **x:** is the drive letter of the device containing the Windows 98 CD, and press **Enter.**

4. A message appears that ScanDisk is going to be performed. After the ScanDisk check, press the **X** key.

5. The Welcome to Windows 98 Setup screen appears. Click on the **Continue** button.

6. The licensing agreement appears on the screen. Read the agreement and click on the **I accept the Agreement** radio button and click the **Next** button. The Product Key window appears.

7. Enter the product key provided by the instructor or lab assistant, or located on the registration certificate or CD case. Click the **Next** button to continue. The Select Directory window appears.

8. The default Windows 98 folder is C:\WINDOWS. Contact the instructor or lab assistant to see if the default or a different drive/folder is to be used. If a different drive/folder is to be used, click on the **Other directory** radio button and click **Next.** If the default directory is to be used, simply click on the **Next** button. The Preparing Directory screen appears.

9. What is the path where Windows 98 is to be installed?

10. Setup checks the computers hardware and then displays the Setup Options screen. The default is Typical. Ensure the **Typical** radio button is selected and click on the **Next** button.

11. The User Information window appears. Type a name and a company in the text boxes and click on the **Next** button.

12. The Windows Components window appears. Ensure the **Install the most common components** radio button is selected and click on the **Next** button.

13. When the Identification screen appears, contact an instructor or lab assistant for the Name and Workgroup. The computer name must be unique and can be up to 15 characters. Spaces are not allowed. The workgroup name must be the same for all computers on the same network (if the computer is networked) and can be up to 15 characters; spaces are not allowed.

14. What computer name and workgroup name is to be used on this computer? Fill in the information below:

    Computer name _____

    Workgroup name _____

15. The Establishing Your Location window appears. Select the appropriate country and click on the **Next** button.

16. The Startup Disk screen appears. Click on the **Next** button. A prompt appears; insert the blank floppy disk into the A: drive and click on the **OK** button.

17. The Start Copying Files window appears. Click on the **Next** button and the Windows 98 files are copied to the computer hard drive. After copying, you are prompted to restart the computer.

18. Click on the **Restart Now** button. After the computer reboots, Windows 98 configures hardware and restarts again. The Welcome to Windows 98 screen appears.

Internet Discovery

Internet Discovery Lab 1

The goal of this lab is to obtain specific information on the Internet regarding Windows 98.

1. What is the URL for the Microsoft Windows 98 on-line help?

2. Locate a magazine article on the Internet that describes how to automatically log a user into a Windows 98 computer.

3. Locate a description of IRQ steering on the Internet and write the URL and basic description in the space below.

4. Find three different locations on the Internet that have the *Microsoft Windows 98 Resource Kit* book for sale. Find the lowest price on these three sites. Write the cost and the site in the space below.

_____

_____

_____

_____

_____

_____

_____

_____

_____

_____

_____

_____

_____

_____

_____

_____

_____

_____

5. Locate a site on the Internet that has Windows 98 troubleshooting tips that is not the Microsoft web site. Write the URL of this location in the space below.

6. At the Frank Condron's World O'Windows web site, he has Easter eggs listed for Windows 98. What panel is used to discover the Windows 98 Easter egg?

Soft Skills: Help Desk Support

1. A customer calls you in the middle of the night and is attempting to install Windows 98 on a PC that crashed. The PC had been running Windows 98. It seems that the boot sector is infected with a virus. What do you tell the customer?
2. A user from Accounting e-mails you about installing a printer. What steps do you give them to install this printer?
3. You receive a call from a user interested in setting up a dual-boot PC. How do you tell the user to proceed?

Critical Thinking

1. How do viruses infect Windows 98?
2. Compare and contrast a DLL file, a DRV file, and a CAB file.
3. When would you use System Monitor?
4. Explain how dual-booting works.

 *Study Skills*

Have You Done Your Homework?

Homework is work but it can be fun because you are learning valuable skills. The only way you can really learn a technical topic such as operating systems is by doing the homework assignments and lab projects.

- Are you doing all of your homework? You should complete all homework assignments as soon as you can once they are assigned. This way, the material is still fresh in you brain and you'll tend to remember it better.
- Are you doing your homework in a quiet place? You should complete your homework where you can be free of outside noises. Don't play the radio, television, or your favorite music CD while you study. If necessary, go to the library because it'll be quiet there.
- Do you have a complex problem that you can't answer? If you get stuck on a problem, look for a similar situation in the textbook or your notes. Instructors will often choose problems for homework assignments that are similar in nature to ones that have already been discussed.
- Try to work through problems before asking another person. If you figure it out, it will build your confidence in the material. If you still run into trouble, ask your study partner or your instructor.
- Store you homework in a separate three-ring binder for later access. This will be a good source of information for test preparation.

Self-Study Question(s)

1. Have you done your homework today?
2. Identify at least one problem you had trouble answering and list the steps as they relate to the above Study Skills that helped you figure out the problem.

5

Chapter 5
Introduction to Microsoft Windows NT

OBJECTIVES

The goal of this chapter is twofold:
- To introduce you to Windows NT.
- To help you prepare and pass the following sections of the A+ Operating System Technologies Exam:

A+ Operating System Technologies Exam Objectives
covered in this chapter (and corresponding page numbers)

1.2 Identify the names, locations, purposes, and contents of major system files.

Domain 2 Installation, Configuration and Upgrading

2.1 Identify the procedure for installing Windows 9.x/Me, Windows NT 4.0 Workstation, Windows 2000 Professional, and Windows XP and bringing the operating system to a basic operational level.

2.2 Identify steps to perform an operating system upgrade from Windows 9.x/Me, Windows NT 4.0 Workstation, Windows 2000 Professional, and Windows XP. Given an upgrade scenario, choose the appropriate next steps.

2.3 Identify the basic system boot sequences and boot methods, including the steps to create an emergency boot disk with utilities installed fro Windows 9.x/Me, Windows NT 4.0 Workstation, Windows 2000 Professional, and Windows XP.

2.4 Identify procedures for installing/adding a device, including loading, adding, and configuration device drivers, and required software.

2.5 Identify procedures necessary to optimize the operating system and major operating system subsystems.

Domain 3 Diagnosing and Troubleshooting

3.1 Recognize and interpret the meaning of common error codes and startup messages from boot sequence, and identify steps to correct the problems.

3.3 Recognize common operational and usability problems and determine how to resolve them.

In this chapter, you will complete the following sections:
- 5.1 Understanding Microsoft Windows NT
- 5.2 Managing Files and Folders
- 5.3 Understanding NTFS Compression
- 5.4 Understanding the Windows NT Registry
- 5.5 Understanding the Pre-Installation Steps of Windows NT
- 5.6 Understanding How to Install and Upgrade Windows NT Workstation
- 5.7 Troubleshooting the Windows NT Installation
- 5.8 Configuring Windows NT Workstation
- 5.9 Understanding the Boot Process
- 5.10 Understanding Task Manager, Event Viewer, and Windows NT Diagnostics
- 5.11 Troubleshooting a Service that Does Not Start
- 5.12 Monitoring System Performance

5.1 Understanding Microsoft Windows NT

Microsoft created NT Workstation for business computers operating in a networked environment. There are actually two major versions of NT: NT Server and NT Workstation. This chapter covers NT Workstation because it is on the A+ exam.

NT Workstation has two versions: Windows NT 3.5.1 and Windows NT 4.0. NT Workstation 4.0 offers many enhancements to prior operating systems such as Window NT Workstation 3.5.1, Windows 3.x, Windows 95, and Windows 98. The NT Workstation enhancements are listed below.

- More efficient use of memory than NT Workstation 3.5.1.
- Improved graphics performance.
- True 32-bit operating system.
- Supports SMP (Symmetric Multiprocessing), which is the ability to run multiple processes on multiple CPUs.
- Better network security.
- Support of long file names.
- More reliable than Windows 3.x and Windows 95.
- Better application reliability.
- NT can be installed on **RISC (Reduced Instruction Set Computer)** systems as well as **CISC (Complex Instruction Set Computer)** systems.

RISC computers have fewer instructions (or reduced) in their instruction set than CISC computers. Because RISC computers have fewer instructions, they rely on additional hardware to perform the operations lacking from the instruction set. This makes them faster yet more expensive. Your PC is a CISC system. IBM makes a RISC system called an IBM RS (RISC System) 6000. It operates a version of the UNIX operating system. The UNIX operating system will be discussed in Chapter 8.

The file systems NT Workstation 4 supports are FAT16 (File Allocation Table) and NTFS (New Technology File System). For CD media, NT Workstation supports **CDFS (Compact Disk File System)**. NT Workstation 4 does not support DVD (Digital Versatile Disk) media.

When NT Workstation boots, a log-in screen appears. A user can log in to a network server or the user can log in locally to the local workstation. After logging in, the desktop appears. The **desktop** is the area where all work is performed. The desktop is the interface between you and the applications, files, and computer hardware. The desktop is part of a **GUI (Graphical User Interface)** environment. Figure 5.1 shows the NT Workstation desktop.

Figure 5.1: Windows NT Workstation Desktop

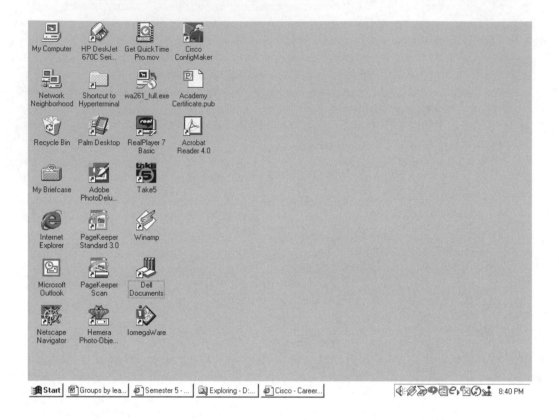

On the desktop are icons. **Icons** are graphics that can represent applications, files, the computer and its hardware devices, and shared network resources. In Figure 5.1, examples of icons include the My Computer icon, the Network Neighborhood icon, and the Recycle Bin icon. The desktop can be modified to have a background or desktop pattern. This can include a graphic or a specific color.

Let's go through a short exercise to help you understand how to maintain an uncluttered Windows NT desktop. In this exercise, you will work with the **Auto Arrange** option to force your icons on your desktop to be automatically arranged towards the left hand side of the desktop. When you attempt to relocate an icon, it will automatically be arranged back to its original position—on the left hand side of the desktop. It is assumed that Windows NT is installed and running on a computer with the Windows NT desktop displayed on your screen and you are logged in.

1. Right-click on a blank desktop space. Once a menu appears, point to the **Arrange Icons** option. If the **Auto Arrange** option does have a check mark beside it, click on an **empty portion of the desktop** to cancel because Auto Arrange is already in use. If

the **Auto Arrange** option does not have a check mark beside it, click on the **Auto Arrange** option to add the check mark and enable **Auto Arrange** capability.

2. Note the position of the icons on the desktop.
3. Click on the **My Documents** icon and while continuing to hold the mouse button down, drag the icon to the center of the screen. You will see that the **My Documents** icon moves to its original position.
4. Click on the **Recycle Bin** icon and while continuing to hold the mouse button down, drag the icon to the top right portion of the screen. This icon moves to its original position too.
5. Right-click on a **blank desktop space.** A menu appears.
6. Using step 1 as a guide, set Auto Arrange back to its original setting.

Notice in Figure 5.1 how part of the desktop includes the Start button and the taskbar. The **Start button,** located in the lower left corner of the desktop, is the most commonly used desktop item. The Start button is used to access applications, files, help, and utilities.

The **taskbar** is the bar that runs across the bottom of the desktop. The taskbar holds buttons that represent applications or files currently loaded into memory. The taskbar also holds icons that allow access to system tools. These tools can include an icon for changing or viewing the date and time, a speaker icon for adjusting speaker volume control, and an icon for a virus utility. In Figure 5.1, the files currently loaded into memory and displayed on the taskbar as a button are a Word document called **Groups by lea...**, an Internet Explorer session opened to **Semester5...**, the Explorer application as shown by the **Exploring-D...** button, and another Internet Explorer session opened to **Cisco-Career....** On the right portion of the taskbar in Figure 5.1 is the speaker icon, which is used to adjust speaker volume.

Other common desktop icons include My Computer, Internet Explorer, Network Neighborhood, and Recycle Bin. The **My Computer** desktop icon is used to access hardware, software, applications, and files located on or in the computer. To use the My Computer icon, simply move the mouse pointer to the icon and double-click on it. The same is true for all desktop icons.

The **Internet Explorer** icon is used to start the Internet Explorer application. The **Network Neighborhood** icon only appears if the computer has a network card installed and is used to display and access all networked computers and networked devices in your workgroup or domain. The **Recycle Bin** is used to hold files and folders that the user deletes. When a file or folder is deleted, it is not immediately discarded; instead, it goes to the Recycle Bin. Once a file or folder is in the Recycle Bin, it can be removed. This is similar to the fact that a piece of trash can be retrieved from an office trash can.

Let's go through an exercise to help you understand how to modify or view your desktop appearance on Windows NT. It is assumed that Windows NT is installed and running on a computer with the Windows NT desktop displayed on your screen and you are logged in.

1. Right-click on an **empty desktop area.** A menu appears.

2. Click on the **Properties** option. The Display Properties window appears. (Note: this screen can also be accessed by clicking on the **Start** button, pointing to the **Settings** option, clicking on the **Control Panel** option, and double-clicking on the **Display** control panel icon.)

3. Click on the **Background** tab. This tab controls how the background color and pattern appears.

4. To change your wallpaper to the **WINNT** option, click on the **WINNT** option.

5. Click on the **Apply** button.

6. To view the **Appearance** tab settings, click on the **Appearance** tab at the top of the window. This tab controls the color scheme and size of the letters. This tab is good to use when people with vision problems need adjustments.

7. Click Cancel to close the window.

An operating system specialist must remember that the files and folders in the Recycle Bin take up hard drive space and that users often forget to empty the files and folders from the Recycle Bin.

5.2 Managing Files and Folders

Managing files and folders is an important part of NT Workstation. You must be able to create, delete, and move files and folders regularly. The easiest way to learn about files and folders is to start with the drive letter. A drive letter followed by a colon represents every drive in the computer. The first floppy drive in a system receives the drive letter A:. The first hard drive partition gets the drive letter C:. Other partitions receive consecutive drive letters (D:, E:, F:, etc.), usually before other devices.

Computer drives hold files and folders. A **file** is an electronic container that holds data or computer code. An example of a file is NT CHAPTER.DOC. A **folder** holds files and can also contain other folders. This is similar to an office folder that has memos and paper documents as well as other folders. A folder was called a **directory** in older DOS operating systems.

In DOS, in addition to naming a file or a folder, you frequently would have to add an extension to a file name. An **extension** is part of the file name and it can be up to three characters in length. The extension is separated from the first part of the file name by a period (.). Most people consider the extension to be part of the file name. An example of a file name (with an extension) is NTCHAP.DOC where *.DOC* is the extension. The good part about extensions today is that the applications normally add an extension automatically to the end of a file name. In most windows, NT Workstation does not automatically show the extensions.

Next, you will perform an exercise to help you understand how to view extensions on Windows NT. It is assumed that Windows NT is installed and running on a computer with the Windows NT desktop displayed on your screen and you are logged in.

1. Right-click Start and click Explore. Windows Explorer opens.

2. To view the extensions in Windows NT Explorer, click on the **View** menu option.
3. Click on the **Folder Options** menu option. Click on the **View** tab and click on the **Hide file extensions for known file types** check box (which will remove the check from the box).
4. Click on the **Apply** button. Windows Explorer displays the extension for all files.
5. Close Explorer.

When NT Workstation recognizes an extension, it associates that extension with a particular application. This means that NT Workstation assigns a known graphical icon to that file. Extensions work the same as they do in Windows 98 so you can refer to Chapter 4 for a discussion of extensions.

File names in NT Workstation can be up to 255 characters. These extended file names are commonly called **long file names**. Folder names and file names can have all characters, numbers, letters, and spaces except for the following:

/ (forward slash) " (quotation marks) \ (backslash)
| (vertical bar) ? (question mark) : (colon)
***** (asterisk)**

An example of an NT Workstation long file name is NT WORKSTATION CHAPTER.DOC. Any time a document has been saved with one of these long file names and is taken to an older computer with an operating system that does not support long file names, the file name is shortened to a maximum of eight characters. Windows does this by using the first six characters of the file name, deleting any spaces, and using two special characters—a tilde (~) and a number. For example, NT WORKSTATION CHAPTER.DOC would be shortened to NTWORK~1.DOC. If there were two files named NT WORKSTATION CHAPTER.DOC and NT WORKSTATION INDEX.DOC, the two files would be saved as NTWORK~1.DOC and NTWORK~2.DOC, respectively.

Any time a file is saved on a disk, the reference for what drive and folder the file is saved to is known as the file's **path.** A path is like a roadmap of how to get to the file. An example of an NT path is as follows: C:\MY Documents\A+ Book\NT WORKSTATION CHAPTER.DOC

To understand a path statement, look at the items in small chunks. In the previous example, C: is the drive letter where the document is stored. The C: represents the first hard drive partition. The name of the document is always at the very end of the path. In the example given, the name of the document is **NT WORKSTATION CHAPTER.DOC**. Everything in the middle of these two items is the name of the folders one must go through to find the NT WORKSTATION CHAPTER.DOC file.

For example, the first folder listed in the example is **My Documents**. The My Documents folder is on the C: drive and is separated from the drive letter by a backslash (\). The next folder (which is a subfolder within the My Documents folder) is called **A+**

Book. The **A+ Book** folder is separated from the parent folder (My Documents) by another backslash (\). The A+ Book folder is also separated from the name of the document by a backslash (\).

NT Workstation comes with an application called Windows NT Explorer. **Windows NT Explorer** is the most common application used to copy or move files and folders. Figure 5.2 shows Windows NT Explorer. You can see how the document A+ Book hardware from Pagemaker.doc looks.

Figure 5.2: Windows NT Explorer

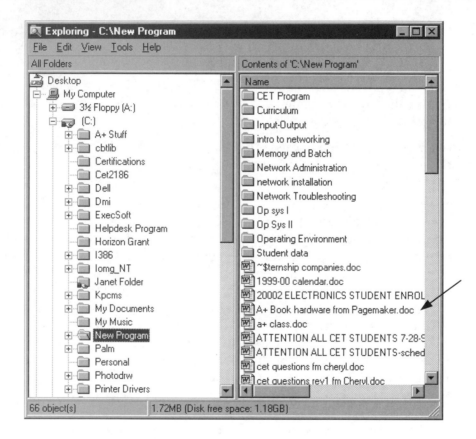

The My Computer desktop icon can be used in a similar fashion to copy or move files and folders. Even though the graphical form is nice, you must thoroughly understand a file's path written in long format. Frequently, technical directions, advisories, support documents, etc., have the path written in long format, such as the example given earlier.

My Computer and NT Explorer can be used for setting attributes for a file or folder. The file and folder attributes are read-only, hidden, archive, and system. The **read-only attribute** marks a file or folder so that it cannot be changed. The **hidden attribute** marks a file or folder so that it is not visible through My Computer or NT Explorer, unless someone changes the default view for the window. Some applications use the **archive attribute** to control which files or folders are backed up. The **system attribute** is placed on certain files used to boot NT Workstation.

Let's perform an exercise to help you understand a file's attributes. It is assumed that Windows NT is installed and running on a computer with the Windows NT desktop displayed on your screen and you are logged in.

1. Insert your floppy in the floppy drive.
2. Click the **Start** button, point to the **Programs** option, and click on the **Windows NT Explorer**.
3. Click the icon associated with the floppy.
4. To create a file, click **File**, click **New,** and then click **New Text Document**. The name of the text document appears as **New Text Document**. Press Enter to keep this name.
5. To set the file's attributes, right-click the new text document and click **Properties**. The Properties window for the file opens.
6. In the **Attributes** section of the page, check **Read-only** and click **OK**. The Properties window closes.
7. Next, you will attempt to modify the file. Remember it is set to read-only. Double-click the document. The document opens.
8. On the document screen, type your name and press Enter. This is considered text.
9. Attempt to save the document. Click **File** and then click **Save**. The **Save As** window opens. Normally, this window does not appear—the file is simply saved. However, because the file has the read-only attribute, you cannot save the text to this file. Windows NT wants you to enter a new file name.
10. To see the error message, click **Save**. A message is displayed indicating you cannot save this file.
11. Click **OK** and then click **Cancel**. The document screen remains.
12. Close the document. Do not save changes if prompted.
13. Close Windows Explorer.

5.3 Understanding NTFS Compression

NT Workstation can use the NTFS and FAT16 file systems. In the NTFS file system, file and folder compression are supported. In NT Workstation, when you click on an NTFS compressed file, the file automatically uncompresses. After all work on the file is complete, the file recompresses when it is saved or the document is closed. Even though a compressed file or folder takes less space on the hard drive, the file or folder takes longer to access when it is compressed. Each file and folder in NTFS has a compression state of either compressed or uncompressed. Just because a folder is in a compressed state, does not mean that all files in the folder are compressed.

Think of compression as "squeezing" redundant data from file and creating a new smaller file. The smaller file will take up less room so you save disk space.

Let's run through a short exercise to help you understand compression. It is assumed that Windows NT is installed and running on a computer with the Windows NT desktop displayed on your screen and you are logged in.

1. Insert your floppy in the floppy drive.
2. Click the **Start** button, point to the **Programs** option, and click on the **Windows NT Explorer**.
3. Right-click **Start** and then click **Explore**. Windows Explorer opens with left and right windowpanes.
4. Click the icon associated with the floppy.
5. Right-click the document you created earlier: **New Text Document**. The Properties window for the file opens.
6. Click once in the **Compressed** check box.
7. Close Windows NT Explorer.

The compression on Windows NT Workstation can only be accomplished if the partition has been formatted as NTFS—not FAT16.

The operating system version is important to you when troubleshooting a computer. With NT Workstation, upgrades or patches to the operating system are provided with service packs. A **service pack** contains fixes for known operating system problems. Some application installation requirements list the minimum service pack that has to be installed before the software will install and/or operate properly. There are several ways to determine what version and service pack are installed with NT Workstation. They are listed below.

• When NT Workstation boots and the blue screen appears, the version and service pack level display.
• Click on the **Start** button. Click on the **Run** option. In the **Open** text box, type **WINVER** and press **Enter**.
• Open Windows NT Explorer. Click on the **Help** menu item. Click on the **About** option. The version appears.

5.4 Understanding the Windows NT Registry

With NT Workstation, every hardware and software configuration is stored in a database called the **registry.** The registry contains such things as folder and file property settings, application preferences and settings, driver files, environment settings, and user profiles. A **user profile** is all settings associated with a specific user including what application the user has access to, desktop settings, and the user's network configuration. The registry loads into RAM during the boot process. As changes are made to the computer's hardware and software settings, the registry updates continuously.

Be careful when modifying the Windows NT registry. You could make a change that could render your system useless. Because most of the things you do on the system ultimately change the registry, when in doubt, simply use the graphical method to change a setting.

The registry consists of five subtrees. Subtrees are also called branches or hives. In Chapter 4, which covered Windows 98, you learned that **subtrees** are like folders and are also sometimes called hives. Think of the hive as a collection of folders. The five standard subtrees are HKEY_LOCAL_MACHINE, HKEY_USERS, HKEY_CURRENT _USER, KEY_CURRENT_CONFIG, and HKEY_CLASSES_ROOT. Each of the subtrees has keys and subkeys containing values.

The registry can contain other subtrees depending on what software (applications, device drivers, or services) are added to the computer. A **service** is a process running on NT that provides a specific function to the computer. Examples of services include: DHCP (Dynamic Host Configuration Protocol) client, computer browser, event log, net logon, and remote access connection manager.

DHCP is a service that allows computers on a network to obtain an IP address. The computers need an IP address to access the Internet.

The registry is located in the %SYSTEMROOT%\SYSTEM32\CONFIG folder, where %SYSTEMROOT% is the boot partition and the name of the folder under the folder where NT Workstation is installed (normally C:\WINNT). Table 5.1 lists the five major subtrees and the function of each.

Table 5.1: Windows NT Workstation Registry Subtrees

Registry Subtree	Function
HKEY_LOCAL_MACHINE	Holds global hardware configuration. Included in the branch is a list of hardware components installed in the computer, the software drivers that handle each component, and the settings for each device. This information is not user-specific.
HKEY_USERS	Keeps track of individual users and their preferences.
HKEY_CURRENT_USER	Holds a specific user's configuration such as software settings, how the desktop appears, and what folders the user has created.

| HKEY_CURRENT_CONFIG | Holds information about the hardware profile that is used when the computer first boots. |
| HKEY_CLASSES_ROOT | Holds file associations and file links. The information held here is what allows the correct application to start when you double-click on a file name in Explorer or My Computer (provided the file extension is registered). |

Editing the NT Workstation Registry

Most changes to the registry are accomplished through the various control panels, but some changes can only be done through the registry editor. By default, only users who log on as the Administrator user can edit the registry, but all users can view the registry. Before making changes to the registry, make sure you make a backup of the registry.

With Windows NT, users log on using a user account. The user account with the greatest control over the computer is called **Administrator**. The Administrator user can do just about anything. For example, if you are logged on as this user, you can add other users, add groups, add and remove hardware, and perform other system administration functions. Microsoft distinguishes between a local Administrator user account and a network Administrator user account. A local user account exists on the local system (or the one you are logging on to). A network user account exists on a server. An example of Microsoft server is a computer that is installed with Microsoft Windows NT Server. Its purpose it to "serve" resources to client computers; Windows NT Workstation could be the client it serves.

NT Workstation ships with two registry editors—**REGEDIT** and **REGEDT32**. Either editor can be used to change the registry. For most technicians, it is simply a matter of which view he or she prefers. However, there are some differences between the two. Table 5.2 shows some of the more important differences.

Table 5.2: REGEDIT and REGEDT32 Differences

REGEDIT	*REGEDT32*
Provides more powerful search capabilities.	Can display and edit values larger than 256 characters.
All of the subtrees are shown in one window.	The subtrees are shown in individual windows.
Allows exporting the registry to a text file.	Can look at and apply access permissions to the subtrees, keys, and subkeys.
Allows importing the registry from the command line.	Can work with multiple registry files simultaneously.

Backing Up and Restoring the NT Workstation Registry

The registry should be backed up once a computer is initially configured and operating properly. The registry should also be backed up when any software or hardware changes are made.

> The registry should be backed up and restored on a different working computer *before* disaster strikes.

There are several ways to back up the registry; they are listed below.
• Use the NT Backup program.
• Use the RDISK (Repair Disk) utility.
• Use the REGEDIT registry editor.

The **NT Backup** program is accessed by first clicking on the **Start** button. Point to **Programs**, point to **Administrative Tools**, and click on the **Backup** option. The Backup dialog box appears. Select the **Windows** menu option and then click on the **Drives** option. Double-click on the drive letter that represents the boot drive (normally C:). In the bottom window, click on a check box to enable at least one file on this drive. At least one file must be chosen in order to back up the registry. Click on the **Backup** button. In the Backup Information dialog box, click once in the **Backup local registry** check box to enable the option. Click on the **OK** button.

The **RDISK (Repair Disk)** utility is used to create an **ERD (Emergency Repair Disk)** after NT is installed. When using the RDISK utility, two options are available: **Update Repair Info** and **Create Repair Disk**. The **Update Repair Info** button updates the emergency repair directory, which is a folder called Repair. The system then prompts you to create an ERD (Emergency Repair Disk). After NT is installed for the first time, the emergency repair folder is updated only using this **Update Repair Info** option. It is especially important to use the **Update Repair Info** option when making an updated ERD.

The **Create Repair Disk** button is used to create an ERD and use the information stored in the Repair folder. This option does not back up the current options to the Repair folder. Since the **Update Repair Info** option creates an ERD anyway, it is best to use the **Update Repair Info** option. An exercise at the end of the chapter details how to create an ERD.

One limitation of the RDISK program is that it does not update the default, security, or SAM files in the Repair folder. This means none of the user account information or changes are backed up. To do a complete update, use the **RDISK /S** command from the command prompt. This takes a bit of time. That is why most people prefer using some type of backup program to back up the registry. The BACKUP utility that ships with NT Workstation is a better method once the Repair folder has been updated using the RDISK program.

The REGEDIT program can also be used to back up or export the registry to back up media. To start the REGEDIT utility, click on the **Start** button. Click on the **Run** option and in the text box type in **REGEDIT** and press **Enter**. In the REGEDIT window, click on the **Registry** menu item. Click on the **Export Registry File** option from the drop-down menu. Select the drive location where the registry will be saved. In the **File name** text box, type in the name for the registry file backup. Click on the **Save** button. Figure 5.3 illustrates the REGEDIT utility.

Figure 5.3: REDEDIT Window

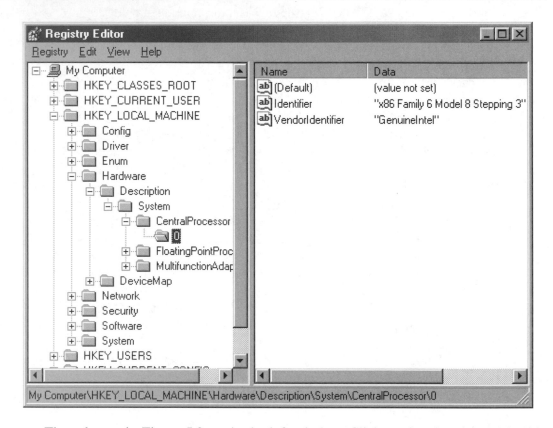

The subtrees in Figure 5.3 are in the left window. Click on the plus sign (+) beside each subtree and more subkeys appear. Click on a folder and the folder values appear in the right window. For example, in Figure 5.3, the Hardware folder has a subfolder called Description. The Description folder has a subfolder called System. The System folder has a subfolder called CentralProcessor. The CentralProcessor folder has a subfolder called 0. Once the 0 subfolder is double-clicked, the values appear in the right window. The value names for the 0 subfolder are (Default), Identifier, and VendorIdentifier. The value data for the 0 subfolder is on the far right side of the right window. These values must be changed sometimes to repair technical problems.

5.5 Understanding the Pre-Installation Steps of Windows NT

Skipping the installation planning is a bad habit for you to get into, especially when installing an operating system. The planning stages save time (and time is money to any business). If you skip the pre-installation steps, you many find yourself troubleshooting the installation process unnecessarily. The following steps outline the various stages of NT Workstation pre-installation.

1. Decide whether the installation will be an upgrade or clean install.
2. Decide whether the computer will have more than one operating system installed.
3. Decide how you want to partition the hard drive.
4. Determine which type of file system NT Workstation will use.
5. Determine if the hardware is compatible.
6. Obtain any drivers, upgrades, or hardware replacements.
7. Determine if the software applications are compatible.
8. Obtain any patches, upgrades, or software replacements.
9. Scan the computer for viruses.
10. Remove any power management or disk management tools.
11. Delete any unwanted files and uninstall any unwanted applications.
12. Back up any data files necessary.
13. Determine the local administrator password.

A **clean install** places an operating system on a computer that does not already have an operating system installed. An operating system **upgrade** is when an operating system already exists on the computer and NT Workstation 4 will be installed on top of this operating system. Windows 3.x and a previous version of NT Workstation (such as 3.5.1) can be upgraded to NT Workstation 4; however, a computer that already has Windows 3.x will probably not have enough hardware to load NT Workstation 4.

When NT Workstation 4 is installed to a different folder than the existing operating system, the computer will automatically be configured to be a dual-boot system. **Dual-boot** means that the computer has two operating systems installed and you can boot from either one. If Windows 95 or 98 is already installed, remove the operating system or install NT to a different folder. The default installation folder for NT is **C:\WINNT**.

If you decide that you are going to install NT on a machine that already has Windows 95/98, first check to see that all applications are supported by NT Workstation 4.0. Then, install NT into a separate directory on the hard drive. The system now has two operating systems and can boot from either of them. Reinstall all Windows 9x applications through NT. If Windows 95/98 is no longer desired, manually delete the folder that contains the Windows 95 or 98 operating system files.

There are two types of NT Workstation partitions: system partition and boot partition. The **system partition** is the active hard drive partition that contains hardware-specific files used to load the operating system. The system partition is normally located on the C: hard drive partition. The **boot partition** is the hard drive partition that holds the majority of the NT Workstation operating system files. What is confusing is that the two types of

NT partitions can be located on the same hard drive partition, or they can be on different partitions. For example, take a hard drive that has one partition, C:. The files needed to boot the system are on C:, and the NT files, located in the folder, are also on the C: partition. Therefore, in this example, both the system partition and the boot partition are on the same hard drive partition (C:).

Another example is a computer that has two partitions, C: and D:. The active partition that the computer boots from is C:. An extended partition with one logical drive is D:. The C: partition already has Windows 95 installed. This computer will be able to dual-boot from either Windows 95 or NT Workstation. When NT Workstation is installed, the files needed to load and boot NT are put on the C: partition (the NT system partition). The NT files are loaded to the folder on the D: partition (the NT boot partition).

NT Workstation can use either the FAT16 or the NTFS file system. The FAT16 file system should only be used if NT is to be dual-booted with an older operating system and there is only one hard drive partition formatted as FAT16. With the FAT16 file system, there is no file or folder compression. Nor are there as many permissions for individual files and folders. Also, the partition does not support large hard drive volumes greater than 2 GB. The NTFS file system supports security options and long file names.

The FAT16 partition can be converted to NTFS after the installation by using the **CONVERT program**. CONVERT *partition letter*: /FS:NTFS will convert a FAT16 partition to NTFS with no data loss. Take the example of the C: hard drive partition that is currently FAT16. The command used to convert the partition to NTFS is CONVERT C: / FS:NTFS. This command would be executed from a command prompt or by using the Run option from the Start menu.

The hardware requirements for NT Workstation are very important in the pre-installation checklist. If you omit this, you may find that you have to troubleshoot when actually there is nothing wrong except that the hardware requirements are not met. Table 5.3 lists the minimum requirements for NT Workstation.

Table 5.3: NT Workstation Minimum Hardware Requirements

Component	Minimum Requirements
CPU	Intel 486 (or compatible) 33 MHz
RAM	12 MB
Free hard drive space	120 MB
Input Device	Keyboard, mouse, or other pointing device
Multimedia Drive	CD-ROM
Video	VGA

On the NT Workstation CD in the Support folder, there is a file called HCL.HLP. This is the hardware compatibility list that contains hardware devices compatible with NT Workstation. If a computer device is not listed here, check Microsoft's web site at www.microsoft.com for the latest listing. Also check the device manufacturer's web site to see if it offers an NT driver for the device or if the driver is compatible with NT.

Microsoft provides a utility called **NTHQ** (**NT Hardware Qualifier**) that identifies what hardware is installed in the system. The NTHQ program is executed from a special floppy disk that you must make. To make this disk, insert a blank floppy disk into the computer. Run the **MAKEDISK.BAT** command, (which is located on the NT Workstation CD in the SUPPORT\HQTOOL folder). After this program executes, reboot the computer from this special disk. If your computer will not boot from the disk, check BIOS settings to make sure the A: drive is the first boot device. The NTHQ program automatically executes. Once hardware has been verified, obtain the appropriate NT driver for the device. If the device is incompatible with NT, replace the hardware device with a compatible one _before_ installing NT.

DOS applications, 16-bit, and 32-bit Windows-based applications can all operate under NT. However, any older application that tries to access hardware directly will not operate properly in the NT environment. One way you can know whether an application is compatible with NT is to try it. Some application manufacturers provide software upgrades (for a fee) so that the application can be run on NT. Contact the software manufacturer for any software compatibility issues. Microsoft also has some application compatibility notes on their web site.

When upgrading an operating system to NT Workstation, it is wise to free up hard drive space and clean off unwanted files and applications. Hard drive space is an important commodity to an operating system. As applications and operating systems increase in size, it is very important to have enough free hard drive space and enough RAM installed.

The last preparation steps before installing or upgrading NT Workstation are to determine what the name of the computer will be and what the local administrator passwords will be. NT Workstation was designed for a corporate networked environment. In a network, every computer must have a unique name. The company may have a standard for naming computers. Gather this information before starting an NT upgrade or installation.

A local administrator has full power over the NT Workstation computer. When someone logs in as the local administrator, he or she can create and delete user accounts, create and delete hard drive partitions, and use all of the administration tools that ship with NT Workstation. Some companies have standards for the local administrator account password.

Check with the network administrator or desktop support supervisor to see if this is the case. Otherwise, determine what password will be set during the installation process. The password can be blank (not advised), or it can be up to 14 characters in length. Also, the password is case-sensitive (unlike the user name, such as Administrator).

Don't forget about checking for viruses. Viruses were discussed in the Virus section in Chapter 4. Review that section if necessary.

5.6 Understanding How to Install and Upgrade Windows NT Workstation

After all pre-installation steps are completed, you can start the installation process. NT Workstation uses the **Setup** program to install the operating system files. There are three ways to start the Setup program: (1) from an NT Workstation CD, (2) by launching the installation program from a local hard drive partition, and (3) across a network. When installing NT across a network, one of two files is used to start the Setup program: WINNT.EXE or WINNT32.EXE. **WINNT.EXE** is used to install NT Workstation to a computer that currently has DOS, Windows 3.x, Windows 95, or Windows 98 installed. The **WINNT32.EXE** file is used to upgrade from a previous version of NT Workstation.

There are two major parts of the installation process—the text mode (otherwise known as DOS mode) and the GUI mode (also known as Windows mode). In **text mode**, characters are shown on a plain blue background. During text mode, the hard drive is partitioned and formatted for either FAT16 or NTFS, the location of where to install the NT files is chosen, hardware is checked for minimum requirements, hard drives are detected, and some of the installation files are copied. During the **GUI mode**, the setup logs are created, the computer is named, the administrator password is entered, and the rest of the operating system files are copied. You may also create an Emergency Repair Disk, as well as install networking components.

5.7 Troubleshooting the Windows NT Installation

Various problems can cause the NT Workstation installation process to halt. If the computer halts during text mode and displays a STOP message, there is probably an incompatible piece of hardware or software installed in the computer, or a virus is present. If no error messages appear, but the computer halts during text mode, check the BIOS settings, especially on older ISA devices. If an error message appears during text mode that indicates the HAL.DLL is missing or corrupt or a similar HAL.DLL message, then the incorrect HAL (Hardware Abstraction Layer) is being loaded. This is a layer between the operating system and hardware devices. The **HAL** allows NT to run with different hardware configurations and components without affecting (or crashing) the operating system. To correct this, restart the Setup program. When the message appears that NT is examining your hardware configuration, press the **F5** key. Select the correct computer type from the list that appears. If the computer type is not listed there, obtain a HAL from the computer manufacturer. Then, select **Other** from the list and load the HAL provided by the computer manufacturer. Another indication that the HAL is incorrect is if the Setup program hangs while copying files to the hard drive.

If the computer halts during the GUI portion of Setup, restart the computer. The installation process attempts to continue from the place it left off. Incompatible hardware devices normally cause this. You can troubleshoot the device once Windows is installed. Also, if the system hangs at random intervals, an IRQ (Interrupt Request) conflict, I/O (Input/Output) port address conflict, or video setting is probably the culprit. IRQ and I/O ports were discussed in Chapter 2 entitled *Basic Operating System Theory*. Incompatible hardware devices are the most common problem during the installation process because they can cause both the text and GUI modes to halt. If the system hangs after the final reboot, the problem is most likely caused by incorrect information in the BOOT.INI file or an incorrect hardware configuration. The **BOOT.INI** file is a Windows NT file that instructs the operating system as to which partition to boot.

Installation problems can be caused by a number of factors. The following list shows the most common causes and their associated solution during the installation process.

- **Incompatible BIOS**—Obtain compatible BIOS, upgrade to a compatible BIOS, replace motherboard with one that has a compatible BIOS, or do not upgrade/ install NT Workstation.
- **BIOS needs to be upgraded**—Upgrade the BIOS.
- **Incompatible hardware**—Replace the hardware or do not upgrade/install NT Workstation.
- **Incompatible drivers**—Obtain NT drivers from the hardware manufacturer.
- **Existing drivers are incompatible or halt the installation/upgrade process**—Obtain NT drivers from the hardware manufacturer.
- **Incompatible TSRs**—Remove TSRs or obtain updated ones from the software manufacturer; otherwise, disable the TSR until after NT has been installed and then try re-enabling the TSR.
- **Incompatible applications**—Obtain upgrades from software manufacturer.
- **Minimum hardware requirements have not been met**—Upgrade the hardware. The number one thing to check is the CPU (486 33 MHz minimum) and RAM (12 MB minimum).
- **A virus is on the hard drive**—Run a virus-checking program to remove the virus.
- **Pre-installation steps have not been completed**—Go back through the list!
- **The installation floppy disks or CD is corrupted** (not as likely as the other causes)—Try the disk in another machine and see if you can see the contents. For the CD, check to see if any scratches or dirt are on the surface. Clean the CD as necessary.
- **Incorrect CD key**—Type in the correct CD key to complete the NT Workstation installation. The key is located on the CD case.
- **Hard drives are not configured correctly**—If a message appears that Setup did not find any hard drives on your computer and this is a new computer, check that the cable(s) are properly connected and that power connects to the drive.

- **Existing FAT32 partition**—If a message appears that there is no valid partition on the hard drive and a previous operating system has been (or is) on the hard drive, there is a good possibility that the drive has been partitioned to FAT32. NT Workstation does not support FAT32. Back up the data and create a new partition (FAT16 or NTFS) if NT Workstation is to be installed.
- If Setup hangs while files are copying to the hard drive, the wrong HAL is installed (see previous section that describes the HAL file) or some BIOS settings are interrupting the copy process. Go into the computer's BIOS and disable video shadow RAM and the 32-bit enhanced file throughput settings.

You need to know the installation steps for the A+ Operating System exam.

5.8 Configuring Windows NT Workstation

Technicians must frequently add new hardware and software using the operating system. NT Workstation has specific tools for these functions. Using the correct procedure is essential for success. The following sections handle many of the tasks a technician must perform.

- Adding and Removing Hardware Components
- Installing and Removing Software
- Adding a Printer

Adding and Removing Hardware Components

All hardware devices must have NT drivers in order to operate with NT Workstation. An important thing to remember is that the only type of user who can install hardware components by default is the Administrator. The Administrator uses various control panels to add hardware components in NT Workstation.

To access the control panels, click on the **Start** button. Point to the **Settings** option and then click on the **Control Panel** menu option. The number of control panels is determined by what hardware is installed in the computer. Table 5.4 lists a few of the control panels and their functions.

Table 5.4: A Few NT Control Panel Functions

Control Panel	Function
Devices	Used to start and stop device drivers; used to control how the device driver loads.
Display	Used to install a monitor driver and configure monitor settings.
Keyboard	Used to install a keyboard driver and configure keyboard.

Network	Used to install NIC drivers, add network software, and configure network connections.
SCSI Adapters	Used to add and configure SCSI device drivers.
Tape Devices	Used to add tape device driver and configure tape device parameters.

Knowing what control panel to use and what specific control panel tab to use to install a hardware device driver is sometimes confusing in NT. The following procedures are provided to help with the most common NT hardware installations.

- To load a monitor device driver, use the **Display** control panel, click on the **Settings** tab, and click on the **Change** button.
- To load a keyboard driver, use the **Keyboard** control panel, click on the **General** tab, and click on the **Change** button.
- To load a modem driver, use the **Modem** control panel, click on the **Add** button.
- To load a multimedia device driver such as a CD-ROM, or joystick driver, use the **Multimedia** control panel, click on the **Devices** tab, and click on the **Add** button.
- To install a NIC driver, use the **Network** control panel, click on the **Adapters** tab, and click on the **Add** button.
- To install a SCSI adapter device driver, use the **SCSI Adapter** control panel, click on the **Drivers** tab, and click on the **Add** button.
- To install a tape drive device driver, use the **Tape Devices** control panel, click on the **Detect** button.

Two important things to remember about NT Workstation: (1) you must have Administrator rights to install a hardware driver and (2) the driver needs to be compatible with NT. Administrator rights means you log on the Windows NT system as the Administrator user account.

Installing and Removing Software

No computer is complete without software. Various types of applications can be used with NT including DOS applications, 16-bit Windows applications, and 32-bit Windows applications. Not all older applications are compatible with NT, but the only way you can know is to load the application and try it. 16-bit applications are installed using the directions from the software manufacturer. 32-bit applications are installed using the **Add/Remove Programs** control panel. To access this control panel, click on the **Start** button. Point to the **Settings** option and click on the **Control Panel** menu option. When the Control Panel window appears, click on the **Add/Remove Programs** control panel. Follow the directions on the screen or the directions from the software manufacturer.

DOS applications run in a special environment called **NTVDM** (NT Virtual DOS Machine). NTVDM simulates a DOS environment inside NT Workstation. Each DOS application loaded into RAM (started), loads into one NTVDM. The environment the DOS application runs in can be customized through NT Workstation. Using My Computer or Explorer, right-click on the EXE file or shortcut that starts the DOS application. Select the **Properties** option from the drop-down menu. A Properties screen appears with tabs across the top. All of the settings made through this window are collectively known as the DOS application's **PIF** (**Program Information File**).

The two most common tabs used with DOS applications are Memory and Screen. The **Memory tab** allows setting memory parameters for the NTVDM that the DOS application is using. With this tab, you can specify a specific amount of expanded and extended memory. Figure 5.4 shows the Memory tab for the DOS application executable file called SHERLOCK.EXE.

Figure 5.4: Memory Tab for a DOS Application

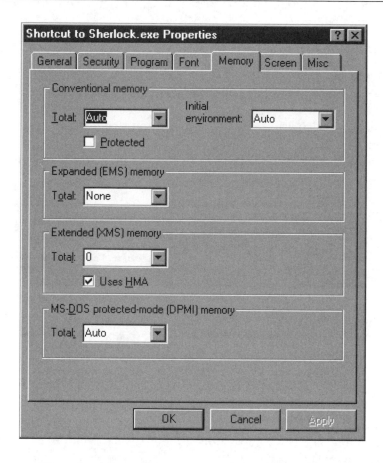

The other commonly used tab for a DOS application is the Screen tab. The **Screen tab** allows the DOS application to run in full-screen mode (the application takes up the entire screen and has no common Windows buttons) or in a window mode (the application runs inside a window that has common Windows buttons to control the window).

Not all DOS applications are compatible with NT Workstation. The only way to know is to install the software and attempt to execute it. If it doesn't work, try adjusting the memory settings described above for the specific amount and type of memory the DOS application requires. DOS applications frequently try to directly access hardware and NT has a built-in feature called HAL (Hardware Abstraction Layer) that prevents direct access to hardware.

A 16-bit Windows application runs in a **Windows On Windows** (or **WOW**) environment. WOW is a WIN16 environment simulator running inside a NTVDM. All 16-bit Windows applications run in a single NTVDM, which means that when one 16-bit Windows application fails, they all fail. You can configure each 16-bit Windows application to operate in its own NTVDM. Create a shortcut for the icon that starts the program. (Use Explorer to create the shortcut. Click and drag the executable file that starts the program to the desktop and a shortcut is automatically created.) Right-click on the shortcut icon. Click on the **Shortcut** tab. Click on the **Run in Separate Memory Space** check box to enable starting the application in its own NTVDM and preventing other 16-bit applications from crashing if this application fails.

Not all 16-bit Windows applications are compatible with NT Workstation. This is because some 16-bit Windows applications use a VxD (virtual device driver) that accesses hardware directly. NT Workstation has a built-in feature called HAL (Hardware Abstraction Layer) that prevents direct access to hardware.

A 32-bit Windows application has its own memory space allocated—2 GB for the operating system files that all applications share and 2 GB for the application. This means that if a 32-bit application crashes, no other 32-bit application will fail because it does not use the same memory space.

Adding a Printer
Printers normally connect to a computer through the parallel port, a USB port, or through a local area network. Only local printers (printers directly connected to the computer) will be covered in this chapter.

To install a printer under NT, use the **Add Printer wizard**. To access the Add Printer wizard, click on the **Start** button. Point to the **Settings** option and then click on the **Printers** option. Double-click on the **Add Printer** icon. Figure 5.5 shows this concept.

Figure 5.5: NT Workstation Printers Option

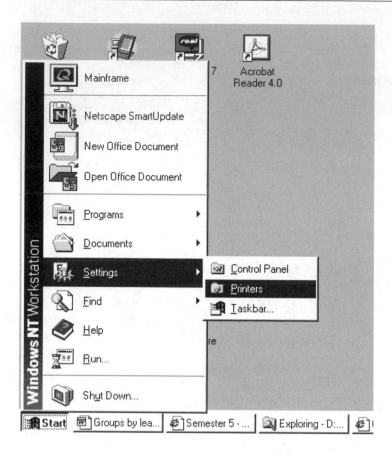

If multiple printers are available to an NT Workstation computer, one printer is marked as the default printer. A default printer is the one to which a computer prints unless a different one is chosen. To set the default printer, click on the **Start** button, point to the **Settings** option, and click on the **Printers** drop-down menu option. The Printers folder opens. Right-click on the printer that will be the default printer. From the drop-down menu, click on the **Set As Default** option.

5.9 Understanding the Boot Process

With NT Workstation and Windows 2000, two types of partitions are important during the boot process—the system partition and the boot partition. The **system partition** is the active drive partition that has the files needed to load the operating system. The system partition is normally the C: drive (the active partition). The **boot partition** is the partition or logical drive where the NT operating system files are located. One thing that people sometimes forget is that the system partition and the boot partition can be on the same partition. These partitions are where certain files needed to boot are located.

Every operating system needs specific files that allow the computer to boot. These files are known as **system files** or startup files. The system files and their specific location on the hard drive are listed in Table 5.5.

Table 5.5: Windows NT Boot Files

Startup File name	File Location
BOOT.INI	Root directory of system partition.
BOOTSECT.DOS (needed if dual or multi-boot system	Root directory of system partition.
HAL.DLL	*%systemroot%*\SYSTEM32 (*%systemroot%* is a variable representing the folder where Windows NT is installed. This is normally C:\WINNT.)
NTBOOTDD.SYS (used with SCSI drives that have the SCSI BIOS disabled)	Root directory of system partition.
NTDETECT.COM	Root directory of system partition.
NTLDR	Root directory of system partition.
NTOSKRNL.EXE	*%systemroot%*\SYSTEM32 (*%systemroot%* is a variable representing the folder where Windows NT is installed. This is normally C:\WINNT.)

Information from Microsoft can be confusing at times because of the *%systemroot%* and *%systemdrive%* entries. This is because computers can be partitioned differently. If you install Windows NT Workstation onto a drive letter (a partition or logical drive) other than the active partition (normally C:), the startup files can be on two different drive letters. Also, you do not have to take the default folder name to install NT. To account for these different scenarios, Microsoft uses the *%systemroot%* to represent the boot partition—the partition and folder that contains the majority of the NT Workstation files. On a computer with a single operating system, this would be C:\WINNT. The *%systemdrive%* represents the root directory of the same drive letter. On a computer with a single operating system, this would be C:\.

The boot process is actually quite involved, but the major steps are as follows:

1. The computer is powered on.
2. POST executes.
3. BIOS searches for an active partition on the hard drive.
4. BIOS reads the Master Boot Record, then locates and loads the information into sector 0 of the system partition. The contents of sector 0 define the type of file system

and the location of the bootstrap loader file, then start the bootstrap loader. With NT Workstation, this file is NTLDR.

5. NTLDR starts the file system.

6. NTLDR reads the BOOT.INI file and displays the various operating system choices. If something other than NT Workstation is chosen, the BOOTSECT.DOS file takes over. If NT Workstation is chosen, the NTDETECT.COM file executes.

7. NTDETECT.COM detects the computer's hardware. A message appears on the screen saying "**NTDETECT V4.0 Checking Hardware...**"

8. NTLDR passes the hardware information to the NTOSKRNL.EXE file and displays the Startup screen. The message on the screen is "**OS Loader V4.0.**" Press **Spacebar** now to invoke Hardware Profile/Last Known Good menu. This menu stays on the screen three to five seconds.

9. The operating system kernel, NTOSKRNL.EXE, executes and the HAL.DLL file loads. The message on the screen displays "**Microsoft® Windows NT™ Version 4.0 (Build 1381) 1 System Processor (*x* MB Memory).**" This is where *x* is the amount of memory on your system.

10. The registry key HKEY_LOCAL_MACHINE\SYSTEM loads. This registry key is located in the *%SYSTEMROOT%*\SYSTEM32\CONFIG\SYSTEM folder. This key has information found during the hardware detection process.

11. The WINLOGON.EXE file executes and the log-on screen appears.

Troubleshooting the Boot Process

Various problems exist that cause NT Workstation to not boot, but boot problems can usually be narrowed down to two main areas: (1) missing or corrupted boot files, or (2) configuration problems. When boot files are missing or corrupted, different error messages can appear. Table 5.6 shows some of the most common error messages seen with missing or corrupt boot files.

Table 5.6: Windows NT Workstation Boot Problems

Symptom	Cause
Message "Boot: Couldn't find NTLDR. Please insert another disk."	NTLDR file is missing or corrupt.
Message "Error opening NTDETECT. Press any key to continue."	Timeout option for the boot loader located in the BOOT.INI file is set to 0 or the path of the operating system in the boot loader section of the BOOT.INI file is not the same as a path listed in the [operating systems] section of BOOT.INI.

Message "The system did not load because of a computer disk hardware configuration problem. Could not read from selected boot disk. Check the boot path and disk hardware. Please check the Windows NT documentation about hardware disk configuration and your hardware reference manuals for additional information. Boot failed."	The device or partition information regarding a path in the [operating systems] section of BOOT.INI is wrong.
Message "multi(0)disk(0)rdisk(0)partition(1)\winnt\system\ntoskrnl.exe The system did not load because it cannot find the following file: C:\\SYSTEM32\NTOSKRNL.EXE. Pease reinstall a copy of the above file."	The path information in the BOOT.INI file's [operating systems] section is wrong or NTLDR and/or BOOT.INI files are not in the root directory of the system partition.
Message "I/O error accessing boot sector file multi(0)disk(0)rdisk(0)partition(1) \bootsect.dos."	The BOOTSECT.DOS file is missing or corrupted. Restore the file from a backup copy.
When you select an operating system from the operating system selection screen, the selected operating system does not load.	Corrupted BOOT.INI file. The path information in the [operating systems] section of BOOT.INI is wrong.
Message "Invalid Partition Table."	A virus is on the hard drive or there is an error in the BOOT.INI file.
Blue screen appears after a power failure.	The boot files are missing or corrupt.

When any of these error messages appear, the most common tool used is the ERD (Emergency Repair Disk). The **ERD** is a disk that can be made during the installation process and after NT Workstation is installed. An entire section is devoted to the ERD immediately following this section.

Other problems that can occur during the boot process are POST errors, STOP errors, and blue screens. **POST errors** are normally caused by an invalid/incorrect hardware configuration or a faulty piece of hardware. If NT Workstation has booted correctly prior to this POST error, press the **spacebar** on the keyboard when the message, "**OS Loader V4.0. Press spacebar NOW to invoke Hardware Profile/Last Known Good menu**" appears on the screen. By default, this message stays on the screen for approximately five seconds. Once you press the spacebar, a menu appears. Press the letter **L** to access the **Last Known Good Configuration** option. This allows you to change the configuration of the new device or disable it until you can determine the exact problem or get an NT driver for the device.

The same process of accessing the **Last Known Good Configuration** is used if you get a **blue screen** (commonly called the "**Blue Screen of Death**") after a configuration change. If a blue screen appears after a power failure, there is a good chance that the boot files are missing or corrupt. Use the ERD or copy over the boot files to the correct location on the hard drive. If a **STOP message** appears, the registry may be corrupt. The registry can be reinstalled using a current ERD for the computer or you can reinstall the registry from a backup.

In older operating systems, a startup disk was used to begin the troubleshooting process. With NT Workstation, the startup disk is actually the set of three installation disks that came with the NT Workstation CD. If you cannot find these, you can go to any computer that has the same version of Windows NT Workstation loaded and make a set. To do this, insert the NT Workstation CD, click on the **Start** button, and click on the **Run** option. In the dialog box, type *drive***:\i386\winnt32.exe /ox** where *drive* is the drive letter that represents the CD-ROM. Click on the **OK** button and follow the prompts on the screen. You will need three floppy disks to complete this procedure. The installation disks are used when repairing a system using the ERD.

Creating an Emergency Repair Disk

The **ERD** (**Emergency Repair Disk**) is used to fix system file problems, the partition boot sector, and startup environment settings, all of which can cause a computer to not boot properly. The ERD is unique to a specific computer. The ERD should be recreated any time changes are made to the computer's configuration. An ERD can also be created during the installation process but it still needs to be recreated whenever changes are made.

To create an ERD, click on the **Start** button and click on the **Run** menu option. In the dialog box, type **RDISK /S**. If an error occurs, type the correct path for the **RDISK** program. By default, RDISK is located in the SYSTEM32 subfolder under C:\WINNT. The RDISK program creates a disk that can be used for emergency repairs. The **/S** switch causes a backup of the Security Accounts Manager (SAM) database as well as the entire registry in a subfolder under called **Repair**. When prompted to create the ERD, click on the **Yes** button if you are simply updating the existing ERD.

If the computer system will not work and you want to use the ERD, you must actually start the NT Workstation installation process and get to the Welcome to Setup screen. Use the NT Workstation setup disks. Once presented with the Welcome to Setup screen, press the **R** option, which is the To repair a damaged Windows NT version 4.0 installation. Once you press **R**, a menu appears. The menu has four selections that are all enabled. Press **Enter** to start the process. A prompt appears telling you to insert the ERD. The CHKDSK program runs and verifies the hard disk clusters. Each operating system file is checked and reinstalled if the file is missing or corrupt. The system and security portion of the registry is replaced (if you confirm that you want them replaced), and the boot sector and boot files are replaced. An exercise at the end of the chapter illustrates how to create an ERD.

Windows Scripting

A Windows **script** can automate desktop shortcuts for users and set or restrict access for the desktop, Start menu, network share mapping, network printer mapping, setting default printer, and launching applications. In Windows scripts can be executed three different ways:

- By double-clicking on an icon associated with a particular file.
- By running a script from the Run dialog box.
- By typing WSCRIPT.EXE *script_name* (where *script_name* is the path and name of the previously created script) from the Run dialog box.

In a DOS environment, a batch file can provide how the operating system loads, how a particular application loads, a menu-driven environment, and other environment settings.

With Windows 98, NT, and Windows 2000, Microsoft provides a Windows-based scripting tool called WSCRIPT.EXE. When the WSCRIPT tool is executed, a Windows-based dialog box appears and allows you to configure script properties.

The Windows scripting host is sometimes known as WSH. The scripting host is integrated into Windows 2000 and the NT option pack and can be downloaded from Microsoft's web site. Microsoft also provides sample scripts and a tutorial.

5.10 Understanding Task Manager, Event Viewer, and Windows NT Diagnostics

Task Manager is a Windows-based utility that displays applications currently loaded into memory, processes that are currently running, microprocessor usage information, and memory usage data. To activate the Task Manager utility, press the **Ctrl + Alt + Delete** keys. From the window options that appear, click on the **Task Manager** button. Two other ways to access Task Manager are:

• Press **Ctrl + Alt + Esc.**

• Right-click on the **taskbar** and then click on the **Task Manager** option.

One of the common uses of Task Manager is to exit from an application that is "hung" or not responding. Task Manager can help with exiting the program. To exit from a 32-bit application, access the Task Manager window and click on the **Applications** tab. Locate the name of the troublesome application and click on it. Normally, the status shows the application as "not responding." Click on the **End Task** button. Close the Task Manager window.

To exit from a DOS or 16-bit Windows applications, access the Task Manager window and click on the **Processes** tab. Locate the appropriate NTVDM.EXE file that contains the program. Note that there may be more than one application running within the NTVDM if they are 16-bit Windows programs that have not been configured to run in separate NTVDMs. Click on the **End Process** button. Close the Task Manager window.

Event Viewer is a Windows tool used to monitor various events in your computer, such as when a driver or service does not start properly. One of the most common reasons a technician uses the Event Viewer is when he or she encounters the message "One or more services failed to start. Please see the Event Viewer for details." (As a side note, the individual administrative tool that controls the service or the Services control panel is used to manage the event. The Event Viewer is used to determine which service had the problem.)

The **EventLog** service starts automatically every time a computer boots. This service is what allows events to be logged. Event Viewer is then used to see the log. Event Viewer is a great troubleshooting tool because, even when a user cannot remember exactly what happens when a problem occurs, Event Viewer tracks the problems.

Event Viewer tracks three different categories of events: system events, security events, and application events. **System events** log events regarding system components. An example would be a driver that does not load during startup. **Security events** are only accessible to system administrators and contain information such as valid and invalid log on attempts and whether someone tried to access a protected file. An administrator can set what security events are logged and tracked by Event Viewer. **Application events** are associated with a specific program. The programmers that design the software decide which events to display in the Event Viewer's application log. All users can view the system log and the application log, but only a member of Administrators can view or enable security log information. The most commonly used log is the system log.

Let's go through an exercise to help you understand Event Viewer.

1. Access the Event Viewer by clicking on the **Start** button, pointing to the **Programs** option, pointing to the **Administrative Tools** option, and clicking on **Event Viewer**.
   Event Viewer opens and the system events appear by default in the Event Viewer window.

2. Double-click on any event that is shown in Event Viewer to see more information.

3. Click the **Close** button to return to the full Event Viewer screen.

4. Select the **Security** option. All Security events display.

5. To access application events, click on the **Log** menu option.

6. Select the **Applications** option. All Application events display.
7. To clear the System log, right-click **System** and click **Clear All Events**. You are prompted to save the log file.
8. Click on the **No** button when asked if you want to save the log. Note: Once events are cleared from the log, they cannot be retrieved. Save the events when prompted if you want to keep the events that are currently logged before clearing the log. You will need to supply a file name.
9. To exit Event Viewer, click on the **Close** box in the upper right corner or click on the **Log** menu option and select the **Exit** option.
10. Click on the **Yes** button when asked if you are sure that you want to clear the log. This process clears all prior logged events in the System log and will only displays events that occur from this point forward.
11. Press the **F5** key to refresh the Event Viewer.

If an error message appears stating that the Event Viewer log is full, start the **Event Viewer.** Click on the **Log** menu option and then click on the **Log Settings** selection. The Event Log Settings dialog box appears. Locate the **Change Settings For** option. Click on the **down arrow** and select the **log** to modify. This will normally be the system log or security log. To change the maximum amount of hard disk space allocated to the specific log, use the arrow in the **Maximum Log Size** text box to select up to a maximum size of 512 K. You can also change how long events are kept. Locate the **Event Log** Wrapping section. Click on one of the following: **Overwrite events as needed**, **Overwrite events older than 0 days**, or **Do not overwrite events**. Figure 5.6 shows what the settings screen looks like.

Figure 5.6: Event Viewer Full Settings

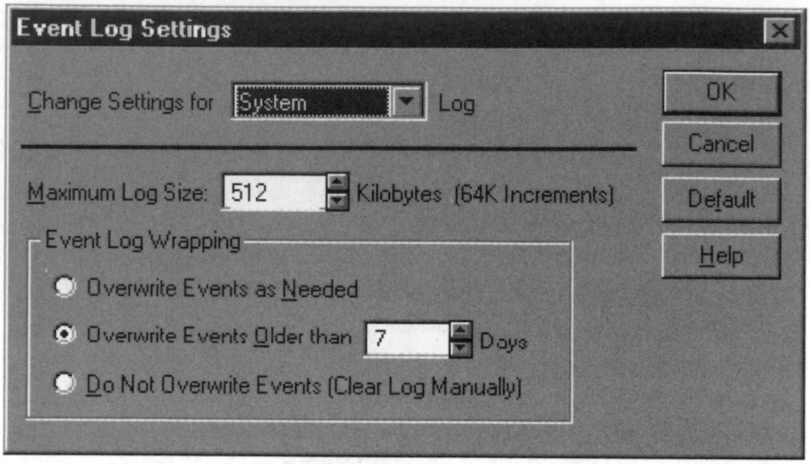

NT Diagnostics is a utility that allows viewing configuration information about the computer's hardware, installed device drivers, and installed services. NT Diagnostics can help when troubleshooting configuration problems. The utility does not really perform any diagnostics; it simply displays information. To access NT Diagnostics, click on the **Start** button, point to **Programs**, point to **Administrative Tools (Common)**, and click on the **Windows NT Diagnostics** option. The tabs across the top of the window include Version, System, Display, Drives, Memory, Services, Resources, Environment, and Network.

The **Version NT Diagnostics tab** displays the version, service pack version, and NT serial number. A Print button on this screen allows a comprehensive listing of all NT Diagnostic information (from all of the various tabs). The **System NT Diagnostics tab** displays the installed HAL type, BIOS manufacturer and date, and the type of processor(s) installed in the computer. The **Display tab** lists video adapter information including chip type, BIOS date, driver, driver version, and amount of video memory. The **Drives NT Diagnostics tab** displays information about any drives connected to the computer, including hard drive partitions. The window also displays network connected drives (shares). Double-click on any drive and the amount of used space, number of bytes per sector, number of sectors per cluster, and available space is shown. The **Memory NT Diagnostics tab** displays physical memory and information about the paging file. This tab is useful in determining if the computer has enough physical memory or if the paging file needs to be adjusted.

The back row of NT Diagnostic tabs includes Services, Resources, Environment, and Network. The **Services tab** displays the current status of all services and device services. Double-click on any service to see detailed information such as the path name to the service or device and any dependencies. There are two types of dependencies, service and group. **Service dependencies** are the services or drivers that must run before the particular service can start. **Group dependencies** are groups of services that must be running before the particular service can start.

The **Resources tab** is probably the most common tab used by technicians. On this display, IRQs, I/O port addresses, DMA channels, memory addresses, and devices can be shown by clicking on the appropriate button at the bottom of the tab. Figure 5.7 illustrates the Resources tab.

The **Environment tab** lists the path to the command interpreter, how many processors are installed, and the directory where the majority of NT files are located. The **Network tab** shows the domain or workgroup name and what user is currently logged onto NT. Other buttons located on the bottom of the tab are Transports, Settings, and Statistics. The Statistics button is helpful when troubleshooting network problems because you can see the number of bytes transmitted, bytes received, network errors, failed sessions, server disconnects, hung sessions, etc.

Figure 5.7: NT Diagnostics Screen

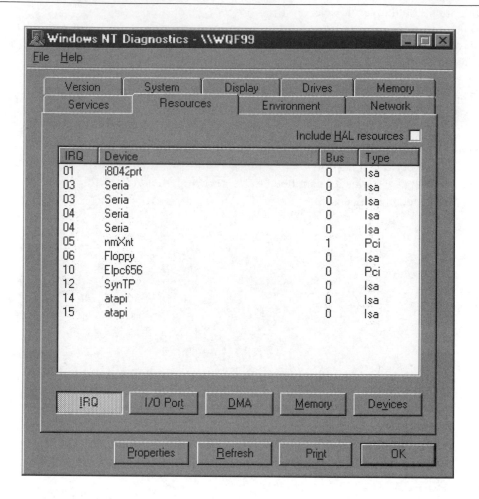

5.11 Troubleshooting a Service that Does Not Start

Some Windows NT services start automatically each time the computer boots. If one of these services has a problem, an error message normally appears during the boot sequence. You can use Event Viewer to see what particular service did not start. To control the service, use the individual administrative tool or the Services control panel.

To access the Services control panel, click on the **Start** button. Point to the **Settings** menu option. Click on the **Control Panel** option. A window appears with all the control panels displayed. Double-click on the **Services** icon. To start or stop a service manually, click on its name. Click on the **Start** or **Stop** button as appropriate.

Shutdown Problems

To shut down Windows NT Workstation properly, click on the **Start** button, click on the **Shut Down** option, click on the **Shut down the computer** radio button, and click on the **Yes** button. A shortcut is to press the **Alt + F4** keys after all applications are closed. If applications are open when you try to shut down NT, the operating system will attempt to close the applications and, if successful, it will shut down. If any documents have not been saved, you are prompted to save changes.

Before NT Workstation can shut down, the operating system sends a message to all devices, services, and applications. Each device that is running sends a message back saying it is okay to shut down. Any active application saves data that has not been previously saved and sends a message back to the operating system. Active system services also respond that it is okay to shut down. If the system has trouble shutting down, it is due to one of these three things. The most common problem is an application that is not responding. When this happens, press Ctrl + Alt + Del to access Task Manager. Click on the **Task Manager** button. Click on the **Applications** tab. Click on the application that has the words "not responding" in the Status column. Click on the **End Task** button. If a single application continually prevents NT Workstation from shutting down, contact the software manufacturer to see if there is a fix.

For services problems, boot the computer into Safe Mode and then shut down the computer. Take note as to whether or not the computer had any problems shutting down. If the process works, access the BOOTLOG.TXT file that is located in the root directory of the drive containing NT Workstation. Inside the file, take note of each service that is disabled because of booting into Safe Mode. Boot the computer normally. Stop each service one at a time to see which service is causing the problem. Before troubleshooting non-responding devices, eliminate services and applications. A device usually does not cause a shutdown problem. While working on the computer, take notice of which devices you are using. Common ones are video, hard drive, CD-ROM, keyboard, and mouse. Verify that all of your devices have the most up-to-date drive loaded and that the driver is compatible with Windows NT.

5.12 Monitoring System Performance

Another utility used to monitor the computer is the Performance Monitor tool. **Performance Monitor** allows creation of graphs, bar charts, and text reports. Specific resources such as memory and CPU usage can be tracked through Performance Monitor. An exercise at the end of the chapter shows how to use the Performance Monitor utility.

To access the Performance Monitor utility, click on the **Start** button, point to **Programs**, point to **Administrative Tools (Common)**, and click on the **Performance Monitor** option. The utility can be customized to show different counters. The button with a + (plus sign) is used to add various counters on the display. Some of the most important memory counters are Available Bytes and Pages/sec. The Available Bytes counter shows the amount of RAM available for running program processes. The Pages/sec counter shows the number of times per second that the information requested could not be found in RAM, and the data had to be retrieved from the hard drive. Since memory is a potential bottleneck for many computers, a technician should familiarize himself or herself with this technique. Sometimes you must adjust the paging file size for optimum computer performance. Note that the paging file is known by various terms including swap file, paging file, or virtual memory.

The **System** control panel is used to set the virtual memory size. Once in the System control panel, click on the **Performance** tab. Click on the **Change** button and the Virtual Memory window appears. Two values are selectable: Initial size and Maximum size. Both of these values should be the same for maximum computer performance. Once you change the values, click on the **Set** button. The Virtual Memory window may also be used to change the amount of space reserved for the registry.

Another potential bottleneck is the hard drive. The hard drive and memory work together because NT makes use of the paging file. The Performance Monitor charts you should watch are the Page Writes/sec and Pages Output/sec for memory and the Disk Writes/sec, Disk Write Bytes/sec, and Avg. Disk Write Queue Length for the logical disk option. You should practice working with the Performance Monitor utility before a problem occurs or the computer slows down.

The Task Manager utility may also be used to monitor your current systems performance. Sometimes a computer starts slowing down. A baseline is needed before the slowdown occurs. A **baseline** is a snapshot of your computer's performance during normal operations (before it has problems).

Start the Task Manager utility and click on the **Performance** tab to see the CPU usage and memory usage statistics. The first window on the left shows the CPU usage percentage. It is actually a percentage of time the processor is running a thread. A thread is a type of Windows object that runs application instructions. This percentage relates directly to the System Monitor's (Processor) %Processor Time counter. The first window on the right displays the CPU usage history, which is a graph of how busy the processor has been over a period of time.

The second window on the left shows the amount of virtual memory being used. The amount shown is in kilobytes as evidenced by the K after the number. The number displayed directly relates to the System Monitor's (Memory) Committed Bytes counter. The second window on the right is a graph of the virtual memory used over time. Memory is a frequent bottleneck for computer performance issues. Task Manager can also be used to see the total amount of RAM installed and how much RAM is available. Task Manager is an invaluable tool for technicians when a computer is slowing down.

Chapter Summary

- Windows NT is a true 32-bit operating system that has better security due to the support of the NTFS file system. It also supports FAT. Its GUI is similar to Windows 98.
- Windows NT provides a hierarchical file system allowing you to create folders and store files within them.
- Windows NT supports compression when using the NTFS. Compression saves disk space by "squeezing" redundant data from a file.
- The Registry is a database, which stores the hardware and software information installed on your computer.
- As part of the pre-installation steps of Windows NT, you should decide whether or not to perform a clean installation or an upgrade. You should consider installing virus protection software, which can detect and remove common software viruses.
- When installing Windows NT, the setup program named WINNT.EXE is used to install Windows NT on a system that currently has DOS, Windows 3.1, Windows 95/98 loaded. The WINNT32.EXE is used to upgrade from a previous version of Windows NT Workstation.
- Troubleshooting your Windows NT installation involves verifying BIOS and driver compatibility, inserting the correct CD key, ensuring you have a minimum CPU that is a 486 33 MHz, and having at least 12 MB of RAM.
- Configuring Windows NT involves using Control Panel to add and remove hardware components such as a mouse or a printer. It also includes adding and removing additional software.
- You need to understand the boot process of Windows NT so you know how to troubleshoot when problems occur. For example, if the NTLDR file is missing or corrupt, Windows NT will not boot.
- Windows Task Manager allows you view the applications that are running on your system. You get to Task Manager by pressing Ctrl + Alt + Del. Event Viewer contains a log of events that have occurred on your system. The Windows NT Diagnostic utility aids in troubleshooting by displaying system information.
- When Windows NT boots, various services are started. You can see the status of a service by checking the Services program in Control Panel. If a service fails to start, consider pressing F8 and booting to Safe Mode during the boot sequence.
- You can use Performance Monitor to monitor the performance of your Windows NT system through the use of objects and counters. It is useful to take a baseline of your system for comparative purposes.

Review Questions

1. Which type of computer system contains a relatively small number of instructions, and is fast and expensive?
   a) CISC
   b) IRQ
   c) RISC
   d) I/O port

2. What icon allows you to view networked computers when a network card is installed?
   a) Internet Connection Sharing
   b) Internet Explorer
   c) Network Explorer
   d) Network Neighborhood

3. The _____ is the area where all of your work in Windows NT begins. This area contains icons and shortcuts.
   a) Desktop
   b) Window
   c) Control Panel
   d) Long File Names

4. What Windows feature prevents you from moving icons on your desktop?
   a) Auto Unclutter
   b) Auto Arrange
   c) Arrange Order
   d) Automatically Detect

5. Where are files held when they are deleted from the hard disk drive?
   a) Trash Can
   b) They are not held anywhere because they are permanently deleted.
   c) Recycle Bin
   d) Quick Launch Bar

6. Where do you view file extensions?
   a) Under the Folder Options menu, which is beneath the View menu.
   b) Under the Folder View menu, which is beneath the Options menu.
   c) Under the View Options menu, which is beneath the Folder menu.
   d) Quick Launch Bar

7. How many characters can you have in a Windows NT file name?
   a) 128
   b) 255
   c) 256
   d) 8

8. Which of the following are invalid Windows NT long file names? Choose all that apply.
   a) PayrollQ.txt
   b) Fo/Le:100.tx*t
   c) Salary.dat
   d) TEXAS.DOC

9. Which NTFS feature allows you to save file disk space by reducing a file's size?
   a) Clustering
   b) Compression
   c) Partitioning
   d) Attributes

10. Which feature allows you make a file hidden?
    a) Clustering
    b) Compression
    c) Partitioning
    d) Attributes

11. Which feature allows you make a file read-only?
    a) Clustering
    b) Compression
    c) Partitioning
    d) Attributes

12. Which partition does Windows NT Workstation compress require?
    a) NFS
    b) NTFS
    c) FAT16
    d) FAT32

13. What is the database of folder and file property settings, and application and user preferences?
    a) REGEDIT
    b) Registry
    c) Backup
    d) DHCP

14. A _____ is a process running on Windows NT.
    a) Function
    b) Process
    c) DLL
    d) CAB file

15. Which registry subtree contains global hardware configuration?
    a) HKEY_USERS
    b) HKEY_CURRENT_USER
    c) HKEY_CLASSES_ROOT
    d) HKEY_LOCAL_MACHINE

16. Which registry subtree contains user preferences?
    a) HKEY_USERS
    b) HKEY_CURRENT_USER
    c) HKEY_CLASSES_ROOT
    d) HKEY_LOCAL_MACHINE

17. The _____ command is used to change the file system type from FAT to NTFS.
    a) CONVERT
    b) COMPRESS
    c) COMPACT
    d) FAT_TO_NTFS

18. What is the name of the special environment where a DOS application runs on a Windows NT computer?
    a) Partition
    b) PIF
    c) NTVDM
    d) Control Panel

19. Which Control Panel function allows you to install a new software application?
    a) Add New Hardware
    b) Add/Remove Programs
    c) Regional Settings
    d) Printers

20. The _____ directory is the default location where Windows NT is installed.
    a) C:\WINNT
    b) C:/WINNT
    c) C:\SETUP
    d) C:\SYSTEM32\WINNT

Lab Projects

Unless otherwise indicated, all projects assume the computer is powered on, you are logged in, and your desktop appears on the screen. Windows NT is up and running.

Lab Project 1

The goal of this project is to help you understand how to create a tree structure from a screenshot.

1. Using Windows Explorer, create the tree structure shown in Figure 5.8 on your floppy. To get to Explorer, right-click Start and then click Explore.

Figure 5.8: A Sample Tree Hierarchy

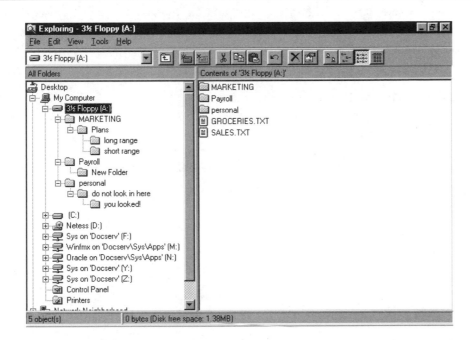

2. Using Notepad, create the following files as Text Documents in the "long range" folder. Place your name in each file.

   Monday.txt
   Tuesday.txt
   Wednesday.txt
   Thursday.txt
   Friday.txt

3. Record the steps you took to create them.

4. Create a "Medium Range" folder in "Plans."

5. Record the steps you took.

_____

6. In the "personal" folder, create a file called "resume.txt."

7. Hide this folder.

8. Record the steps you took.

_____

9. View all files.

10. Record the steps you took.

_____

11. Make a complete copy of the files in "Plans" and put them in a folder called "BACKUP of PLANS."

12. Record the steps you took.

_____

13. Make a disk copy of this floppy to another floppy.

14. Record the steps you took.

_____

15. Create another folder in the root directory of the floppy that has your name in it.

16. Create another folder in the root directory of the floppy named "Windows NT Project 1."

17. Show your instructor.

18. Close all windows.

Lab Project 2

The goal of this project is to reinforce your understanding of how to create a tree structure. You will also work with attributes to help you understand how they work.

1. Using Windows Explorer, create the tree structure shown in Figure 5.9 on your floppy.

Figure 5.9: A Sample Tree Hierarchy

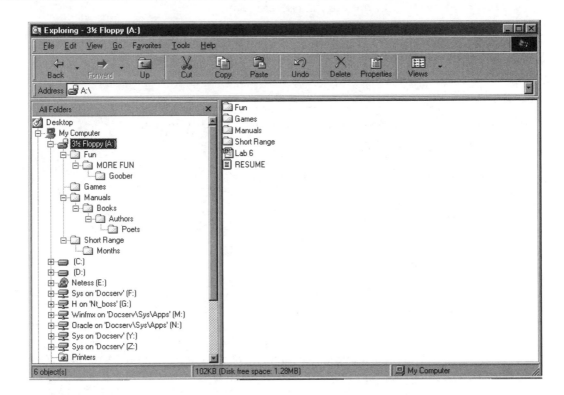

2. Using Notepad, place these files in the "Poets" folder:

   Poe.txt
   Walker.txt
   Smith.txt

3. Activate the setting so hidden and system files are not displayed.

4. Record the steps you took.

5. Hide "Smith.txt."

6. Record the steps you took.

7. Record whether or not you can now see "Smith.txt."

   _____

8. Activate the setting to show all files.

9. Record the steps you took.

   _____

10. Record whether or not you can now see "Smith.txt."

    _____

11. Make the file "Walker.txt" read-only.

12. Record the steps you took.

    _____

13. Open the file named "Walker.txt" in Notepad and attempt to add data.

14. Record what occurred.

    _____

15. Close Notepad.

16. Make "Walker.txt" readable.

17. Open the file named "Walker.txt" in Notepad and attempt to add data.

18. Record what occurred.

    _____

19. Close Notepad.

20. Create another folder in the root directory of the floppy that has your name in it.

21. Create another folder in the root directory of the floppy named "Windows NT Project 2."

22. Show your instructor.

23. Close all windows.

Lab Project 3

The goal of this project is to help you understand how to use the Windows NT Recycle Bin.

1. Turn the computer on and verify that Windows NT Workstation loads.

2. Log on to Windows NT Workstation using the userid and password provided by the instructor or lab assistant.

3. Double-click the **My Computer** desktop icon and then double-click the **C: Drive**. The C: Drive directory structure displays in the My Computer window.

4. Right-click on an empty space in the window, choose New, and then select Text Document. A new text document file appears in the window with the name highlighted.

5. Type **TESTDOC** and press **Enter.** You have now created a blank text file on the C: drive.

6. Record the file extension of TESTDOC.

_____

7. Right-click the **TESTDOC** file and choose **Delete**. Confirm the file deletion by clicking **Yes**. The TESTDOC file has now been sent to the Recycle Bin.

8. Double-click the **Recycle Bin** desktop icon. The Recycle Bin window opens with the **TESTDOC** file appearing in the window.

9. Highlight the **TESTDOC** file and select the **File** menu option.

10. Record which option you would select to return the TESTDOC file to its previous location.

_____

11. Choose the **Empty Recycle Bin** option, and then click **Yes** to confirm the deletion. The TESTDOC file has been permanently deleted.

Lab Project 4

In this lab, you will use REGEDIT to modify how Windows NT Workstation participates in the Browser process.

CAUTION: Editing the registry can cause your computer to run erratically, or not run at all! When performing any registry editing, follow ALL directions carefully including spelling, syntax use, etc. Failure to do so may cause your computer to fail!

1. Turn on the computer and verify that the NT Workstation loads.

2. Log on to NT Workstation using the userid and password provided by the instructor or lab assistant.

3. Click on the **Start** button. Click on the **Run** option.

4. In the text box, type in **REGEDIT** and press **Enter.** The REGEDIT utility appears.

5. In the left window, click on the **+ (plus sign)** by the **HKEY_LOCAL_MACHINE** folder.

6. In the left window, click on the **+ (plus sign)** by the **System** folder.

7. In the left window, click on the **+ (plus sign)** by the **CurrentControlSet** folder.

8. In the left window, click on the **+ (plus sign)** by the **Services** folder.

9. In the left window, click on the **+ (plus sign)** by the **Browser** folder.

10. In the left window, click on the **Parameters** folder. The values contained in the Parameters folder list in the window on the right side.

11. In the right window, locate the MaintainServerList value.
    Note: The MaintainServerList value controls how the local computer participates in Browser selections. It has three possible DATA values:

    A. Yes: Will always participate as a Browser
    B. No: Will not participate as a Browser
    C. Auto: Can be a Browser if necessary

    Record the DATA value for your computer's **MaintainServerList** value.

    _____

12. Double-click the **MaintainServerList** value. The **Edit String** window opens.

13. If the current DATA value is **Auto**, change the value to **Yes**. If the current DATA value is **Yes**, change the value to **Auto**. If the current DATA value is **No**, change the value to **Auto**.

    What effect will the change made have on the computer's Browser status?

    _____

14. Click on the **OK** button to close the **Edit String** window. The new DATA value appears in the REGEDIT window.

    When will the new registry setting take effect?

    _____

15. Close the **REGEDIT** utility and reboot the computer.

Lab Project 5

In this lab, you will use REGEDT32 to create and configure a new registry setting to resize the desktop icons from the default of 32 pixels to 20 pixels.

CAUTION: Editing the registry can cause your computer to run erratically, or not run at all! When performing any registry editing, follow ALL directions carefully including spelling, syntax use, etc. Failure to do so may cause your computer to fail!

1. Turn on the computer and verify that the NT Workstation loads.

2. Log on to NT Workstation using the userid and password provided by the instructor or lab assistant.

3. Click on the **Start** button. Click on the **Run** option.

4. In the text box, type in **REGEDT32** and press **Enter.** The REGEDT32 utility appears.

5. Click on the **HKEY_CURRENT_USER** window.

6. Expand the **Control Panel** folder by clicking on the **+ (plus sign)** beside it.

7. Expand the **Desktop** folder.

8. Select the **WindowMetrics** option.

9. Form the **Edit** menu item, click on the **Add Value** option. The Add Value window opens.

10. In the Value Name field, type **Shell Icon Size**.

11. In the Data Type field, choose **REG_SZ** and then click on the **OK** button. The String Editor opens.

12. From the String Editor window, enter the value of **20**, (the default number of pixels in a desktop icon is 32) and click on the **OK** button. The new value displays.

    How is the new value displayed in REGEDT32?

    _____

13. Close the **REGEDT32** utility and reboot the computer for the new registry setting to take effect.

    After the computer reboots, are your desktop icons smaller or larger than before? If not, perform the exercise again.

    _____

Lab Project 6

The goal of this project is to help you understand the proper use of the Windows NT Backup utility to back up and restore files. This project assumes you will back up all files from the C: drive to a tape backup unit. If you don't have one, then choose three files to back up to floppy instead.

To back up files, follow these steps:

1. Turn the computer on and verify that Windows NT Workstation loads.

2. Log on to Windows NT Workstation using the userid and password provided by the instructor or lab assistant.

3. From the **Start** menu, point to the **Programs** selection, then the **Administrative Tools** option, and double-click on the **Backup** menu selection. The Windows NT Backup utility starts.

4. From the **Drives** window, select the **C: drive check box** to enable it. This action selects all files and folders on the C: drive for back up.
   Note: To select specific files only, double-click the drive, then browse to and select specific files and/or folders.

   How many drives list in the Drives window?

   _____

5. After choosing the files to be backed up, click on the **Backup** button to begin the backup process. The Backup Information window opens.

6. From the Backup Information window, type **TESTTAPE** in the Tape Name field, choose **Verify After Backup**, select **Normal** for the backup type, and click on the **OK** button.

   What types of backups are available?

   _____

   What does the setting Verify After Backup mean?

   _____

7. Windows NT Backup begins backing up the selected files and folders. The status of the backup can be monitored from the Backup Status window. When Windows NT Backup completes the backup process, a summary report displays. Click on the **OK** button to close the Backup Status window.

   To restore files and/or folders, follow these steps:

8. Verify that the tape with the needed files is inserted into the tape drive.

9. From within the Windows NT Backup utility, open the **Tapes** window. Browse to and select the files to be restored, and then click on the **Restore** button. The Restore Information window opens.

10. Click on the **OK** button to begin the restore process.

    Can the files be restored to an alternate location?

    _____

11. Confirm any file replacement messages by selecting **Yes to all**.

12. When the restore process is completed, a summary report displays in the Restore Status window.

13. Click on the **OK** button to close the Restore Status window and then exit the Windows NT Backup utility.

Lab Project 7

The goal of this project is to help you understand the proper installation of devices using Windows NT Workstation.

Most devices installed on a Windows NT Workstation are installed using the device icons within Control Panel. Each device type has its own specific installation steps, and you must follow the device-specific on-screen prompts and instructions for proper installation.

To install a device on a Windows NT Workstation, follow these general steps:

1. With the computer turned off and following proper ESD precautions, install the device into the appropriate BUS slot or external port.

2. Turn the computer on and verify that Windows NT Workstation loads.

3. Log on to Windows NT Workstation using the userid and password provided by the instructor or lab assistant.

4. From the **Start** menu, point to the **Settings** option, and then select **Control Panel**.

   How else can you open the Control Panel?

   _____

5. From the Control Panel window, double-click on the **Control Panel** icon that represents the device you are installing. Follow the appropriate device-specific instructions and steps to install the device.

Modems

6. Double-click the **Modems** control panel icon.

7. Click on the **Next** button to allow NT to detect the modem.

8. When the modem is detected, click on the **Next** button and follow the on-screen prompts to complete the installation.

Sound Cards

9. Double-click the **Multimedia** control panel icon.

10. Select the **Devices** tab.

11. Click on and highlight **Audio Devices**, click on the **Add** button, and then follow the on-screen prompts to complete the device installation.

Network Adapters

12. Double-click the **Network** control panel icon and select **Adapters**.

13. Click **Add** and then follow the on-screen prompts to complete the device installation.

Ports

14. Double-click the **Ports** control panel icon.

15. Click on the **Add** button and follow the on-screen prompts to complete the device installation.

Printers

16. Double-click the **Printers** control panel icon.

17. Double-click **Add Printer** icon and follow the on-screen prompts to complete the device installation.

SCSI Adapters

18. Double-click the **SCSI Adapters** control panel icon and select **Drivers**.

19. Click on the **Add** button and follow the on-screen prompts to complete the device installation.

Tape Devices

20. Double-click the **Tape Devices** control panel icon and select **Drivers**.

21. Click on the **Add** button and follow the on-screen prompts to complete the device installation.

Note: If the device cannot be installed through a control panel, follow the device manufacturer's specific installation instructions.

Lab Project 8

The goal of this project is to help you understand how to view memory usage statistics in Windows NT Workstation.

At times, it is beneficial to view memory usage statistics to troubleshoot resource conflicts. NT provides two utilities that can be used for this: Task Manager and NT Diagnostics. Task Manager allows you to view the memory usage for individual processes whereas NT Diagnostics allows you to view overall memory usage statistics.

To use Task Manager, follow these steps:

1. Turn the computer on and verify that Windows NT Workstation loads.

2. Log on to Windows NT Workstation using the userid and password provided by the instructor or lab assistant.

3. Press **Ctrl , Alt, and Delete** simultaneously. The Windows Security window opens.

4. Choose **Task Manager** and then select the **Processes** tab.

5. The currently running processes display along with each process's CPU and Memory usage statistics.

   How much memory are the Winlogon.exe and the Taskmgr.exe processes using?

   _____

   To use the Windows NT Diagnostics utility, follow these steps:

6. Click on the **Start** button, point to **Programs, Administration Tools,** and then click on the **Windows NT Diagnostics** option.

   What other method can be used to start Windows NT Diagnostics?

   _____

7. When the Windows NT Diagnostics window opens, select the **Memory** tab. The memory statistics display.

   How much memory is set aside for file caching?

   _____

Lab Project 9

The goal of this project is to help you understand the proper use of the NT Diagnostics utility.

Windows NT provides a diagnostic utility that can be used to view system information. This is especially useful for diagnosing and troubleshooting system errors and conflicts.

1. Turn on the computer and verify that the Windows NT Workstation loads.

2. Log on to NT Workstation using the userid and password provided by the instructor or lab assistant.

3. From the **Start** menu, choose **Programs, Administrative Tools**, and then select **Windows NT Diagnostics**. The Windows NT Diagnostics utility starts.

   How else can you start Windows NT Diagnostics?

   _____

4. Click on the **Version** tab. From the Version tab, you can view the version, service pack, registration key, and registration information for the computer.

5. Select the **System** tab. From the System window, you can view system, HAL, BIOS, and CPU information.

   What type of processor is installed in the computer?

   _____

6. Select the **Display** tab. From the Display window, you can view video adapter and display information.

   What type of video adapter is installed in the computer?

   _____

7. Select the **Drives** tab. From the Drives window, you can view the hard drives and CD-ROM drives.

8. Select the **Memory** tab. From the Memory window, you can view memory statistics for the system.

   How much total physical memory is installed in the system?

   _____

9. Select the **Services** tab. From the Services window, you can view installed services and their status, as well as installed devices and their operational state.

10. Select the **Resources** tab. From the Resources window, you can view IRQ, I/O port, DMA, memory, and device statistics for your system.

    Are any devices using DMA channels? If so, which devices?

    _____

11. Select the **Environment** tab. From the Environment window, you can view the system and local user environment variables.

12. Select the **Network** tab. From the Network window, you can view the network environment settings and statistics.

    Is your workstation a member of a domain, or a workgroup?

    _____

13. Close the Windows NT Diagnostics utility.

Lab Project 10

The goal of this project is to use the Windows NT Diagnostics tool to troubleshoot device conflicts.

1. Turn on the computer and verify that the Windows NT Workstation loads.

2. Log on to NT Workstation using the userid and password provided by the instructor or lab assistant.

3. Click on the **Start** button.

4. Point to the **Programs** option.

5. Point to the **Administrative Tools (Common)** option.

6. Click on the **Windows NT Diagnostics** option. The Windows NT Diagnostics window opens. The tab that opens by default is the Version tab.

   What version of NT Workstation is running and what, if any, service pack is installed?

   _____

7. Click on the **Resources** tab.

   What IRQs are not used?

   _____

8. Click on the **I/O Port** button.

   What device uses 03F7 I/O address space?

   _____

9. Click on the **DMA** button.

   What device uses DMA channel 2?

   _____

10. Click on the **Memory** button.

    What device uses the memory range 000A0000—000AFFFF?

    _____

11. Close the **NT Diagnostics** window.

Lab Project 11

The goal of this project is to help you understand the installation of a local printer on a Windows NT Workstation. You will need a computer with NT Workstation loaded and a printer physically attached to a printer port. Before an NT Workstation can send a print job to a local printer, the driver for that printer must be installed and configured.

1. Turn the computer on and verify that NT Workstation loads.

2. Log on to NT Workstation using the userid and password provided by the instructor or lab assistant.

3. Double-click on the **My Computer** desktop icon and then double-click on the **Printers** folder. The Printers folder opens.

4. Double-click the **Add Printer** icon, and the Add Printer wizard starts.

5. Select **My Computer** and then click on the **Next** button and the Ports window opens.

   How many LPT and COM ports are listed in the Ports window?

   _____

6. Choose the printer **Manufacturer** of your printer and then select the printer **Model**. If the attached printer is not listed, click on the **Have Disk** button, insert the print driver disk or CD, enter a path to the driver files (such as A: or the drive letter for the CD-ROM), and click on the **OK** button.

7. Select the appropriate printer mode and click on the **Next** button.

8. Enter a **name** for the printer in the **Printer Name** field and click on the **Next** button.

   The Sharing and Additional Drivers window opens.

   What name did you assign to the printer?

   _____

9. Select the **Not Shared** option, leave the Share Name field blank, and click on the **Next** button.

10. Select the **Yes** option to print a test page and click on the **Finish** button.

11. If prompted, insert the Windows NT Workstation CD-ROM or enter a path to the installation files.

12. The printer installation process finishes and returns to the Printers Folder. If the installation is successful, a printer test page prints.

    Did a test page print successfully? If not, redo the exercise. Take special precautions when selecting the appropriate print driver.

    _____

Lab Project 12

The goal of this project is to create an ERD on an NT Workstation computer.

1. Turn on the computer and verify that the NT Workstation loads.

2. Log on to NT Workstation using the userid and password provided by the instructor or lab assistant.

3. Click on the **Start** button.

4. Click on the **Run** option.

5. In the Open: text box, type, **RDISK** and press **Enter.** The Repair Disk Utility appears on the screen.

6. Click on the **Update Repair Info** button. A prompt may appear that states that the repair information that was previously saved will be deleted. If this prompt appears, click on the **Yes** button.

   What is the difference between using the Update Repair Info button and the Create Repair Disk options?

   _____

7. A prompt appears asking if you want to create an Emergency Repair Disk. Click on the **Yes** button to create an Emergency Repair Disk.

8. Insert a floppy disk into the A: drive and click on the **OK** button.

9. When finished, remove the disk, label it with the current date and store in a safe location.

10. Click on the **Exit** button to close the Repair Disk program.

Lab Project 13

The goal of this project is to help you understand the proper use of the Performance Monitor utility in monitoring system performance in Windows NT Workstation. Windows NT provides the Performance Monitor utility that can be used to monitor system performance.

1. Turn the computer on and verify that Windows NT Workstation loads.

2. Log on to Windows NT Workstation using the userid and password provided by the instructor or lab assistant.

3. From the **Start** menu, point to **Programs, Administrative Tools**, and then select **Performance Monitor**. The Performance Monitor utility starts.

4. Before Performance Monitor can be used to monitor system performance, you must configure the system counters to monitor. To add counters to Performance Monitor, click the **+ (plus)** button on the Performance Monitor toolbar. The Add To Chart window opens.

5. From the Object drop-down menu, select **Processor**.

   How many objects are available for adding system counters?

   _____

6. From the Counter window, hold the **Ctrl** key down while selecting the following:

   %Processor Time, %User Time, %Interrupt Time, and **%Privileged Time**. Click on the **Add** button, and then click on the **Done** button to return to the Performance Monitor window.

7. Each of the selected counters will be represented on the Performance Monitor chart by different color. Minimize the **Performance Monitor utility** window by clicking on the icon with the line symbol in the top right corner of the window.

8. From the **Start** menu, choose **Programs, Accessories,** and then select **Notepad**.

9. Maximize the **Performance Monitor utility** window by clicking on the Performance Monitor icon located on the taskbar.

   Did starting the Notepad application cause any activity with the selected counters in Performance Monitor?

   _____

10. Move the mouse cursor across the Performance Monitor screen.

    Did moving the mouse cursor cause any counter activity?

    _____

    If moving the mouse cursor caused counter activity, which of the selected counters show the most activity?

    _____

11. When finished with monitoring the selected counters, close the **Performance Monitor** utility.

Lab Project 14

The goal of this project is to create a computer baseline report that can be used when the computer does not function properly.

1. Turn on the computer and verify that the Windows NT Workstation loads.

2. Log on to NT Workstation using the userid and password provided by the instructor or lab assistant.

3. Click on the **Start** button.

4. Point to the **Programs** option.

5. Point to the **Administrative Tools (Common)** option.

6. Click on the **Event Viewer** option.

   What log is opened by default?

   _____

   If any events list on the screen, write the two most recent events in the space below.

   _____

7. Scroll down through the events and note the different symbols that precede events.

   How many different event symbols show in the Event Viewer system log?

   _____

   What does each symbol mean?

   _____

8. Double-click on an **event** in the system log.

   What information does the Event Detail window give you that was not available on the original screen?

   _____

9. Click on the Event Detail **Help** button.

   What happens to event data if you archive the event log in a TXT file?

   _____

10. Click on the **Close** button in the Event Viewer Help window.

11. Click on the **Close** button in the Event Detail window.

12. Click on the **Log** menu item and select the **Security** option.

    What types of events are kept in the security log?

    _____

13. Click on the **View** menu option.

What is an alternative to viewing the newest information first?

_____

14. Click on the **Log** menu item and select the **Log Settings** option.

    What is the current setting for the Maximum Log Size option?

    _____

15. Click on the **Change Settings for Security** down arrow. The Maximum Log Size option is applicable to each type of event log.

    How many types of logs are available in the drop-down menu and what are the names of the logs?

    _____

16. Click on the **Cancel** button.

17. Click on the **Log** menu item and select the **System** option.

18. Click on the **View** menu item and select the **Filter Events** option. The Filter Events option is a dialog box that allows you to define the event time period, event type, and event category. When filtering is enabled, the word Filtered appears in the title bar.

    What types of events are filtered by default?

    _____

19. Click on the **Cancel** button in the Filter window.

20. Click on the **Close** button in the Event Viewer window.

Lab Project 15

The goal of this project is to use Task Manager to halt an application.

At times, it may become necessary to halt a hung or stalled application. Windows NT Workstation provides a method to accomplish this through the Task Manager utility.

1. Turn on the computer and verify that Windows NT Workstation loads.

2. Log on to Windows NT Workstation using the userid and password provided by the instructor or lab assistant.

3. From the **Start** menu, point to **Programs, Accessories,** and then click on the **Notepad** option. The Notepad utility runs.

4. To access Task Manager, simultaneously press the **Ctrl, Alt**, and **Del** keys and then select **Task Manager**.

   What things can you view from Task Manager?

   _____

5. Select the **Applications** tab.

   What applications are listed as open?

   _____

6. Highlight the **Notepad.exe** application and select **End Task**. The Notepad.exe application closes.

7. Close the **Task Manager** utility.

Lab Project 16 Challenge

The goal of this challenge project is to install the Windows NT Workstation operating system on a computer. For this project, you will need a computer without an operating system and a Windows NT Workstation installation CD-ROM along with the setup disks.

The method used to start the Windows NT Workstation installation process depends on whether or not your system supports booting from a CD-ROM.

If your computer supports booting from a CD-ROM, follow these steps:

1. Insert the Windows NT Workstation CD-ROM into the CD-ROM drive and start the computer. The computer boots from the CD and begins the installation process.

2. At the Welcome to Setup screen, press **Enter** to begin Windows NT Setup.

3. Setup detects the mass storage devices installed on your computer and displays a list of the detected devices. Press **Enter** to continue. Continue with Step 5 below.

If your computer does NOT support booting from a CD-ROM, follow these steps:

1. Insert the Windows NT Setup Disk 1 into the floppy drive and start the computer.

   Note: If you do not have the Windows NT Setup floppy disks, you can create them from the installation CD. Insert the installation CD in a computer that has been booted with CD-ROM support. From a command prompt, change to the I386 directory on the CD.

   Type /OX and press Enter when upgrading from DOS, WIN 3.x, or WIN9x, or type 32 /OX and press Enter to upgrade an older version of NT or install a new version of NT. You are prompted to label and insert three floppy disks.

2. When prompted, change the floppy disks.

3. At the Welcome to Setup screen, press **Enter.** Windows prompts to detect the mass storage devices. Press **Enter** to continue.

4. Setup detects the mass storage devices installed and displays a list of the detected devices. Press **Enter** to continue.

From this point on, setup for both types of installations is identical:

5. Page down through the Licensing Agreement and press **F8** to agree.

6. At the Hardware Components page, verify that the listed components match what is installed in the computer and press **Enter.**

7. Highlight the **un-partitioned disk space** where you want to install Windows NT and press **Enter**.

8. Choose to format the partition as **NTFS** and press **Enter**.

9. Press **Enter** to install Windows NT into the default directory.

10. Press **Enter** to have Windows Setup examine the hard disks. After the examination is complete, Setup begins copying files to the hard drive.

11. When prompted, remove all floppy disks and CDs from the drives and press **Enter** to restart the computer.

12. The computer restarts and Setup enters the graphical (GUI) Setup mode. If you chose to use the NTFS file system format, Setup converts the partition to NTFS and restarts the computer a second time.

13. When prompted, re-insert the **Windows NT Installation CD-ROM** and click **OK**.

14. At the Gathering Information about your Computer window, click on the **Next** button.

15. Choose **Typical** installation and click **Next**.

16. Enter the **Name** and **Organization** information provided by the instructor or lab assistant, and click **Next**.

17. Enter the **CD-key** (found on the back of the CD case or provided by the instructor or lab assistant) and click **Next**.

18. Enter the **computer name** (provided by the instructor or lab assistant) and click **Next**.

19. Enter and confirm the **Password** (provided by the instructor or lab assistant) and click **Next**.

20. Select **No, Do not create an Emergency Repair Disk** and click on the **Next** button.

21. Select **Install the most common components**, and click on the **Next** button.

22. Click on the **Next** button to begin installing Windows NT Networking.

23. Choose **This computer will participate on a network**, select **Wired to the Network**, and click **Next**.

24. Select **Start Search**, and Setup searches for network adapters. When the network adapter is detected, click on the **Next** button.

25. Select the **TCP/IP Protocol**, click on the **Next** button, and click on the following **Next** button.

26. If directed by the instructor or lab assistant to use DHCP, choose **Yes** at the TCP/IP Setup window. If not using DHCP, choose **No**. Setup installs the selected networking components.

27. If you are NOT using DHCP, enter an IP address and subnet mask provided by the instructor or lab assistant and click **OK**.

28. Click on the **Next** button to start the network.

29. At the Make this computer a member of window, enter the workgroup or domain information (provided by the instructor or lab assistant) and click on the **Next** button.

30. Click on the **Finish** button to complete Windows NT setup.

31. Select the appropriate **Time Zone** information and click on the **Close** button.

32. Test the selected video settings by selecting the **Test** button. If the settings are correct, click **Yes**, and then click **OK** twice.

33. When prompted, remove all floppy disks and CDs from the drives and select **Restart Computer**. The computer restarts using the newly installed Windows NT Workstation operating system.

Internet Discovery

Internet Discovery Lab 1

The goal of this project is to access the Internet to obtain specific information regarding a computer or its associated parts.

1. List two web sites that have information about troubleshooting the NT Workstation installation.

2. You have just loaded Norton AntiVirus and now NT Workstation does not boot. Find a web site that details what to do.

3. How do you create boot floppies for NT Workstation? Find a web site that describes how and write the URL in the space below.

4. What is Paged Pool Memory as it relates to NT Workstation? Find a definition for this term on the Internet and write the URL and the definition in the space below.

5. On the Microsoft Internet site, find a description of how to install NT Workstation unattended. Write the URL in the space below.

6. An internal 100 MB Iomega Zip Drive is installed in an NT Workstation computer. The computer displays the error, "ASPI for Win32 not initialized." Find a URL that details the resolution process. Write the URL in the space below.

Soft Skills: Help Desk Support

1. A customer calls and asks you how to install a Network Interface Card (NIC). What do you tell them?

2. You receive a page from a Technical Support Engineer. She is receiving this message when booting Windows NT: "Couldn't find NTLDR. Please insert another disk." What is the problem and what suggestions do you offer her for correcting the problem?

3. You attempt to install Windows NT Workstation on a computer that is an Intel 80386 running at 33 MHz with 16 MB of RAM and 100 MB of hard disk space free. What potential issues do you see with this configuration before you install the operating system?

Critical Thinking

1. What is the difference between the system partition and the boot partition?
2. Explain the terms NTVDM and WOW.
3. Define the purpose of the BOOT.INI file.

Study Skills

The "Do" in "Hear, See, Do, Say"

Key to learning are these four points: hear, see, do, say. You need to <u>hear</u> the topic, <u>see</u> it, <u>do</u> and then <u>say</u> it. The "hear" is the easy part. You just simply need to listen to the instructor's lecture in class and take good notes. You can "see" a topic by the instructor writing on the board, showing a video. or presenting PowerPoint slides to you. The "do" part is doing the work and in this case, it means doing the Review Questions and Lab Projects. You complete the "say" point when you are asked a question in class or when you must give answers on a test. Instructors may use traditional pen and paper tests combined with hands-on lab tests. This Study Skills section focuses on the "Do" which translates into doing the Lab Projects."

- Before you begin a Lab Project, read the lab over quickly to get a sense of what is being asked.

- Once you have read it quickly, carefully read it over again. This time, you are reading for content. At this point, you may need to go through the chapter again to help you understand some specific point in the Lab Project.

- Make sure you have the necessary materials to do the Lab Project. For example, if you need a floppy disk, then make sure you have one. Always be prepared. If you need additional paper to complete additional tasks, bring it.

- If you have a hands-on test, practice the Lab Projects, practice the Lab Projects, and then practice the Lab Projects again. Ask the instructor for additional Lab Projects if you want additional help.

- Always bring your textbook and notebook to class. Sometimes instructors will give open book hands-on tests but time the test to simulate a real-world business situation. By all means, bring your material for that type of test. Otherwise, you could fail the test.

- If you are having trouble doing a Lab Project, do a little research first. Dig into your textbook, search the Internet, use the online help available on the computer, and ask your partner. Of course, you can ask your instructor but they will be impressed at the amount of research you've done.

Self-Study Question(s)

1. What Lab Projects did you complete this week?
2. Which Study Skill(s) listed above helped you this week in completing your Lab Projects?

6

Chapter 6
Introduction to
Windows 2000

OBJECTIVES

The goal of this chapter is twofold:
- To introduce you to Windows 2000 Professional.
- To help you prepare and pass the following sections of the A+ Operating System Technologies Exam:

A+ Operating System Technologies Exam Objectives
covered in this chapter (and corresponding page numbers)

1.1 Identify the major desktop components and interfaces, and their functions. Differentiate the characteristics of Windows 9x/Me, Windows NT 4.0 Workstation, Windows 2000 Professional, and Windows XP.

1.2 Identify the names, locations, purposes, and contents of major system files.

1.4 Identify basic concepts and procedures for creating, viewing, and managing disks, directories and files. This includes procedures for changing file attributes and the ramifications of those changes (for example, security issues).

1.5 Identify the major operating system utilities, their purpose, location, and available switches.

Domain 2 Installation, Configuration and Upgrading

2.1 Identify the procedure for installing Windows 9.x/Me, Windows NT 4.0 Workstation, Windows 2000 Professional, and Windows XP and bringing the operating system to a basic operational level.

2.3 Identify the basic system boot sequences and boot methods, including the steps to create an emergency boot disk with utilities installed fro Windows 9.x/Me, Windows NT 4.0 Workstation, Windows 2000 Professional, and Windows XP.

2.4 Identify procedures for installing/adding a device, including loading, adding, and configuration device drivers, and required software.

In this chapter, you will complete the following sections:
- 6.1 Understanding Windows 2000 Professional
- 6.2 Managing Files and Folders on Windows 2000
- 6.3 Understanding Compression and Encryption on Windows 2000 Professional
- 6.4 Understanding the Windows 2000 Registry
- 6.5 Understanding the Pre-Installation Steps of Windows 2000
- 6.6 Understanding How to Install and Upgrade Windows 2000 Professional
- 6.7 Troubleshooting the Windows 2000 Professional Installation
- 6.8 Understanding How to Dual-Boot Windows 2000 Professional and Windows NT
- 6.9 Configuring Windows 2000 Professional
- 6.10 Understanding the Windows 2000 Professional Boot Process
- 6.11 Understanding Task Manager, Dr. Watson, and Event Viewer
- 6.12 Troubleshooting a Service that Does Not Start
- 6.13 Monitoring System Performance

6.1 Understanding Windows 2000 Professional

There are several versions of Windows 2000—Windows 2000 Professional, Windows 2000 Server, Windows 2000 Advanced Server, and Windows 2000 Data Center Server. Windows 2000 Professional is an operating system designed for a computer workstation.

Windows 2000 Professional is an operating system designed for business workstations —computers in the workplace that are connected to a network. Windows 2000 is easy to install and you can upgrade from Windows 95, Windows 98, or NT Workstation. The installation can also be done across a network. In addition, Windows 2000 is more robust and more stable than previous workstation operating systems. Windows 2000 supports plug and play. The operating system supports many new hardware devices including DVD, video capture devices, speakers, USB, multiple monitor support, removable storage drives, infrared devices, and digital cameras. One important change is that with Windows 2000, the computer does not have to be restarted after every change, as it did with Windows 95/98. This feature is only available if the drivers being installed are Windows 2000 certified.

Windows 2000 Professional is based on a 32-bit architecture. This operating system provides better performance than NT Workstation and Windows 95/98. Every 32-bit application under Windows 2000 and NT runs in its own memory space. When an application freezes or crashes, other 32-bit applications loaded into memory are not affected. Like NT Workstation, all 16-bit applications run in a single process called **NTVDM (NT Virtual DOS Machine)**. If one 16-bit application crashes, any other 16-bit applications loaded into memory crash too. However, Windows 2000 Professional offers the option of running a 16-bit application in its own memory space to prevent this from happening. To allow a 16-bit application to run in its own memory space, go to a command prompt and type **start /separate** *process_name*, where *process_name* is the name of the 16-bit application process.

Windows 2000 Professional can use up to 4 GB of RAM and support **SMP (Symmetric Multiprocessing)**. SMP is the ability to support two processors that operate simultaneously. Each process being run by an application or the operating system is distributed equally across two microprocessors and makes sure that one processor does not become a bottleneck for the system.

A new Windows 2000 Professional feature is **WFP (Windows File Protection)**, which is a program that protects system files. WFP is a program that runs in the background. It detects whether a system file has been altered or deleted. In previous operating systems, applications or users changed the system files and thus made the operating system unstable. WFP detects when a system file has been altered, deleted, or overwritten.

The file systems that Windows 2000 Professional supports include FAT16, FAT32, and NTFS.

After a user logs on to Windows 2000, the desktop appears. The log-in screen cannot be bypassed as it can in Windows 95/98. A user can log in to a network server, log in to a network workgroup (peer-to-peer network), or log in locally (as a stand-alone computer not participating in a network). Figure 6.1 shows the Windows 2000 desktop.

Figure 6.1: The Windows 2000 Professional Desktop

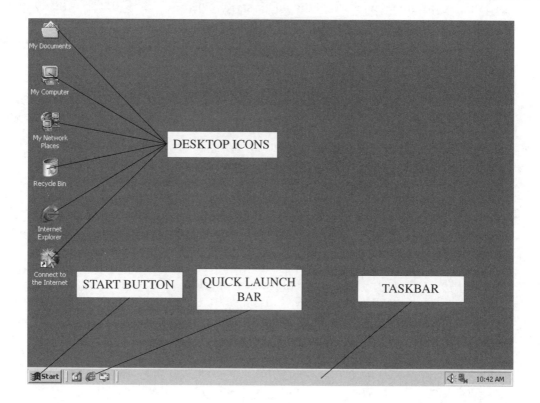

The desktop consists of many icons. Icons are pictures on the desktop. When you double-click on them, you interface with various devices, files, and applications on your computer. The desktop sometimes is cluttered with things the user puts on it.

Let's go through a brief exercise to help you understand how to maintain your Windows 2000 desktop. In this exercise, you will work with the **Auto Arrange** option to force your icons on your desktop to be automatically arranged towards the left hand side of the desktop. When you attempt to relocate an icon, it will automatically be arranged back to its original position—on the left hand side of the desktop. It is assumed that Windows 2000 is installed and running on a computer with the Windows 2000 desktop displayed on your screen.

1. Right-click on a blank desktop space. Once a menu appears, point to the **Arrange Icons** option. If the **Auto Arrange** option does have a check mark beside it, click on an **empty portion of the desktop** to cancel because Auto Arrange is already in use. If the **Auto Arrange** option does not have a check mark beside it, click on the **Auto Arrange** option to add the check mark and enable **Auto Arrange** capability.
2. Note the position of the icons on the desktop.
3. Click on the **My Documents** icon and while continuing to hold the mouse button down, drag the icon to the center of the screen. You will see that the **My Documents** icon moves to its original position.
4. Click on the **Recycle Bin** icon and while continuing to hold the mouse button down, drag the icon to the top right portion of the screen. This icon moves to its original position too.
5. Right-click on a **blank desktop space.** A menu appears.
6. Using step 1 as a guide, set Auto Arrange back to its original setting.

The desktop can also be modified to have a wallpaper scheme (which is a background picture, pattern, or color), a color scheme (which is the color scheme used in displaying folders), and a screen saver (which is the picture, color, or pattern that displays when the computer is inactive).

Let's perform an exercise on the desktop properties. Again, it is assumed that Windows 2000 is running on a computer.

1. Right-click on an **empty desktop area.** A menu appears.
2. Click on the **Properties** option. The Display Properties window appears. (Note: this screen can also be accessed by clicking on the **Start** button, pointing to the **Settings** option, clicking on the **Control Panel** option, and double-clicking on the **Display** control panel icon.)
3. Click on the **Background** tab. This tab controls how the background color and pattern appears.
4. To change your wallpaper to the **Chateau** option, click on the **Chateau** option.
5. Click on the **Apply** button.
6. To view the **Appearance** tab settings, click on the **Appearance** tab at the top of the window. This tab controls the color scheme and size of the letters. This tab is good to use when people with vision problems need adjustments.
7. Click Cancel to close the window.

Whenever you open an application or utility, a window appears. Windows are a normal part of the desktop. The desktop consists of icons, the Start button, the Quick Launch bar, and the taskbar. Common icons include: My Documents, My Computer, My Network Places, and Recycle Bin.

The **My Documents** icon is used to quickly access the My Documents folder (directory) on the hard drive. The My Documents folder is the default location for files the user saves. The **My Computer** icon is used to access the hardware, software, and files located on the computer. The **My Network Places** icon is used to access network resources such as computers, printers, scanners, fax machines, and files. The **Recycle Bin** is used to hold files and folders the user deletes. When a file is deleted from a folder, it is not immediately discarded; instead, it goes into the Recycle Bin. Once a file or folder is in the Recycle Bin, it can be removed. This is similar to a piece of trash being retrieved from an office trash can. You must remember that these files and folders in the Recycle Bin take up hard drive space and that users frequently forget to empty the files and folders from the Recycle Bin.

The **Internet Explorer** icon is used to start the Internet Explorer application. This application allows Internet connectivity. The **Connect to the Internet** icon is a Windows 2000 wizard that steps you through setting up an Internet connection. The Connect to the Internet icon is actually a common type of desktop icon called a shortcut. A **shortcut** is an icon that looks like any other icon except that it has a bent arrow in the lower left corner. A shortcut represents a path (location on the drive) to a file, folder, or program. It is a link (a pointer) to where the file or application resides on a disk. When you double-click on a shortcut icon, the shortcut icon tells Windows where to find the specific file that the icon represents. If the shortcut represents an application, the application opens. If the shortcut represents a document, the application used to create the document appears along with the document. A shortcut offers faster access to an application or file than going through the Start button or through the My Computer icon. Users frequently place shortcuts on the desktop and it is important that you know how to create one. Users create shortcuts to their favorite applications or to documents used frequently.

The **Start button** is located in the lower left corner of the desktop and is used to launch applications and utilities, search for files and other computers, get help, and add or remove hardware and software. Figure 6.2 shows the Start button.

Figure 6.2: The Windows 2000 Start Button

The **Shut Down Start button** option is used to shut down the computer, log off from the network, restart the computer, and possibly put the computer in standby or hibernate. The standby and hibernate options are available on computers that support power saving features. Standby is helpful on laptop computers to save on battery life. Hibernate is available on computers that support Windows 2000 power options and is similar to the shutdown option except that it can be scheduled. At a specific time, the computer is shut down and when the computer restarts, the active components on the desktop at the time of hibernation are still there.

The **Run Start button** option is used for starting an application or bringing up a command prompt window. The Help option is used for Windows 2000 general usage and troubleshooting assistance. The **Search Start button** option is used to locate files, remote network devices, web sites on the Internet, and people in the Windows address book. The Settings option is used to access various suboptions that allow computer configuration. The Documents selection contains the 15 most recently used files (provided the application supports this option). The Programs choice allows access to applications installed in the computer. The Windows Update option is used to connect to the Microsoft web site to access the most recent drivers and operating system updates.

The **Quick Launch bar** is a set of icons to the right of the Start button and allows you to launch applications with one click on a Quick Launch icon. An important icon on the Quick Launch bar is the **Show Desktop** icon. This icon looks like a desk with a pencil touching a piece of paper. Single-click on the Show Desktop icon to reduce all windows on the screen and display the desktop. If you click on the icon a second time, the original document (that was on the screen when you clicked the Show Desktop icon) reappears. Another Quick Launch icon is the Internet Explorer icon. Click once on this icon and the Internet Explorer application opens.

The **taskbar** is the bar that runs across the bottom of the screen. The taskbar holds buttons that represent applications or files currently loaded into memory. The taskbar also holds icons that allow access to system utilities. These utilities can include a clock icon for the date and time, an icon of a speaker for volume control, and an icon for a virus utility. Look back to Figure 6.1 to identify the taskbar.

One type of window used with Windows operating systems is a dialog box. A **dialog box** allows you to set application or operating system preferences. The most common features found in a dialog box are a check box, a text box, tabs, a drop-drown menu, a Help button, a Close button, an OK button, and a Cancel button.

A **text box** is an area where you can type a specific parameter. When you click inside a text box, a vertical line appears, which is the insertion point. Any typed text is placed to the right of the insertion point. Notice in Figure 6.3 how the Top, Bottom, Inside, Outside, etc. options are text boxes. Text boxes sometimes have up or down arrows that can be used to select a preset option, or you can simply type your own parameter.

Tabs normally appear across the top of a dialog box. Each tab normally holds a group of related options. Click once on the tab to bring that particular major section to the window forefront. The tabs in Figure 6.3 are Margins, Paper Size, Paper Source, and Layout.

The **Help button** is used to provide context-sensitive assistance and is the question mark located in the upper right corner of the dialog box. When you click on the Help button (the question mark), the cursor turns into an arrow with a question mark attached. Click on any item you want basic information on and a pop-up window appears on the screen. To close the pop-up window, click anywhere on the screen.

The **Close button**, which is located to the right of the Help button and is an "X," is used to close the dialog box. When the Close button is used, no changes that have been made in the dialog box are applied.

A **check box** is an option that, when checked, is enabled or turned on. In Figure 6.3, clicking inside the check box enables the Mirror Margins option. When an option is enabled, a check mark appears in the check box.

A **radio button** is a round circle, but it operates the same way a check box does. A radio button is enabled when a solid dot appears inside it. Click once on a blank radio button and a solid dot appears. Click once on a radio button that has a dot in it, and the dot disappears. The option is disabled when no dot appears.

Drop-down menus are presented when you click on a down arrow. In Figure 6.3, the drop-down menu appears when you click on the down arrow in the section marked Apply to. Once presented with the drop-down menu, click on the preferred option and that option appears in the drop-down window.

Figure 6.3: Windows 2000 Dialog Box Components

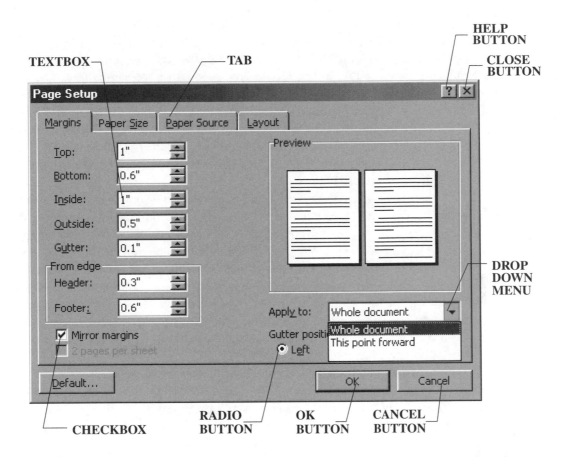

The **OK button** and the **Cancel button** are standard buttons in a dialog box. When you click on the OK button, all options selected or changed within the dialog box are applied. When you click on the Cancel button, all changed options in the dialog box are not applied; the options are left in their current state.

Another related button that can be found in a dialog box, but is not shown in Figure 6.3, is the Apply button. The **Apply button** is used to make changes immediately (before clicking on the OK button). This is useful when you want to see the results of your selection, such as when you are making changes to the desktop's background.

One of the most frequent mistakes users make is clicking on the Close button without first applying or clicking OK for changes. When a dialog box is closed with the Close button, no changes in the dialog box window are saved or applied.

6.2 Managing Files and Folders on Windows 2000

Users are always creating, deleting, and moving files and folders. It is important that you are able to do these tasks quickly and without error. The important thing to remember is to think about what file and folder you want to work with, where the files and folders are located now, and where you want the files or folders to be eventually.

A drive letter followed by a colon represents every drive in a computer. For example, the floppy drive is represented by A: and the first hard drive partition is represented by C:. The CD-ROM drive, or DVD drive, is also represented by a letter followed by a colon. Disks or drives hold files. A **file** is an electronic container that holds computer code or data. Another way of looking at it is thinking of a file as a box of bits. A file is kept on some type of media such as a floppy disk, hard drive, tape, or CD. Each file is given a name called a file name. An example of a file name is 2000CHAP.DOC.

Files are kept in folders. A **folder** holds files and can also contain other folders. In folder operating systems, a folder was called a directory. Every file and folder is given a name. It is easier to understand file and folder names if we look at how older operating systems named files and folders and then look at how Windows 2000 names differ. An **extension** is an addition to the file name and it can be up to three characters in length. The file name and the extension are separated by a period. An example of a file name with an extension is BOOK.DOC where BOOK is the name of the file and DOC is the extension.

Normally with Windows 95/98/NT/2000, the application automatically adds an extension to the end of the file name. In most windows, Windows 2000 does not automatically show the extensions.

In this next exercise, you will learn view the extensions of files. Be default, Windows 2000 does not display the extension of files.

1. Right-click **Start** and then click **Explore**. Windows Explorer opens.
2. To view the extensions in Windows NT Explorer, click on the **View** menu option.
3. Click on the **Folder Options** menu option. Click on the **View** tab and click on the
4. **Hide file extensions for known file types** check box (which will remove the check from the box).
5. Click on the **Apply** button. Windows Explorer displays the extension for all files.
6. Close Explorer.

Windows 2000 Files

File names in Windows 2000 Professional can be up to 255 characters. These extended file names are commonly called long file names. Folders and file names can have all characters, numbers, letters, and spaces _except_ the following:

/ (forward slash) " (quotation marks) \ (backslash)
| (vertical bar) ? (question mark) : (colon)
* (asterisk)

An example of a long file name is WINDOWS 2000 CHAPTER.DOC. Any time a document has been saved with one of these long file names and is taken to a computer with an older operating system that does not support long file names, the file name is shortened to a maximum of eight characters. Windows does this by using the first six characters of the file name, deleting any spaces, and using two special characters—a tilde (~) and a number. For example, 2000 PROFESSIONAL CHAPTER.DOC would be shortened to 2000PR~1.DOC. If there were two files named 2000 PROFESSIONAL CHAPTER.DOC and 2000 PROFESSIONAL INDEX.DOC, the two files would be saved as 2000PR~1.DOC and 2000PR~2.DOC, respectively.

When a file is saved in a Windows application, it automatically goes to a specific folder. This is known as the default folder. With Windows NT and 2000, this folder is the **My Documents** folder. In documentation, installation instructions, and when writing down the exact location of a file, the full path should be used. A file's **path** is like a road map to the file and includes the drive letter plus all folders and subfolders as well as the file name and extension. For example, if the CHAP1.DOC file is in the MY DOCU-MENTS folder on the first hard drive partition, the full path is C:\MY DOCU-MENTS\CHAP1.DOC. The first part is the drive letter where the document is stored, C:. The C: represents the first hard drive partition. The name of the document is always at the very end of the path. In the example given, CHAP1.DOC is the name of the file. Everything in between the drive letter and the file name is the name of one or more folders where the CHAP1.DOC file is located. The folder in this example is the MY DOCU-MENTS folder.

If the CHAP1.DOC file is located in a subfolder called COMPUTER BOOK, which is located in the folder called MY DOCUMENTS, then the full path is C:\MY DOCU-MENTS\COMPUTER BOOK\CHAP1.DOC. Notice how the backslashes in the path are always used to separate the folder names as well as separate the drive letter from the first folder name.

Explorer is the most common application used to create, copy, and move files or folders; however, the My Computer window can be used in a similar fashion. When you are copying a file or folder, use the Copy/Paste functions. When you are moving a file or folder, use the Cut/Paste functions.

Let's go through an exercise to help you understand how to manipulate files and folders. It is assumed that Windows 2000 is installed and running on a computer with the Windows 2000 desktop displayed on your screen.

1. Insert your floppy in the floppy drive.
2. Right-click **Start** and then click **Explore**. Windows Explorer opens with left and right windowpanes.
3. Click the icon associated with the floppy.
4. Click **File**, click **New,** and then click **Folder**. The name of the folder appears as **New Folder**. Type **My Folder** and press Enter.
5. In the left windowpane, expand the floppy by clicking the plus sign to the right of it. If you see a minus sign to the right, it already is expanded.
6. Click **My Folder**.

7. To create a folder within **My Folder**, click **File**, click **New,** and then click **Folder**. The name of the folder appears as **New Folder**. Enter your own name for the name of the folder and press Enter.

8. To create a file within **My Folder**, click **File**, click **New,** and then click **New Text Document**. The name of the text document appears as **New Text Document**. Press Enter to keep this name.

9. To rename the new text document, right-click the document and click **Rename**. The document name is highlighted. Type **My Document** and press Enter. Note if the extension (.txt) appears, then the folder is set to show extensions. In this case, leave it otherwise you will receive a message that changing the file extension may cause the file to become unusable.

10. To delete the text document, right-click the document (note its name has changed) and click **Delete**. A confirmation message will appear.

11. Click **Yes** to delete. The file is deleted.

12. Close Windows Explorer.

When you delete a file or folder from a floppy disk, the file or folder is permanently deleted. When deleting a file or folder from a hard drive, the file or folder is automatically sent to the Recycle Bin. The contents of the Recycle Bin take up hard drive space. Many users do not realize that they are not really deleting the file, but simply moving it to the Recycle Bin. To delete a file permanently from the hard drive, hold down the **Shift** key while pressing the **Delete** key on the keyboard and the file is permanently removed. Otherwise, you will have to remember to empty the Recycle Bin periodically. Figure 6.4 shows how the A+ COMPLETE BOOK.DOC long file name looks in graphical form using the Explorer application.

Figure 6.4: Windows 2000 Explorer

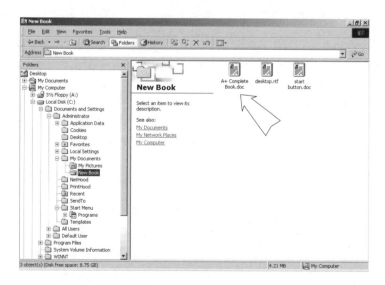

File Attributes

My Computer and Explorer can be used for setting attributes for a file or folder. The file and folder attributes are read-only, hidden, archive, and system. The **read-only attribute** marks a file or folder so that it cannot be changed. The **hidden attribute** marks a file or folder so that it is not visible through My Computer or Explorer unless someone changes the default view for the window. Some applications use the **archive attribute** to control which files or folders are backed up. The **system attribute** is placed on certain files used to boot Windows 2000 Professional.

Let's perform an exercise to help you understand a file's attributes. It is assumed that Windows 2000 is installed and running on a computer with the Windows 2000 desktop displayed on your screen.

1. Insert your floppy in the floppy drive.
2. Right-click **Start** and then click **Explore**. Windows Explorer opens with left and right windowpanes.
3. Click the icon associated with the floppy.
4. To create a file, click **File**, click **New,** and then click **New Text Document**. The name of the text document appears as **New Text Document**. Press Enter to keep this name.
5. To set the file's attributes, right-click the new text document and click **Properties**. The Properties window for the file opens.
6. In the **Attributes** section of the page, check **Read-only** and click **OK**. The Properties window closes.
7. Next, you will attempt to modify the file. Remember it is set to read-only. Double-click the document. The document opens.
8. On the document screen, type your name and press Enter. This is considered text.
9. Attempt to save the document. Click **File** then click **Save**. The **Save As** window opens. Normally, this window does not appear—the file is simply saved. However, because the file has the read-only attribute, you cannot save the text to this file. Windows 2000 wants you to enter a new file name.
10. To see the error message, click **Save**. A message is displayed indicating you cannot save this file.
11. Click **OK** and then click **Cancel**. The document screen remains.
12. Close the document. Do not save changes if prompted.
13. Close Windows Explorer.

Refer to Figure 6.5 for a sample screenshot of a file's attributes.

Figure 6.5: The Attributes of the File named "Todd Meaders.doc"

Determining the Windows 2000 Version

The operating system version is very important to you. With Windows 95, 98, NT, and 2000, upgrades or patches to the operating system are provided through **service packs**, which fix problems within the operating system. You must determine what operating system version is on the computer so that he or she can research whether or not a service pack is needed. Several ways to determine what version of 2000 is loaded on a computer are listed below.

• Right-click on the **Start** button. Click on the **Explore** option. Windows Explorer opens. Click on the **Help** menu option. Click on the **About Windows** option from the drop-down menu.

• Click on the **Start** button. Click on the **Run** option. In the Open text box, type **winver** and press Enter. A window appears with the version.

• Click on the **Start** button. Point to the **Programs** option. Point to the **Administrative Tools** option. Click on the **Computer Management** option. Right-click on the **Computer Management (Local)** option and then click on the **Properties** option.

• Click on the **Start** button. Click on the **Run** option. In the Open text box, type **winmsd** and then press Enter.

Refer to Figure 6.6 for a sample screenshot of the WINVER program.

Figure 6.6: The WINVER Program Displaying the Windows Version

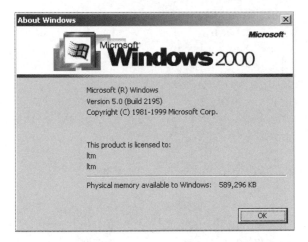

Refer to Figure 6.7 for a sample screenshot of the WINMSD program.

Figure 6.7: The WINMSD Program Displaying the Windows Version and Other Information

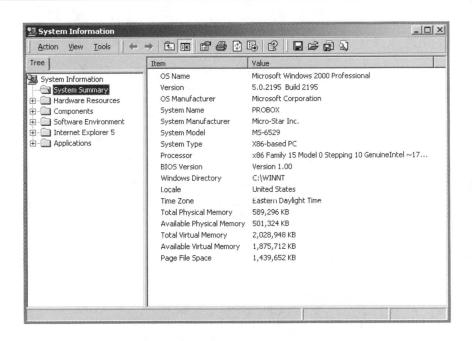

6.3 Understanding Compression and Encryption on Windows 2000 Professional

If the hard drive is partitioned for the NTFS file system, files and folders can be compressed or encrypted with Windows 2000. **Compression** is where a file or folder is compacted to take up less disk space. However, with compression enabled, the computer's performance can degrade. This is because, in order to open a compressed file, that file must be uncompressed, copied, and then recompressed. Degradation can also occur if a compressed file is transferred across a network because the file must be uncompressed before it is transferred. You can enable file compression by using Windows Explorer.

In this next exercise, you will learn how to compress a file on a computer running Windows 2000.

1. Right-click **Start** and then click **Explore**. Windows Explorer opens with left and right windowpanes.
2. Click the local disk (C:) icon. Note that this partition must be NTFS.
3. To create a file, click **File**, click **New,** and then click **New Text Document**. The name of the text document appears as **New Text Document**. Press Enter to keep this name.
4. Right-click the document, and click **Properties**. The Properties window opens.
5. Click **Advanced**. The **Advanced Attributes** window opens.
6. Click **Compress contents to save disk space**.
7. Click **OK** twice to close the windows.
8. Notice the attribute is now set to compressed.
9. Close all windows.

Refer to Figure 6.8 for a sample screenshot of the Compress check box. It is located in the lower half of the screen.

Figure 6.8: Compressing a File

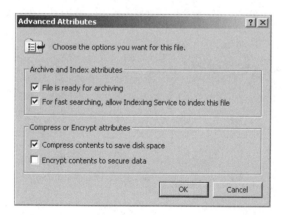

When working with compressed files and folders, users like them to appear in a different color than other files and folders. The color will be blue. To set this option, click on the **View** Windows Explorer menu option. Click on the **Options** selection. Click in the **Display compressed files and folders with alternate color** radio button. Click on the **OK** button. Refer to Figure 6.9 for a sample screenshot. Note this is the first check box selected.

Figure 6.9: Screenshot of Folder Options

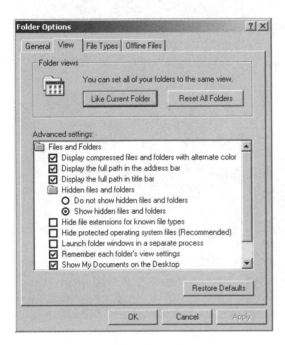

Files and folders can also be compressed or uncompressed from the command line using the **COMPACT** command. The COMPACT command can also be used to view the compression state of folders. The COMPACT command automatically compresses or uncompresses all the files and subfolders when you change a folder's compression state. Refer to Figure 6.10 for the information on the COMPACT command.

Figure 6.10: The COMPACT Command

```
C:\WINNT\System32\cmd.exe                                        _ |&| X|

C:\>
C:\>compact /?
Displays or alters the compression of files on NTFS partitions.

COMPACT [/C | /U] [/S[:dir]] [/A] [/I] [/F] [/Q] [filename [...]]

    /C          Compresses the specified files.  Directories will be marked
                so that files added afterward will be compressed.
    /U          Uncompresses the specified files.  Directories will be marked
                so that files added afterward will not be compressed.
    /S          Performs the specified operation on files in the given
                directory and all subdirectories.  Default "dir" is the
                current directory.
    /A          Displays files with the hidden or system attributes.  These
                files are omitted by default.
    /I          Continues performing the specified operation even after errors
                have occurred.  By default, COMPACT stops when an error is
                encountered.
    /F          Forces the compress operation on all specified files, even
                those which are already compressed.  Already-compressed files
                are skipped by default.
    /Q          Reports only the most essential information.
    filename    Specifies a pattern, file, or directory.

Used without parameters, COMPACT displays the compression state of
the current directory and any files it contains. You may use multiple
filenames and wildcards.  You must put spaces between multiple
parameters.

C:\>
```

 You can compress a file or folder only on an NTFS partition.

Encryption is a method of securing data from unauthorized users. Windows 2000 has a new encryption feature called **EFS (Encrypting File System)**. When a file or folder is encrypted with EFS, only the authorized user can view or change the file. Administrators have the ability to recover encrypted files if necessary. EFS is not compatible with any prior version of Windows.

In this next exercise, you will learn how to compress a file on a computer running Windows 2000. This exercise assumes you performed the previous exercise on compression because that is where you created the file used in this next exercise.

1. Right-click **Start** and then click **Explore**. Windows Explorer opens with left and right windowpanes.
2. Click the local disk (C:) icon.

3. Right-click the **New Text Document** document you created in the previous exercise, and click **Properties**. The Properties window opens.
4. Click **Advanced**. The **Advanced Attributes** window opens.
5. Click **Encrypt contents to secure data**. Notice the check box that was selected for compression is removed.
6. Click **OK** twice to close the windows.
7. Notice the attribute is now set to compressed.
8. Close all windows.

Refer to Figure 6.11 for a sample screenshot.

Figure 6.11: Encrypting a File

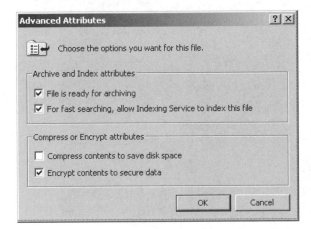

 You can encrypt a file or folder only on an NTFS partition.

The **CIPHER** command line utility can be used to encrypt and decrypt files. The command can also be used to encrypt and decrypt folders. If you type **CIPHER** from the command prompt without any switches, current file/folder encryption states are displayed on the screen. Type **CIPHER /?** to view the various switch options available with this command. Refer to Figure 6.12 for a sample screenshot of this command.

Figure 6.12: The CIPHER Command

```
C:\WINNT\System32\cmd.exe                                            _|_|X|
Microsoft Windows 2000 [Version 5.00.2195]
(C) Copyright 1985-1999 Microsoft Corp.

C:\>cipher /?
Displays or alters the encryption of directories [files] on NTFS partitions.

  CIPHER [/E | /D] [/S:dir] [/A] [/I] [/F] [/Q] [/H] [/K] [pathname [...]]

    /E        Encrypts the specified directories. Directories will be marked
              so that files added afterward will be encrypted.
    /D        Decrypts the specified directories. Directories will be marked
              so that files added afterward will not be encrypted.
    /S        Performs the specified operation on directories in the given
              directory and all subdirectories.
    /A        Operation for files as well as directories. The encrypted file
              could become decrypted when it is modified if the parent directory

              is not encrypted. It is recommended that you encrypt the file and
              the parent directory.
    /I        Continues performing the specified operation even after errors
              have occurred.  By default, CIPHER stops when an error is
              encountered.
    /F        Forces the encryption operation on all specified objects, even
              those which are already encrypted.  Already-encrypted objects
              are skipped by default.
    /Q        Reports only the most essential information.
    /H        Displays files with the hidden or system attributes.  These
              files are omitted by default.
    /K        Create new file encryption key for the user running CIPHER. If thi
s
              option is chosen, all the other options will be ignored.
    pathname  Specifies a pattern, file or directory.

    Used without parameters, CIPHER displays the encryption state of
    the current directory and any files it contains. You may use multiple
    directory names and wildcards.  You must put spaces between multiple
    parameters.

C:\>
```

The compression and encryption attributes cannot both be checked. In other words, you cannot compress and encrypt a folder or file. You can either compress or encrypt, but not both.

6.4 Understanding the Windows 2000 Registry

Every software and hardware configuration is stored in a database called the **registry**. The registry contains such things as folder and file property settings, port configuration, application preferences, and user profiles. A **user profile** contains a user's specific configuration settings such as what applications the user has access to, desktop settings, and the user's network configuration. The registry loads into memory during the boot process. Once in memory, the registry is updated continuously through changes made to software and hardware. The registry is divided into five subtrees. **Subtrees** are sometimes called branches or hives. The five standard subtrees are as follows:

HKEY_LOCAL _MACHINE, HKEY_USERS, HKEY_CURRENT_USER, HKEY_ CURRENT _CONFIG, and HKEY_CLASSES_ROOT. Each of these subtrees has keys and subkeys that contain values related to hardware and software settings.

Table 6.1 lists the five subtrees and each function.

Table 6.1: Windows 2000 Registry Subtrees

Registry Subtree	Function
HKEY_LOCAL_MACHINE	Holds global hardware configuration. Included in the branch is a list of hardware components installed in the computer, the software drivers that handle each component, and the settings for each device. This information is not user-specific.
HKEY_USERS	Keeps track of individual users and their preferences.
HKEY_CURRENT_USER	Holds a specific user's configuration such as software settings, how the desktop appears, and what folders the user has created.
HKEY_CURRENT_CONFIG	Holds information about the hardware profile that is used when the computer first boots.
HKEY_CLASSES_ROOT	Holds file associations and file links. The information held here is what allows the correct application to start when you double-click on a file name in Explorer or My Computer (provided the file extension is registered).

The registry can contain other subtrees that are user-defined or system-defined depending on what hardware and software is installed in the computer.

Most changes to Windows 2000 are made through the various control panels, but sometimes the only way to make a change is to edit the registry directly. However, be careful, because if youmake an incorrect registry change, you could cause your system to become inoperable.

Windows 2000 has two registry editors called **REGEDIT** and **REGEDT32**. Both registry editors can be used to change the registry; however, there are some differences between the two. Table 6.2 shows some of the more important differences.

Table 6.2: REGEDIT and REGEDT32 Differences

REGEDIT	REGEDT32
Provides more powerful search capabilities.	Can display and edit values larger than 256 characters.
All of the subtrees are shown in one window.	The subtrees are shown in individual windows.
Allows exporting the registry to a text file.	Can look at and apply access permissions to the subtrees, keys, and subkeys.
Allows importing the registry from the command line.	Can work with multiple registry files simultaneously.

With the REGEDIT program, subtrees are listed in the left window. Figure 6.13 shows the REGEDIT utility.

Figure 6.13: Windows 2000 REGEDIT Window

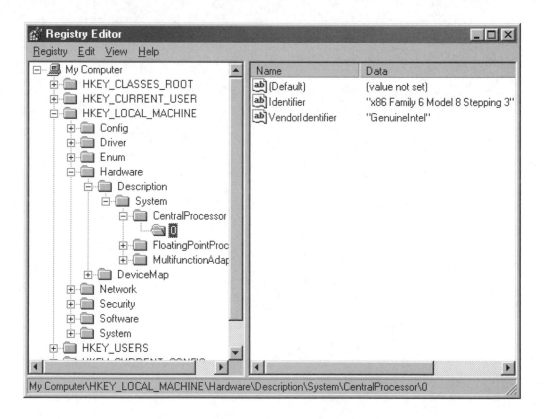

Notice in Figure 6.13 how the subtrees show up in the left window. When you click on the + (plus) symbol beside each subtree, more subkeys appear. After several layers, when you click on a folder in the left window, values appear in the right window. These values are the ones you must sometimes change to fix a problem.

In REGEDT32, each registry hive appears in a separate window inside the Registry Editor window. Each subtree has individual folders with a + (plus) symbol beside them. Click on the plus symbol to view subkeys. Values appear in the right window as they do in the REGEDIT program. Each hive, or even part of a subtree, can be backed up individually.

Backing Up and Restoring the Windows 2000 Registry

The registry should be backed up whenever the computer is fully functional and when any software or hardware changes are made. The registry can be backed up and restored several different ways. The two most common methods used are the REGEDIT and the Backup utilities. The REGEDIT program allows you to export the registry to a file that has the .REG extension. The backed up file can be modified with a text editor if necessary. The file can also be imported back into the computer.

The registry should be backed up and restored on a working computer *before* disaster hits. The time to learn how to restore the registry is *not* when the computer is down.

The Backup utility is accessed through Start button, Programs, Accessories, System Tools, Backup. The Backup utility is the preferred method for backing up the Windows 2000 registry. The Backup option to look for is the System State, which is discussed in the next section. Refer to Figure 6.14 for a sample screenshot of backing up the System State. In the left-hand pane, notice that "System State" is checked (in blue).

Figure 6.14: Screenshot of Backing Up the System State

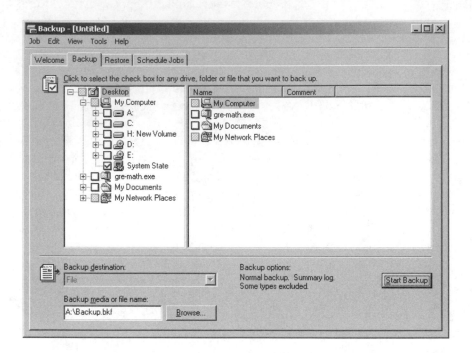

One option available in the Backup utility is the System State. The **System State** is a group of important Windows 2000 files including the registry, the system files, the boot files, and the COM+ Class Registration database. With the Backup utility, you cannot back up or restore these items individually. They are all needed because they depend on one another to operate properly. The Backup utility is accessed through Start button, Programs, Accessories, System Tools, Backup. Click on the **Backup** tab. Click in the box next to **System State**. Select the destination and click on the **Start Backup** button. Once you click on the Start Backup button, the Backup Job Information dialog box appears. The **Advanced** option has a setting: **Automatically backup system protected files with the System State**. This option, when selected, backs up all system files in the *%systemroot%* folder (which is normally C:\WINNT). Click on the **OK** button to start the backup. Refer to Figure 6.15 for a screenshot of selecting the automatic backing up of system protected files.

Figure 6.15: Automatically Back Up System Protected Files with System State

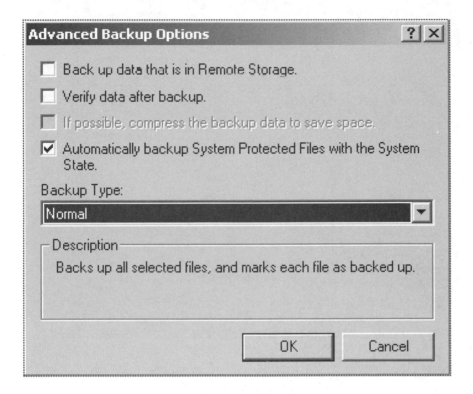

The registry files are located in a folder normally labeled *%systemroot%*\Repair\Reg-backup (which is normally C:\WINNT\REPAIR\REGBACKUP). The registry can be restored without having to restore the other System State files. In order to use the Backup program, you must be an administrator or a member of the Backup Operators group.

Restoring the System State After a Failure
In order to correct a problem with the system files, registry, or Windows 2000 boot failure, you must restore the registry. You may also have to restore the System State files (which include the registry) to make the system operational again. To start the restoration process, install a copy of Windows 2000 to the same folder in which it was installed originally. When you are prompted to format the hard drive volume or leave it, select the **Leave the current file system intact** option. Use the Backup utility to restore the System State and/or the registry, (Programs, Accessories, System Tools, Backup). Click on the **Restore** tab and select the device that holds the backed up files.

Refer to Figure 6.16 for restoring the System State. Notice it is selected in the right-hand pane.

Figure 6.16: Restoring the System State Data

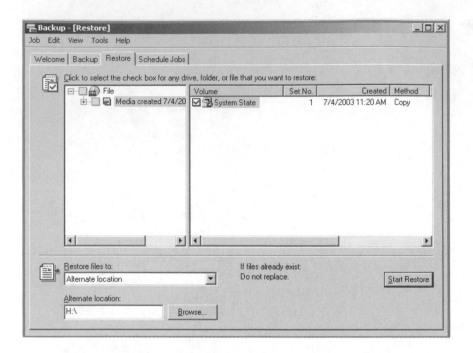

6.5 Understanding the Pre-Installation Steps of Windows 2000

Windows 2000 Professional can be installed from a central location or locally. Because this book focuses on the A+ Certification exam, only local installation is covered. The pre-installation of any operating system is more important than the installation process. The steps to be taken before installing Windows 2000 Professional are outlined below.

1. Decide whether the installation will be an upgrade or a clean install.
2. Determine the file system(s) to be used.
3. Decide whether the computer will have more than one OS installed.
4. Scan for viruses.
5. Determine if the hardware is compatible.
6. Obtain any drivers, upgrades, or hardware replacements.
7. Decide if the software applications are compatible.
8. Obtain any patches, upgrades, or software replacements.
9. Back up any data files necessary.
10. Remove any power management or disk management tools.

The first decision to make when planning to install Windows 2000 Professional is whether to upgrade from another operating system or perform a clean install. A **clean install** puts an operating system on a computer without an operating system, or reformats the hard drive so that the computer's existing operating system is removed.

If the decision is to do an upgrade, then determine which operating system is already installed. Windows 2000 Professional supports upgrading from Windows 95, Windows 98, NT Workstation 3.51, and NT Workstation 4. When Windows 2000 Professional is installed as an upgrade, the user's applications and data are preserved if the operating system is installed in the same folder (directory) as the original operating system. If Windows 2000 Professional is installed in a different folder, then all applications must be reloaded.

Another decision you must make if upgrading to Windows 2000 Professional is whether or not to convert the hard drive partition to NTFS. Once a partition is converted to NTFS, the partition cannot be changed back. If you are unsure whether or not to convert the partition, leave it unchanged and later use the CONVERT.EXE program to upgrade. Most people want to convert the partition to NTFS for the following reasons:
• Security (individual files can be protected with NTFS).
• More efficient use of cluster space (the cluster size can be defined based on the user's needs with NTFS).
• NTFS supports file compression.
• NTFS supports larger hard drive partition sizes.

Actually there are two versions of NTFS: NTFS4 and NTFS5. Windows 2000 supports and uses NTFS5 and NT Workstation supports and uses NTFS4. In NTFS5, a new type of encryption called Encryption File System (EFS) is supported. Additionally, disk quotas can be set to monitor and limit user disk space. If upgrading from a prior version of NTFS, the drive is automatically configured for NTFS5. If any NTFS4 volumes are not powered during the installation process, the volume is automatically upgraded when the drive is mounted. If you want to dual-boot between Windows NT 4 and Windows 2000 Professional (have both operating systems loaded), make sure that NT Workstation Service Pack 4 or higher is installed because some of the features in NTFS5 change the data structure on disks.

In order to take advantage of Windows 2000 Professional's reliability, enhancements, and security features, sometimes a clean installation is the best choice, especially if the current operating system is Windows 95 or Windows 98. Because a clean installation involves formatting the hard drive, the user's data must be backed up and all applications reinstalled once the Windows 2000 Professional installation is complete. Also, all user-defined settings are lost. Another important point to remember is that not all Windows 3.x, 95, and 98 applications are compatible with Windows 2000 Professional.

If the computer already has NT Workstation 3.5 or 4, then a Windows 2000 Professional upgrade is recommended. However, if there are hardware drivers for such devices as a DVD player, power management software, or network utilities loaded on the computer, a clean installation may be a better choice. Whichever is the case, the user's data

Microsoft has a web site that addresses application compatibility and you should check at this site before making your decision: http://windows.microsoft.com/windows2000/reskit/webresources. You can also contact the company that developed your application and see if the application is compatible with Windows 2000.

and applications should be backed up and restored once the Windows 2000 Professional installation is complete.

The third decision that must be made is whether or not Windows 2000 Professional will be installed with one or more other operating systems. This situation is often called a dual-boot scenario. **Dual-boot** means that the computer can boot from more than one operating system. Windows 2000 Professional can be dual-booted with DOS, Windows 3.1 or higher, Windows 95, Windows 98, and Windows NT Workstation. If this is desired, a separate hard disk partition should be created and used for each operating system. When doing a dual- or multi-boot configuration, make sure that Windows 2000 Professional is installed *after* the other operating systems. Multi-booting is beyond the scope of this chapter. See the Microsoft web site for more details.

The fourth step is to scan the system for viruses. Viruses were covered in Chapter 4, *Introduction to Windows 98*.

The fifth thing to do when installing Windows 2000 Professional is to determine what computer hardware is installed. Table 6.3 lists the minimum and preferred hardware requirements for installing Windows 2000 Professional.

Table 6.3: Windows 2000 Professional Hardware Requirements

Component	Minimum	Preferred
CPU	Intel Pentium (or compatible) 33 MHz	Intel Pentium II (or compatible) 300 MHz or higher
RAM	32 MB	64 MB
Free hard drive space	650 MB	2 GB
Input Device	Keyboard, mouse, or other pointing device	Keyboard, mouse, or other pointing device
Multimedia Drive	CD-ROM or DVD	CD-ROM or DVD 12x or faster

The sixth decision relates to obtaining drives, upgrades, or hardware replacements. Once you have verified all of your hardware, you may have to get Windows 2000 device drivers from the hardware device manufacturer or its web site. You may also need to upgrade the hardware device, which usually means replacing it. This is sometimes the

Microsoft has a tool called **Readiness Analyzer** that checks your system for hardware and software compatibility issues. This tool can be downloaded from Microsoft's web site at http://www.microsoft.com/windows2000/upgrade/compat/default.asp. Be aware that the Readiness Analyzer might not be able to detect all hardware devices or software applications.

cost of going to a bigger and better operating system. You may also decide at this point not to upgrade, but to buy a computer with Windows 2000 already installed.

The seventh determination you must make before installing Windows 2000 Professional is whether or not any existing software applications are compatible. The preparation for installing a new operating system is usually more work than the actual installation process, but any omitted step will cost you more time in the long run. Use the Readiness Analyzer or contact the developer of each software application to determine if it is compatible. You may also go to the software developer's web site. The information may be posted there. A list of compatible software is also listed on Microsoft's web site.

The eighth decision is as follows: Once you have determined whether the software is compatible with Windows 2000, you may have to obtain software patches, upgrades, or buy a new version. This is best done before you install Windows 2000. Be proactive, not reactive—solve any problems you can _before_ upgrading or installing any operating system.

The ninth decision deals with backing up your data. As with any upgrade, hardware change, or software change, data must be backed up. It is almost funny that the worst people in the world for backing up data are operating system specialists—the very ones who are entrusted with the clients' data and computer. Since Windows 2000 is really designed for the corporate/small business environment, backing up data is an essential step. Whether you do a clean install or an upgrade, if the user has data on the computer, it must be backed up before starting the installation process. Also, before backing up data, remove any unwanted files and/or applications that are no longer needed in order to free up hard drive space.

The last step in the pre-installation checklist is to remove any power or disk management tools loaded on your computer. Computer manufacturers for older operating systems frequently provide these types of tools. Power or disk management tools can interfere with the new tools provided with Windows 2000. Disable these utilities and applications before attempting an operating system installation or upgrade. One important note about disk drives is that you cannot install Windows 2000 on a compressed hard drive partition. Uncompress the partition before starting the Windows 2000 installation process.

6.6 Understanding How to Install and Upgrade Windows 2000 Professional

After all the pre-installation checklist steps are completed, you are ready to install Windows 2000 Professional. The installation process is easy if you performed the pre-installation steps. An exercise at the end of the chapter guides you through both a clean installation (one where no other operating system is on the machine) and an upgrade to Windows 2000. The number one piece of advice you need to follow when installing any operating system is this: Do your homework first. The number of possible problems will be greatly reduced.

There are two major portions of the installation process the text mode (otherwise known as the DOS mode) and the GUI mode (also known as the Windows mode). In **text mode**, the monitor only shows characters with a blue background. The text mode portion of Setup checks for the proper minimum hardware requirements; detects plug and play devices and adapters; locates hard drives; creates the registry; partitions and formats the hard drive for the file system you select; copies most of the Windows 2000 installation files to begin the installation process; and restarts the computer to begin the Windows mode. During the **GUI mode** portion of Setup, hardware devices are detected, installed, and configured; the Setup logs are created; the operating system starts; and you are allowed to create an Administrator password.

6.7 Troubleshooting the Windows 2000 Professional Installation

Various problems can cause the Windows 2000 installation process to halt. There are two major places the installation stops—during the text mode portion of Setup or during the GUI mode portion of Setup. If the computer halts during text mode and displays a text message, there is probably an incompatible piece of hardware installed in the computer. If the computer halts during the GUI portion of Setup, restart the computer. The installation process attempts to continue from the place it left off. Incompatible hardware devices normally cause this. You can troubleshoot the device once Windows is installed.

An incompatible hardware device is the most common problem during the installation process because it can cause both the text and GUI modes to halt. Installation problems can be caused by a number of factors. The following list shows the most common causes and their associated solution during the installation process.

- **Incompatible BIOS**—Obtain compatible BIOS, replace motherboard with one that has a compatible BIOS, or do not upgrade/install Windows 2000 Professional.
- **BIOS needs to be upgraded**—Upgrade the BIOS.
- **Incompatible hardware**—Replace the hardware or do not upgrade/install Windows 2000 Professional.
- **Incompatible drivers**—Obtain Windows 2000 drivers from the hardware manufacturer.
- **Existing drivers are incompatible or halt the installation/upgrade process** Obtain Windows 2000 drivers from the hardware manufacturer.
- **Incompatible TSRs**—Remove TSRs or obtain updated ones from the software manufacturer.
- **Incompatible applications**—Obtain upgrades from software manufacturer.
- **Minimum hardware requirements have not been met**—Upgrade the hardware. The primary things to check are the CPU (133 MHz minimum) and RAM (32 MB minimum).

- **A virus is on the hard drive**—Run a virus-checking program and remove the virus. One of the tools that comes on the 2000 Professional CD is the InoculateIT anti-virus program (sometimes called the AV boot disk). To make a bootable disk that has the anti-virus program on it, insert the 2000 Professional CD into the CD-ROM drive and a 1.4 MB disk in the floppy drive. Click on the **Start** button and then click on the **Run** option. Click on the **Browse** button; locate and double-click on the drive letter that represents the CD-ROM. Locate the VALUEADD folder, the 3RDPARTY subfolder, and the CA_ANTIV subfolder. Double-click on the CA_ANTIV subfolder. In the right window, double-click on the **MAKEDISK.BAT** file. Click on the **OK** button. To use the disk and run the anti-virus program, make sure the BIOS is set to boot from the floppy drive. Insert the disk into the floppy drive and boot the computer. When the menu appears, press **1** and follow the directions on the screen. Refer to Figure 6.17 for a sample screenshot.

Figure 6.17: Screenshot of the MAKEDISK.BAT Anti-Virus Program

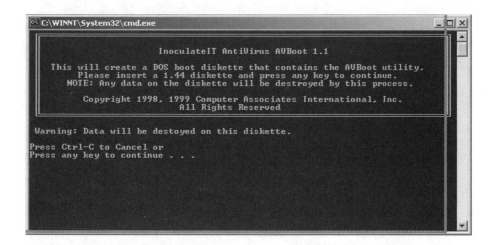

- **Pre-installation steps have not been completed**—Go back through the list!
- **The installation floppy disks or CD are corrupted** (not as likely as the other causes)—Try the disk in another machine and see if you can see the contents. For the CD, check to see if any scratches or dirt are on the surface. Clean the CD as necessary.
- **Incorrect CD key**—Type in the correct CD key to complete the 2000 installation. The key is located on the CD case.

Several text files can be helpful in determining the installation problem. WINNT.LOG and WINNT32.LOG are created during the installation process. SETUPLOG.TXT logs information detected during the text mode portion of Setup and includes device drivers that are copied. SETUPERR.LOG lists errors logged during the installation. SETU-PACT.LOG and SETUPAPI.LOG are located in the folder that contains most of the Windows 2000 files (normally C:\WINNT). SETUPACT.LOG displays information about the files copied during the installation and SETUPAPI.LOG contains information about device driver files copied during installation. The NBTLOG.TXT is used when the Enable Boot Logging boot option is chosen and lists the drivers loaded during the boot process. BOOTLOG.TXT is located in the root directory and lists boot-logging messages when booting from Safe Mode.

Different function keys can also help when troubleshooting the Windows 2000 installation process, as well as troubleshooting boot problems. Table 6.4 shows a list of keystrokes that can be used.

Table 6.4: Windows 2000 Professional Startup Keystrokes

Keystroke	Purpose
F5	System hangs at "Setup is inspecting your computer" screen. Select Standard PC from the list.
F6	Used when you need to go back and load third-party drivers.
F7	Loads the normal HAL instead of ACPI HAL.
F8	Brings up the Advanced Options menu.
Shift + F10	Displays a command prompt during the GUI mode portion of installation.

6.8 Understanding How to Dual-Boot Windows 2000 Professional and Windows NT

Sometimes users would like to try a new operating system, but keep their old operating system loaded as well. If this is the case, two operating systems can reside in one system and it is called **dual-booting**. Because Windows 2000 supports and uses NTFS5, and NT Workstation supports and uses NTFS4, it is best if the two operating systems are placed on different hard drive partitions. As soon as Windows 2000 is installed onto a partition formatted as NTFS, Windows 2000 upgrades the partition to NTFS5 without any prompting.

NT Workstation should be installed first. Service Pack 4 or later should also be installed. Install Windows 2000 to a different hard drive partition by inserting the Windows 2000 Professional CD. The CD normally starts automatically. A dialog box appears, asking if you want to install Windows 2000 Professional. Another dialog box appears asking if you want to install a new operating system or upgrade your old one. For a dual-boot situation, make sure you select that you want to install a new copy.

After installing all files and rebooting, a menu appears with the Microsoft Professional option, NT Workstation option, and NT Workstation (VGA Mode) option. To select which option is the default operating system, right-click on the **My Computer** desktop icon and click on the **Properties** option from the drop-down menu. Click on the **Advanced** tab and select the **Startup and Recovery** button. Select the default boot option. Refer to Figure 6.18 for a sample screenshot of selecting an operating system. In the **Default operating system** drop-down box, you'll notice in the figure there are several operating system choices. This means this system is dual-booted.

Figure 6.18: Selecting an Operating System in the Startup and Recovery Options

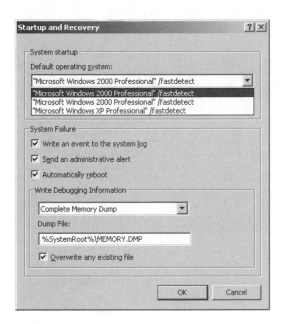

6.9 Configuring Windows 2000 Professional

One of the most common windows used by operating system specialists is the Control Panel window. A **control panel** is a method for configuring various Windows 2000 components. The Add/Remove Hardware, Add/Remove Programs, and Printing control panels are used when installing or removing hardware and software. Each control panel icon represents a Windows utility that allows you to customize a particular part of the Windows 2000 environment. The number of control panels displayed depends on the type of computer and the components contained within that computer. Figure 6.19 shows some of the more common Windows 2000 Professional control panels.

Figure 6.19: Windows 2000 Control Panels

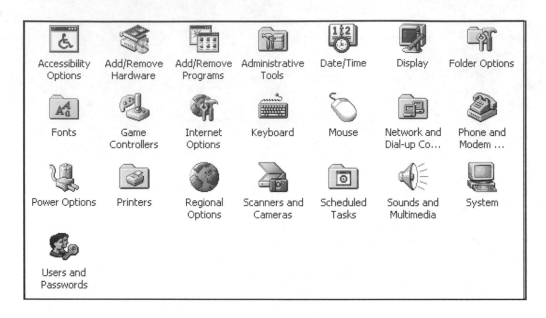

You must know which control panel to use for changing a computer's configuration. Table 6.5 shows a few of the common Windows 2000 control panels and the function of each.

Table 6.5: A Few Windows 2000 Control Panel Functions

Control Panel	Function
Devices	Used to start and stop device drivers; used to control how the device driver loads
Display	Used to install a monitor driver and configure monitor settings
Keyboard	Used to install a keyboard driver and configure keyboard
Network	Used to install NIC drivers, add network software, and configure network connections
Tape Devices	Used to add tape device driver and configure tape device parameters

Operating system specialists frequently must add new hardware and software using the operating system. Windows 2000 Professional has specific tools for these functions. Using the correct procedure is essential for success. Here are some tasks you are likely to perform:

- Adding plug and play devices
- Adding non-plug and play devices
- Adding a printer
- Removing hardware devices
- Installing and removing software

Hardware devices are physical components that connect to the computer. Hardware devices can be either plug and play or non-plug and play. A device driver is a piece of software that allows hardware to work with a specific operating system. Some device drivers are automatically included with Windows 2000 Professional. You must be aware of what hardware is installed into a system so that the latest 2000-compatible drivers can be downloaded and installed.

Adding Plug and Play Devices

Plug and play devices are hardware and software designed to automatically be recognized by the operating system. These include: USB devices, FireWire devices, SCSI devices, PC Card and CardBus devices, VL bus devices, PCI, ISA, and EISA devices, and printers. In order for Windows 2000 to fully support plug and play devices, the computer should have a BIOS that supports ACPI. Successful plug and play device installation involves the following:

- Possessing the most up-to-date device driver
- Following the directions provided by the device manufacturer

Always make sure that the computer is turned off when installing any new hardware component. Some plug and plug devices can be inserted into the computer without restarting. These include a PC card, a laptop into a docking station, and a USB device. However, devices such as internal modems or network cards require that the computer be turned off during installation.

Install the device according to the manufacturer's instructions. Once it is installed, power on the computer. The Windows 2000 Found New Hardware wizard appears. Windows 2000 attempts to find a driver for the new device. Plug and play devices make use of a special .CAB (cabinet) file called DRIVER.CAB, which is located in *%systemroot%*\DriverCache\i386 folder (where *%systemroot%* is normally C:\WINNT). This file is over 50 MB and contains more than 3,000 compressed files. If Windows 2000 detects new hardware, it will automatically search DRIVER.CAB for a driver. If a driver cannot be found, a dialog box appears. The best policy with any operating system is to use the latest driver, even if the operating system detects the device. An exercise at the end of this chapter outlines how to install a new hardware driver.

Remember that if the Windows 2000 Professional operating system cannot configure a plug and play device and prompts for a device driver, you must have administrator rights to install the driver.

Adding Non-Plug and Play Devices

Devices known as **legacy** devices are also called non-plug and play devices. For devices that are not plug and play, Windows 2000 has a tool (wizard) called Add/Remove Hardware. The Add/Remove Hardware wizard allows hardware configuration and is used for hardware that is not automatically recognized by Windows 2000 Professional. It is also used for plug and play devices that don't install properly with Windows 2000's automatic detection. You must have administrator privileges in order to load device drivers for new hardware. Refer to Figure 6.20 for a screenshot of adding/removing hardware.

Figure 6.20: The Add/Remove Hardware Screen

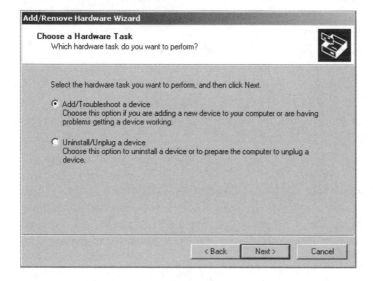

Adding a Printer

Printers can be connected to a computer through the printer's parallel port or through a local area network. Only local printers (printers connected to the computer's parallel port) will be covered in this chapter. Windows 2000 can automatically detect printers. If Windows 2000 detects the printer, the operating system automatically installs the drivers, updates the registry, and allocates system resources to the printer. Automatically detected printers are normally USB, or infrared printers.

To install a printer, connect the printer to the appropriate computer port with the appropriate cable. Power on the computer and the Windows 2000 wizard normally detects and leads you through the installation process. However, if it does not, have the Windows 2000 printer driver ready, click on the **Start** button, point to the **Settings** selection, and double-click on the **Printers** option. When the Printers window opens, click on the **Add Printer** icon. Click on the **Next** button and select the **Local Printer** selection. When prompted for the printer driver, insert the CD or disk that contains the printer driver and use the Browse button to locate the driver. Continue through the Add Printer wizard until the printer is installed.

To configure a printer as a default printer (the printer that applications normally use), locate the printer in the Printers folder. Access the Printers folder by clicking on the **Start** button, pointing to the **Settings** option, and clicking on the **Printers** selection. Once you locate the appropriate printer icon, right-click on the icon. Click on the **Set as Default Printer** option. In the Printers folder, the default printer has a check mark next to (above) the icon. Refer to Figure 6.21 for a screenshot of adding a printer using the Add Printer Wizard.

Figure 6.21: Adding a Printer Using the Add Printer Wizard

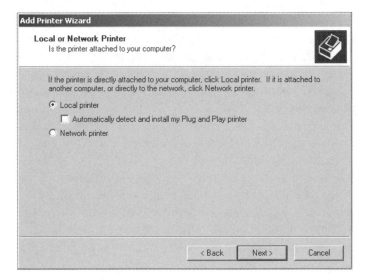

Removing Hardware Devices

When removing most hardware devices (all but printers), use the Add/Remove Hardware tool (wizard). To access this wizard, click on the **Start** button. Point to the **Settings** option and then click on the **Control Panel** option. Start the wizard by double-clicking on the **Add/Remove Hardware** icon located in the Control Panel window. Click on the **Uninstall/Unplug a device** radio button. Click on the **Next** button. Click on the **Uninstall a device** option and click on the **Next** button. Click the **Yes, I want to uninstall this device** option and click on the **Next** button. Click on the **Finish** button to complete the hardware removal.

If you are removing a printer from the system, use the Printers control panel. Access this control panel by clicking on the **Start** button. Point to the **Settings** option and click on the **Printers** option. Right-click on the **printer** you want to delete and choose the **Delete** option.

Be careful when deleting anything! One of the textbook authors was attempting to delete a print job associated with a printer and accidentally deleted the printer itself. Needless to say, it took a few minutes to reinstall the printer drivers.

Installing and Removing Software

No computer is functional without software. One thing you should know about Windows 2000 is that it does not support some of the older 16-bit software. Most software today is 32-bit and comes on CD that includes an Autorun feature. If there is no Autorun feature, an installation wizard steps you through installing the software when the CD is inserted into the drive. If there is not an Autorun feature on the CD or if the software is on a disk, then the Add/Remove Programs control panel is used to install or remove the software.

To access the Add/Remove Programs control panel, click on the **Start** button and point to the **Settings** option. Click on the **Control Panel** option and then double-click on the **Add/ Remove Programs** control panel icon. In the left panel in the window, click on the **Add New Programs** icon. Click on the appropriate **CD** or **Floppy** button depending on the type of media. Make sure the software disk or CD is inserted in the appropriate drive. If Add New Programs cannot find a SETUP.EXE file on the designated disk, it prompts with a dialog box. Use the **Browse** button to locate the installation file. Click on the **Finish** button to complete the process.

To remove a software application, use the same Add/Remove Programs control panel; however, instead of clicking on Add New Programs, click on the **Change or Remove Programs** icon in the left panel. A list of installed applications appears in the right panel. Locate the software to be removed and click on its **name**. Click on the **Change/Remove** button. When asked if you are sure you want to remove this software, click on the **OK** button and close the Control Panel window.

The Add/Remove Programs control panel can also be used to add programs from Microsoft, add programs across your network, and add/remove Windows components. The **Add Programs from Microsoft** icon automatically opens a web browser to the Microsoft web site. There you can locate, download, and install software upgrades, patches, and service releases. The **Add Programs from Your Network** icon is used to install software from a network share on another computer. The **Add/Remove Windows Components** icon is used to add or remove standard Windows applets, games, accessibility options, and communication components. Refer to Figure 6.22 for a screenshot of adding and removing programs.

Figure 6.22: The Add/Remove Programs Screen

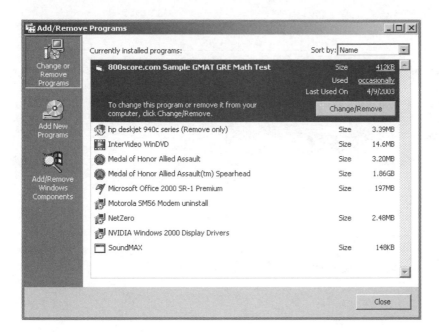

Once an application is installed, launch the application by clicking on the **Start** button and pointing to the **Programs** option. Locate the application name and click on it. If the application does not appear on the list, do not panic. Windows 2000 has a feature that only shows the most commonly used programs in the Programs list. If the application name does not appear, point to the double down arrows at the bottom of the **Programs** submenu. The less frequently used program names appear on the screen.

The Computer Management Console
The **Computer Management Console** is a large group of Windows 2000 tools displayed on one screen. The tools are called snap-ins and the 2000 Professional CD contains additional snap-ins that you can add to your system.

To add the Administrative tools to the Start button Programs option, right-click on an empty space on the taskbar. Click on the Properties option and click on the Advanced tab. Click in the Display Administrative Tools check box and click on the OK button.

The Computer Management Console allows you to manage shared folders, manage disk drives, start and stop services, look at performance logs and system alerts, and access Device Manager to troubleshoot hardware problems.

Figure 6.23 shows a partial listing of a Computer Management Console screen.

Figure 6.23: Computer Management Console

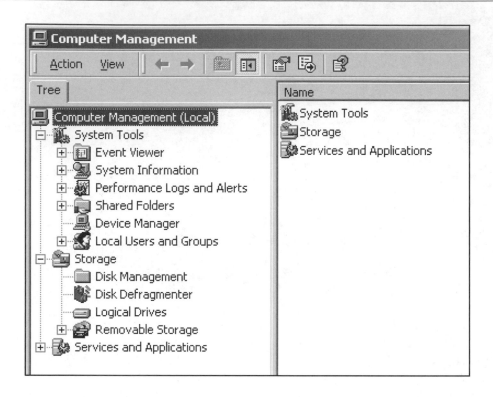

The three major tool categories found in the Computer Management Console include: System Tools, Storage, and Services and Applications. Let's perform an exercise with the Computer Management Console tool. Again, it is assumed that Windows 2000 is running on a computer.

1. Click **Start**, point to the **Settings** option and then click on the **Control Panel**. The Control Panel window opens.
2. Double-click on the **Administrative Tools** icon. The Administrative Tools window opens.
3. Double-click **Computer Management**. The Computer Management window opens.
4. To get an idea of what is in Computer Management, open a few of the tools and then close them.
5. When finished, close Computer Management.

6.10 Understanding the Windows 2000 Professional Boot Process

With NT Workstation and Windows 2000, two types of partitions are important during the boot process—the system partition and the boot partition. The **system partition** is the active drive partition that contains the files needed to load the operating system. The system partition is normally the C: drive (the active partition). The **boot partition** is the partition or logical drive where the Windows 2000 operating system files are located. People sometimes forget that the system and boot partitions can be on the same partition.

Every operating system needs specific files that allow the computer to boot. These files are known as **system files** or startup files. The system files and their specific location on the hard drive are listed in Table 6.6.

Table 6.6: Windows 2000 Boot Files

Startup File Name	File Location
BOOT.INI	Root directory of system partition
BOOTSECT.DOS (needed if dual- or multi-boot system)	Root directory of system partition
CDLDR	Root directory of system partition
HAL.DLL	*%systemroot%*\SYSTEM32 (*%systemroot%* is a variable representing the folder where Windows 2000 is installed. This is normally C:\WINNT.)
HYBERFIL.SYS	*%systemdrive%* (*%systemroot%* is the root directory on the drive where Windows 2000 boot files are located, which is normally C:\.)
NTBOOTDD.SYS (used with SCSI drives that have the SCSI BIOS disabled)	Root directory of system partition
NTDETECT.COM	Root directory of system partition
NTLDR	Root directory of system partition
NTOSKRNL.EXE	*%systemroot%*\SYSTEM32
System Key	*%systemroot%*\SYSTEM32\CONFIG

The table can be confusing because of all of the *%systemroot%* and *%systemdrive%* entries. This is because computers can be partitioned differently. If you install Windows 2000 onto a drive letter (a partition or logical drive) other than the active partition (normally C:), the startup files can be on two different drive letters. Also, you do not have to take the default folder name of WINNT to install Windows 2000. To account for these different scenarios, Microsoft uses the *%systemroot%* to represent the boot partition, the partition and folder that contains the majority of the Windows 2000 files. On a computer with a single operating system, this would be C:\WINNT. The *%systemdrive%* represents the root directory of the same drive letter. On a computer with a single operating system, this would be C:\.

If Windows 2000 is installed onto the C: drive and the C: drive is the active partition, then the BOOT.INI, BOOTSECT.DOS, HYBERFIL.SYS, NTBOOTDD.SYS, NTDE-TECT.COM, and NTLDR files would all be in the root directory of C:. The HAL.DLL and NTOSKRNL.EXE files would be located in the SYSTEM32 folder (that is located under the WINNT folder) on the C: drive. The system key would be in the CONFIG folder (that is located under the SYSTEM32 folder that is located under the WINNT folder) on the C: drive.

Another example: If you installed Windows 2000 onto the D: drive, but the C: drive is the active partition, the BOOT.INI, BOOTSECT.DOS, HYBERFIL.SYS, NTBOOTDD. SYS, NTDETECT.COM, and NTLDR files would all be in the root directory of C:. The HAL.DLL and NTOSKRNL.EXE files would be located in the SYSTEM32 folder (that is located under the WINNT folder) on the D: drive. The system key would be in the CONFIG folder (that is located under the SYSTEM32 folder that is located under the WINNT folder) on the D: drive.

The **system key** is a protection feature for Windows 2000 passwords. By default, the system key is stored on the local computer, but it can also be stored on a floppy disk used to boot Windows 2000. An algorithm secures the system key and it is stored in various locations through the registry. The boot process is actually quite involved, but the major steps are as follows:

1. The computer is powered on.
2. POST executes.
3. BIOS searches for an active partition on the hard drive.
4. BIOS reads the Master Boot Record, then locates and loads the information into sector 0 of the system partition. The contents of sector 0 define the type of file system, the location of the bootstrap loader file, and start the bootstrap loader. With Windows 2000, this file is NTLDR.
5. NTLDR starts the file system.
6. NTLDR reads the BOOT.INI file and displays the various operating system choices contained within the BOOT.INI file. If something other than Windows 2000 is chosen, the BOOTSECT.DOS file takes over. If Windows 2000 is chosen, the NTDE-TECT.COM file executes.
7. NTDETECT.COM detects the computer's hardware.
8. NTLDR passes the hardware information to the NTOSKRNL.EXE file and displays the startup screen.
9. The operating system kernel, NTOSKRNL.EXE, executes and the HAL.DLL file loads. HAL stands for Hardware Abstraction Layer. This is a layer between the operating system and the hardware devices. The HAL allows Windows 2000 to run with different hardware configurations and components, without affecting (or crashing) the operating system.

10. The registry key HKEY_LOCAL_MACHINE\SYSTEM loads from the registry key located in *%systemroot%*\System32\Config\System. This key has information found during the hardware detection process.
11. The Windows 2000 Professional screen appears.
12. The Starting Up process bar displays.
13. The WINLOGON.EXE file executes and the log-on screen appears.

Troubleshooting the Windows 2000 Boot Process

When Windows 2000 has startup problems, the Emergency Repair Disk, Recovery Console, and the Advanced Options menu are used. Many times startup problems are due to a virus. The AVBoot disk can be used to check the computer for a virus. The procedure for creating this is in the Troubleshooting the Windows 2000 Professional Installation section of this chapter. Other utilities that can be used within Windows 2000 to help with MBR, boot sector, and system files are FIXBOOT, FIXMBR, System File Checker, and the Advanced Options menu.

To use FIXBOOT, type **FIXBOOT** *x:* command where *x:* is the drive letter of the volume that has the problem. To use FIXMBR, type **FIXMBR** from a command prompt. Both the FIXBOOT and FIXMBR commands are covered in the Recovery Console section later in the chapter. The System File Checker program can be run from the Run dialog box by typing *x:***\WINNT\SYSTEM32\SFC.EXE /scannow** where *x* is the drive letter where 2000 is installed.

Indications that there is a problem with the Master Boot Record or the system files are as follows:

• Invalid partition table
• Error loading operating system
• Missing operating system
• A disk read error has occurred
• NTLDR is missing
• NTLDR is corrupt

When Windows 2000 has startup problems due to incompatible hardware or software, or a corrupted installation process, the Windows 2000 Advanced Options menu can help. This option can be selected by pressing the **F8** key when the **For troubleshooting and advanced startup options for Windows 2000, press F8** message appears on the screen during the boot process. Also, look back to Table 6.4 for a review of keystrokes that can be used to bring up different start options used in troubleshooting Windows 2000 Professional.

The most commonly used boot option is **Safe Mode**. In prior Windows operating systems, when the system had a problem it automatically booted into Safe Mode. This is not the case with Windows 2000. You must use the Advanced Startup Options to select Safe Mode.

Safe Mode is used when the computer stalls, slows down, does not work right, video is not working properly, intermittent errors appear, or new hardware/software installation causes problems. When the computer boots in Safe Mode, the mouse, keyboard, CD-

ROM, and VGA video device drivers are all that are loaded. After the computer boots to Safe Mode, you can disable or delete a system service; delete, reload, or upgrade a device driver, and disable or delete a shortcut in the Startup folder, any of which can cause the computer to hang during startup. The bottom line is that Safe Mode puts the computer in a "bare bones" (or minimum) mode so you can troubleshoot problems.

Another menu item that is useful when troubleshooting device drivers is Boot Logging. This option creates a file called NTBTLOG.TXT that is placed in the *%systemroot%* folder, (which is normally C:\WINNT). The NTBTLOG.TXT file contains a list of the drivers that load and the drivers that do not load. If you suspect a problem with a driver, use the **Enable Boot Logging** option from the Advanced Options menu to see if Windows 2000 loaded the driver. An exercise at the end of the chapter explains how to take advantage of this feature.

The other menu items are also used for troubleshooting and Table 6.7 shows the function of each menu option.

Table 6.7: Windows 2000 Advanced Options Menu

Menu Option	*Function*
Safe Mode	Loads the bare minimum device drivers needed to boot the system.
Safe Mode with Networking	Loads the bare minimum device drivers needed to boot the system plus the network services and drivers needed for the computer to participate in a network.
Safe Mode with Command Prompt	Loads the bare minimum device drivers needed to boot the system but does not load the GUI interface (EXPLORER.EXE).
Enable Boot Logging	Creates a log file called NTBT in the *%systemroot%* folder. This file contains all drivers that load during startup and shows each driver's status.
Enable VGA Mode	Used when an incompatible or corrupted video driver has been loaded. It boots the system with a generic VGA driver so you can troubleshoot.
Last Known Good Configuration	Used to load a previous configuration that worked (as long as you don't log on).
Debugging Mode	Used to debug the OS kernel.
Boot Normally	Boots Windows 2000 with normal device drivers, registry, and startup folders.

In older operating systems, a startup disk was used to begin the troubleshooting process. With Windows 2000, the startup disk is actually the set of four installation disks that came with the Windows 2000 Professional CD. If you cannot find these, you can go to any computer that has the same version of Windows 2000 Professional loaded and make a set. Insert the Windows 2000 Professional CD, click on the **Start** button, and click on the **Run** option. In the dialog box, type **x:\bootdisk\makeboot.exe a:** where *x* is the drive letter that represents the CD-ROM. Click on the **OK** button and follow the prompts on the screen. You will need four floppy disks to complete this procedure.

Creating an Emergency Repair Disk for Windows 2000 Professional

The **ERD (Emergency Repair Disk)** is used to fix system file problems, the partition boot sector, and startup environment settings, all of which can prevent a computer from booting properly. By default, the ERD does not contain a backup copy of the registry. Use the Backup utility to back up and restore the registry. You should make a new ERD whenever hardware or software changes are made to the computer. Store the disk in a safe place. The ERD contains the following files: AUTOEXEC.NT, CONFIG.NT, SETUP. LOG, NTLDR, NTDETECT.COM, BOOT.INI, NTBOOTDD.SYS, and HAL.DLL. The **AUTOEXEC.NT** and **CONFIG.NT** files are used to initialize the DOS environment. The **SETUP.LOG** lists the files installed by the Setup program. **NTLDR** is the file used to load Windows 2000; **BOOT.INI** is used to tell the computer which hard drive and which hard drive partition to use to boot Windows 2000. The **NTBOOTDD.SYS** is only used when the computer has a SCSI hard drive installed. The **HAL.DLL** file is used by Windows 2000 to keep hardware problems from crashing the operating system. An exercise at the end of this chapter illustrates how to make an ERD.

If the computer system will not work and you want to use the ERD, start the Windows 2000 installation process and get to the Welcome to Setup screen. Use the Windows 2000 Professional Setup disks (you must get through Disk 4 before you see the Welcome to Setup screen) or boot your computer from the Windows 2000 Professional CD. Once presented with the Welcome to Setup screen, press the **R** option, which is the **To repair a Windows 2000 installation by using the emergency repair process** option that allows you to use the ERD to fix the computer. Once you press **R**, you are asked if you want a manual repair or a fast repair. The manual repair allows you to select what portions of the operating system are repaired. The fast repair will check and try to repair system files, the partition boot sector, or the startup environment settings. Most people select the fast repair option. After making a selection, you are prompted to insert the ERD. Follow the instructions on the screen to complete the repair process. The computer must reboot after the repairs have been done.

Using Recovery Console

Another useful tool to use when Windows 2000 crashes or does not boot properly is the Recovery Console. The **Recovery Console** boots the computer to a command prompt and allows access to the hard drive no matter what type of file system is being used (FAT,

FAT32, or NTFS). From the command prompt, the administrator can manipulate files and folders, start and stop services, repair the Master Boot Record, repair the boot sectors, or format the hard drive. You must have the administrator password to access the full potential of this option.

The Recovery Console can be loaded to the hard drive and added to the Start menu, but it is not loaded by default. Refer to Figure 6.24 for a sample screenshot of how to add the Recovery Console. To do so, you must run **winnt32.exe** with the **/cmdcons** option. Note it requires an additional 7 MB of disk space.

Figure 6.24: Adding the Recovery Console

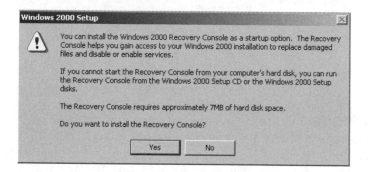

Recovery Console is started from the Windows 2000 CD or the Setup floppy disks. Start the computer from the Setup disks or CD. (If you do not have the Setup disks, go to a working 2000 computer and make some.) When the Setup program begins, press **Enter** at the **Setup Notification** screen. At the **Welcome to Setup** screen, press **R** (Repair a Windows 2000 installation). On the next screen, press **C** to access the Recovery Console. A screen appears that shows all of the Windows 2000 installations that are detected. Press the number corresponding to the Windows 2000 installation you want to work with. You are prompted for the local administrator password. Type the password to continue.

If the registry is corrupt or has been deleted, you are not prompted for an administrator password. Instead, the system boots to a prompt where you can use basic commands like CHKDSK, FIXBOOT, and FIXMBR to repair the system. However, you cannot access any folders on the hard drive.

The drive letters available at the Recovery Console command prompt might not be the same ones you used in the GUI environment. Use the MAP command to see the drive letters (and the volumes that do not have drive letters). The syntax for the MAP command is covered later in this chapter.

In order to work in the Recovery Console, you must be able to work from a command prompt. You may want to go through Chapter 3 entitled *DOS and the DOS Command Line Interface* in this book to understand the process and procedures needed when working from a command prompt. Some of the most frequently used commands at the Recovery Console command prompt are outlined below.

- **DISABLE**—Used to disable a system service or hardware driver.
- **DISKPART**—Used to manage and manipulate the hard drive partitions.
- **ENABLE**—Used to enable a system service or hardware driver.
- **EXIT**—Exits the Recovery Console and restarts the computer.
- **EXPAND**—Used to uncompress a file from the Windows 2000 CD.
- **FIXBOOT**—Used to rewrite the hard drive's boot sector.
- **FIXMBR**—Rewrites the startup partition's Master Boot Record.
- **FORMAT**—Used to format a disk and can be used to format it for a particular file system.
- **LISTSVC**—Lists all of the services, hardware drivers, and their start-types. The **listsvc** command is useful to use before using the **disable** or **enable** command.
- **MAP**—Used to list the computer's drive letters, types of file systems, volume sizes, and physical device mappings.

To receive help on any of the commands listed above, go to the Recovery Console prompt and type the command followed by /?. For example, to find out all about the FIXBOOT command, type FIXBOOT /? (while in Recovery Console mode).

6.11 Understanding Task Manager, Dr. Watson, and Event Viewer

Task Manager is a Windows-based utility that displays applications currently loaded into memory, processes that are currently running, microprocessor usage information, and memory usage data. To activate Task Manager press the **Ctrl + Alt + Del** keys and from the window options that appear, click on the **Task Manager** button. Two other ways of accessing this utility are (1) by pressing **Ctrl + Alt + Esc** and (2) by right-clicking on the **taskbar** and then clicking on the **Task Manager** option.

One of the common uses of Task Manager is to exit from an application that is "hung up" or not responding. Task Manager can help with exiting the program. Once inside the Task Manager window, click on the **Applications** tab. Locate the name of the troublesome application and click on it. Normally, the status shows the application as "not responding." Refer to Figure 6.25 for a screenshot of ending a process in Task Manager.

Figure 6.25: Using Task Manager to End a Process

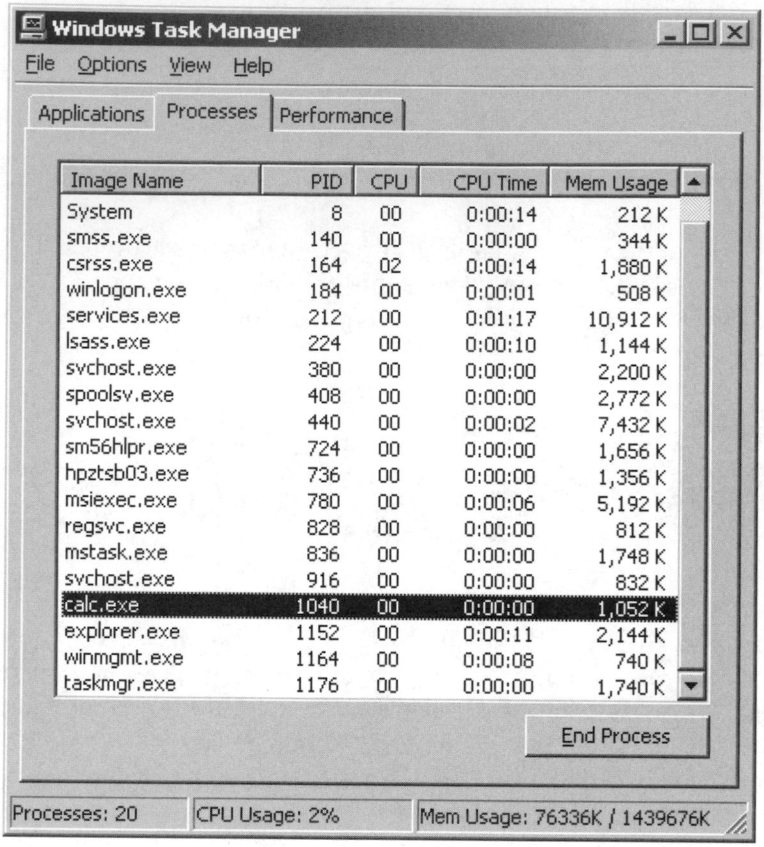

Dr. Watson is a utility that automatically loads when an application starts. Dr. Watson can detect and display troubleshooting information as well as create a text log file when a system or application error occurs. You might need this information when communicating with Microsoft or the application developer's technical support. Make notes of any messages that appear on the screen when any type of problem occurs. To start Dr. Watson in Windows 2000, click on the **Start** button, click on the **Run** option, type **drwtsn32**, and press **Enter**. Click on the **application error** and click on the **View** button. The default location for the log file is C:\Documents and Settings\All Users\Documents\DrWatson. The name of the log file is drwtsn32.log. When an error occurs, Dr. Watson appends information to the end of this log file. Refer to Figure 6.26 for a screenshot of Dr. Watson.

Figure 6.26: Running Dr. Watson

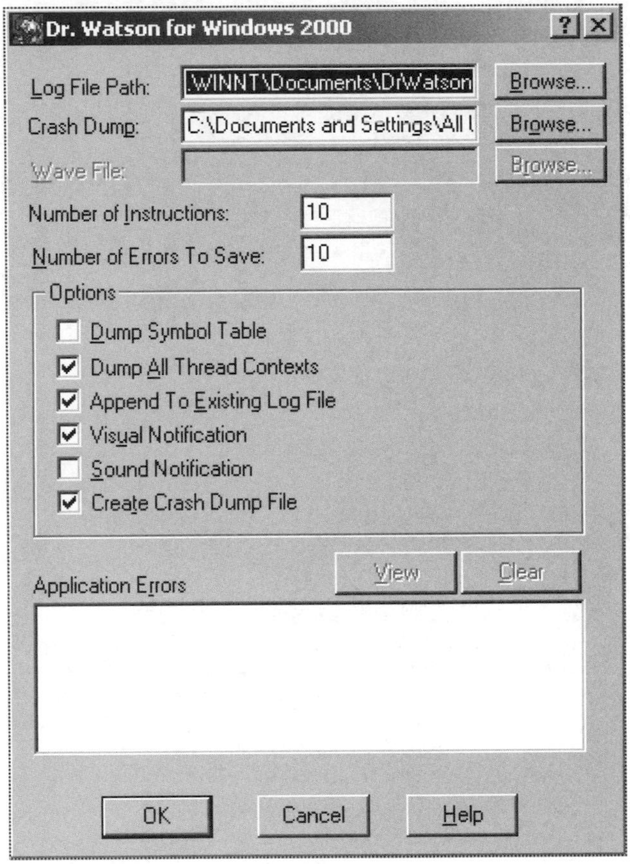

Event Viewer is a Windows tool used to monitor various events in your computer such as when a driver or service does not start properly. The EventLog service starts automatically every time a computer boots to Windows 2000. This service is what allows the events to be logged and then Event Viewer is used to see the log.

Let's perform an exercise to demonstrate the use of Event Viewer.

1. Click **Start**, point to the **Settings** option and then click on the **Control Panel**. The Control Panel window opens.
2. Double-click on the **Administrative Tools** icon. The Administrative Tools window opens.
3. Double-click **Event Viewer**. The Event Viewer window opens.
4. Click on the **System Log** option in the left panel. The system log events are displayed in the right window.
5. Close Event Viewer.

The **System log** displays events that deal with various system components, such as a driver or service that loads during startup. The type of system log events cannot be changed, added, or deleted. The **Security Log** can display events, such as when different users log in to the computer (both valid and invalid log ins). You can pick which events are displayed in the security log. The **Application Log** displays events associated with a specific program. The programmers that design the software decide which events to display in the Event Viewer's application log. All users can view the system log and the application log, but only a member of Administrators can view or enable security log information. The most commonly used log is the system log.

Refer to Figure 6.27 for a screenshot of Event Viewer.

Figure 6.27: Event Viewer

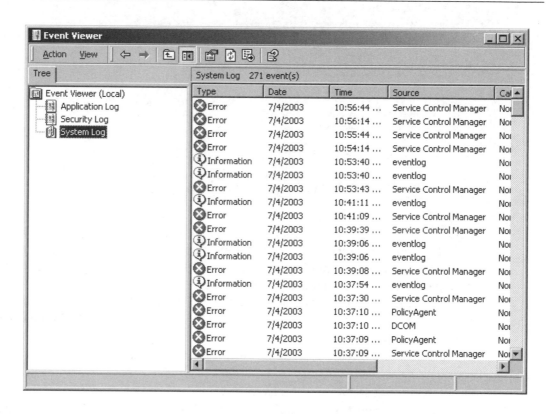

If an error message appears stating that the Event Viewer log is full, start the Event Viewer. Note that you must be an administrator or a member of the administrators group to perform this procedure. Click on the **Action** menu option and then click on the **Properties** selection. Click on the **General** tab. Click in the **Clear log** option. The Log Size option may need to be changed to one of the following: Overwrite events older than 0 days, Maximum log size, or Overwrite events as needed.

6.12 Troubleshooting a Service that Does Not Start

Some Windows 2000 services start automatically each time the computer boots. If one of these services has a problem, an error message normally appears during the boot sequence. You can use Event Viewer also as previously discussed or use the Services and Application tool available through the Computer Management administrative tool.

Before Windows 2000 can shut down, the operating system sends a message to all devices, services, and applications. Each device that is running sends a message back saying it is okay to shut down now. Any active application saves data that has not been previously saved and sends a message back to the operating system. Active system services also respond that it is okay to shut down. If the system has trouble shutting down, it is due to one of the following three things:

- The most common problem is an application that is not responding. When this happens, press **Ctrl + Alt + Del** to access Task Manager. Manually stop any applications from running to see if that is causing the problem. If a single application continually prevents Windows 2000 from shutting down, contact the software manufacturer to see if there is a fix.

- For services problems, boot the computer into Safe Mode and then shut the computer down. Take note as to whether or not the computer had any problems shutting down. If the process works, access the BOOTLOG.TXT file that is located in the root directory of the drive that contains Windows 2000. Once inside the file, take note of each service that is disabled because of booting into Safe Mode. Boot the computer normally. Stop each service one at a time to see which service is causing the problem.

- To troubleshoot devices not responding, eliminate services and applications first. A device frequently does not cause a shutdown problem. Then, while working on the computer, take notice of what devices you are using—common ones are video, hard drive, CD-ROM, keyboard, and mouse. Verify that all of your devices have the most up-to-date driver loaded and that the driver is compatible with Windows 2000.

6.13 Monitoring System Performance

Another utility used to monitor the computer is the Performance Logs and Alerts snap-in tool. **Performance Logs and Alerts** allows creation of graphs, bar charts, and text reports. An exercise at the end of the chapter shows how to use the Performance Logs and Alerts utility. This utility can also be customized to show different counters. The button with a + (plus sign) is used to add various counters to the display. Some of the most important memory counters are Available Bytes, Pages/sec, and Paging file\%Usage. The **Available Bytes** counter shows the amount of RAM available for running program processes. The **Pages/sec** counter shows the number of times per second the information requested could not be found in RAM, and the data had to be retrieved from the hard drive. The **Paging file\%Usage** counter shows what percentage of allocated space for the paging file is in use.

Sometimes you must adjust the paging file size for optimum computer performance. The System control panel is used to set the virtual memory size. Once in the System control panel, click on the **Advanced** tab. Click on the **Performance Options** button and two sections appear in the window: Application Response and Virtual Memory. Click on the **Change** button in the Virtual Memory section and the Virtual Memory window appears. Refer to Figure 6.28 for a screenshot of Performance Monitor. In the figure, the CPU time is being monitored.

Figure 6.28: Using Performance Monitor

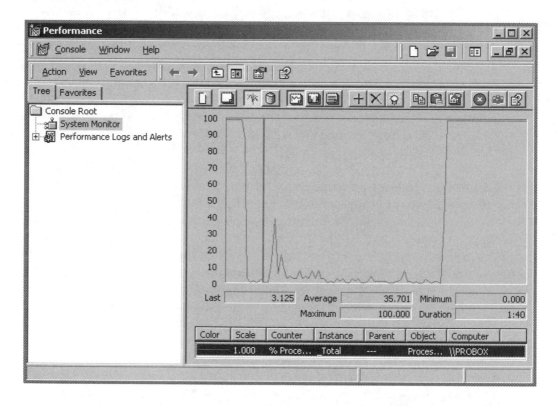

Two values are selectable: Initial size and Maximum size. Both of these values should be the same for maximum computer performance. Once you change the values, click on the **Set** button. The default amount of virtual memory is 1.5 times the amount of RAM installed in the computer. The Virtual Memory window can also be used to change the amount of space reserved for the registry.

The Task Manager utility can be used to monitor your current system's performance. Sometimes a computer can start slowing down. A baseline report is needed before the slowdown occurs. A **baseline** report is a snapshot of your computer's performance during normal operations (when it does not have any problems).

Start the Task Manager utility by pressing **Ctrl + Alt + Del** and clicking on the **Task Manager** button (or right-clicking on an empty space on the taskbar and clicking on the **Task Manager** option). Click on the **Performance** tab to see the CPU usage and memory usage statistics. The first window on the left shows the CPU usage percentage. Actually, it is a percentage of time the processor is running a thread. A thread is a type of Windows object that runs application instructions and was discussed in Chatper Two entitled *Basic Operating System Theory.* The first window on the right displays the CPU usage history, which is a graph of how busy the processor has been over a period of time.

The second window on the left shows the amount of memory being used. The amount shown is in kilobytes as evidenced by the K after the number. The second window on the right is a graph of the memory used over time.

Refer to Figure 6.29 for a screenshot of the Performance tab in Task Manager.

Figure 6.29: The Performance Tab in Task Manager

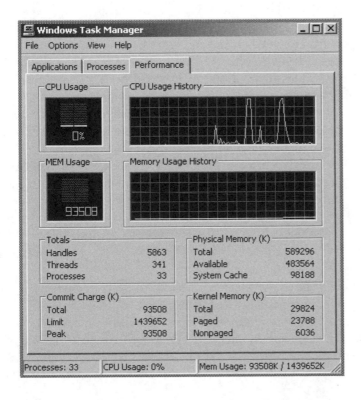

Memory is a frequent bottleneck for computer performance. Task Manager can also be used to see the total amount of RAM installed and how much RAM is available. Task Manager is an invaluable tool when a computer is slowing down.

Chapter Summary

- Windows 2000 Professional is a 32-bit operating system that provides better performance than Windows NT Workstation and Windows 95/98. Every 32-bit application runs in its own memory space meaning that if that application fails, other applications are not affected.
- Windows 2000 Professional provides a hierarchical file system allowing you to create folders and store files within them.
- Windows 2000 Professional supports both compression and encryption on files and folders. Compression saves disk space. Encryption provides security.
- The Windows 2000 Professional Registry is a database, which stores the hardware and software information installed on your computer.
- As part of the pre-installation steps of Windows 2000 Professional, you should decide whether or not to perform a clean installation or an upgrade.
- There are two modes of the Windows 2000 Professional installation—text mode and GUI mode. A few tasks done in text mode are as follows: the installation process checks for minimum hardware requirement and Plug and Play devices are installed. During GUI mode, other components are installed and configured.
- You can dual-boot Windows 2000 with other Microsoft operating system. When dual-booting, you should create a separate partition for each operating system. Install Windows 2000 Professional after installing the other operating systems.
- Troubleshooting your Windows 2000 installation includes verifying BIOS and driver compatibility, inserting the correct CD key, ensuring you have a minimum CPU that is a 486 33 MHz, and having at least 12 MB of RAM.
- Configuring Windows 2000 involves using Control Panel to add and remove hardware components such as a mouse or a printer. It also includes adding and removing additional software.
- You need to understand the boot process of Windows 2000 so you know how to troubleshoot when problems occur. The system partition is the active partition containing the operating system boot files. The boot partition is where the operating system files are located. They can be on the same partition.
- When Windows 2000 boots, various services are started. You can see the status of a service by checking the Services program in Control Panel.
- You can use Performance Monitor to monitor the performance of your Windows 2000 system through the use of objects and counters. It is important to take a baseline of your system for comparative purposes over time.

Review Questions

1. A _____ is an electronic container that holds data; you can open a _____ using an editor.
   a) Folder
   b) File
   c) Tab
   d) Taskbar

2. What is the maximum number of characters in a Windows 2000 file name?
   a) 8
   b) 256
   c) 255
   d) 512

3. What attribute allows you to only view the contents of a file?
   a) Hidden
   b) Archive
   c) Read
   d) System

4. How would you run a 16-bit application named programA in its own separate memory space?
   a) start /separate programA
   b) start programA \separate
   c) start \separate programA
   d) start /memory_space programA

5. What feature of Windows 2000 allows you to run multiple processors to operate on a task at the same time?
   a) WFP
   b) PnP
   c) SMP
   d) XMS

6. What feature of Windows 2000 runs in the background and protects system files from alteration?
   a) WFP
   b) PnP
   c) SMP
   d) XMS

7. With _____ the contents of the file are altered as a means of securing the data.
   a) Compression
   b) Encryption
   c) WFP
   d) XMS

8. What do you select to view the file names of compressed files in blue?
   a) **Select** Display compressed files and folders with alternate color
   b) **Select** Display noncompressed files and folders with alternate color
   c) **Select** Display compressed files and folders in blue color
   d) **Deselect** Display compressed files and folders with alternate color

9. The _____ is a group of important Windows 2000 files including the registry, the system files, the boot files, and the COM+ Class Registration database.
   a) System State
   b) REGEDT32
   c) SMP
   d) WFP

10. _____ is the most common application used to create, copy, or move files or folders.
   a) System State
   b) Registry
   c) Explorer
   d) Internet Explorer

11. Which of the following are considered issues related to a Windows 2000 installation?
   a) Incompatible BIOS
   b) BIOS needs to be upgraded
   c) Incompatible hardware
   d) All of the above

12. The system variable named *%systemroot%* is normally
   a) C:/WINNT
   b) C:/WINNT
   c) C:/Windows
   d) A:/WINNT

13. The _____ command line utility can be used to encrypt and decrypt files.
   a) COMPACT
   b) ENCRYPT
   c) SECRET
   d) CIPHER

14. The _____ command line utility can be used to compress files.
    a) COMPACT
    b) ENCRYPT
    c) COMPRESS
    d) SQUEEZE

15. Which of the following commands will display the Windows version?
    a) COMPACT
    b) COMPRESS
    c) WINDOWS_VERSION
    d) WINVER

16. Which file system does compression require?
    a) NTFS
    b) EXT2
    c) FAT16
    d) FAT32

17. With Windows 95, 98, NT, and 2000, upgrades or patches to the operating system are provided through _____, which fix problems within the operating system.
    a) WFP
    b) SMP
    c) Baseline
    d) Service Packs

18. A _____ report is a snapshot of your computer's performance during normal operations when it does not have any problems.
    a) Baseline
    b) Service Pack
    c) Task manager
    d) Performance

19. Which key do you press during the installation to load third party drivers?
    a) F8
    b) F2
    c) F6
    d) F13

20. Which registry subtree holds file associations and file links?
    a) HKEY_LOCAL_MACHINE
    b) HKEY_USERS
    c) HKEY_CURRENT_USER
    d) HKEY_CLASSES_ROOT

Lab Projects

Unless otherwise indicated, all projects assume the computer is powered on, you are logged in, and your desktop appears on the screen. Windows 2000 is up and running and you are logged on.

Lab Project 1

The goal of this lab is to help you determine system and disk information.

1. To determine CPU type, operating system and version, and RAM capacity, right-click the **My Computer** icon and click **Properties**. The System Properties screen appears with your system information.

2. Record the operating system type and level, the processor type, and RAM.

   _____

3. Click **OK** when complete.

4. In order to determine disk capacity and file system type, double-click **My Computer**. The My Computer screen appears with icons. Note the drive icons.

5. Right-click on the C: drive and then click **Properties**. Make sure the **General** tab is selected. The Properties page for the C: drive appears.

6. Record the used space for drive C:.

   _____

7. Record the free space for drive C:.

   _____

8. Record the file system type for drive C:.

   _____

9. Draw the pie chart representing how the space is used.

10. Insert a floppy in the floppy drive.

11. Repeat steps 6 through 9 for the floppy. Write your answers on a separate piece of paper.

12. Log out.

Lab Project 2

The goal of this lab is to help you understand Task Manager.

1. First, you will start a task. Press **Start** in the lower right-hand corner. Next, click **Run** and in the Open dialog box, type **calc** and press Enter. The calculator program appears on the screen.

2. To see the **calc** task (along with other jobs), press **Ctrl + Alt + Del**.

3. Record three other tasks.

_____

4. You can scroll up and down this list and click End Task to delete a task. Be careful because ending a task will terminate it and you do not want to terminate a system task.

5. Highlight **calc** and press **End Task**.

6. Now, see what happens with you remove a system task. Terminate the **explorer** task. Follow the steps that appear on the screen. _Hint_: The shutdown screen will appear.

7. Record what happens.

_____

8. Log out.

Lab Project 3

The goal of this project is to help you understand how to create a tree structure from a screenshot.

1. Using Windows Explorer, create the tree structure shown in Figure 6.30 on your floppy. To get to Explorer, right-click **Start** and then click **Explore**.

Figure 6.30: A Sample Tree Hierarchy

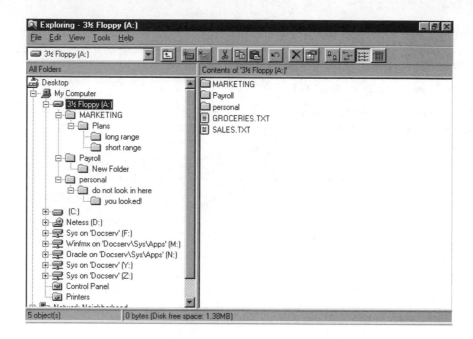

2. Using Notepad, create the following files as Text Documents in the "long range" folder. Place your name in each file.

   Monday.txt
   Tuesday.txt
   Wednesday.txt
   Thursday.txt
   Friday.txt

3. Record the steps you took to create them.

_____

4. Rename the "New Folder" to "Weekly."

5. Record the steps you took.

_____

6. Create a "Medium Range" folder in "Plans."

7. Record the steps you took.

_____

8. In the "personal" folder, create a file called "resume.txt."

9. Hide this folder.

10. Record the steps you took.

_____

11. View all files.

12. Record the steps you took.

_____

13. Make a complete copy of the files in "Plans" and put them in a folder called "BACKUP of PLANS."

14. Record the steps you took.

_____

15. In Explorer, right-click on the floppy to make a disk copy of the floppy. Insert another floppy disk and make a disk copy of this floppy to another floppy.

16. Record the steps you took.

_____

17. Create another folder in the root directory of the floppy that has your name in it.

18. Create another folder in the root directory of the floppy named "Windows 2000 Project."

19. Show your instructor.

20. Close all windows.

Lab Project 4

The goal of this project is to reinforce your understanding of how to create a tree structure. You will also work with attributes to help you understand how they work.

1. Using Windows Explorer, create the tree structure shown in Figure 6.31 on your floppy.

Figure 6.31: A Sample Tree Hierarchy

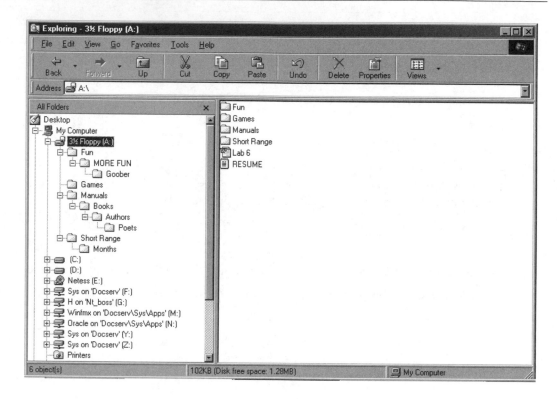

2. Using Notepad, place these files in the "Poets" folder:

   Poe.txt
   Walker.txt
   Smith.txt

3. In Explorer, click **Tools**, then click **Folder Options** and then click **View**. The Folder Options window opens. In the Advanced Settings box, activate the setting so hidden and system files are not displayed.

4. Right-click on "Poe.txt" and click **Properties**. The Properties page appears. Click **Hidden** and then click **OK**. The files are hidden.

5. In Explorer, click **Tools**, then click **Folder Options** and then click **View**. The Folder Options window opens. In the Advanced Settings box, activate the setting to show all files. This will display hidden files.

6. Record whether or not you can now see "Poe.txt."

_____

7. Using the previous step as a guide, make the file "Walker.txt" read-only.

8. Record the steps you took.

_____

9. Open the file named "Walker.txt" in Notepad and attempt to add data.

10. Record what occurred.

_____

11. Close Notepad.

12. Make "Walker.txt" readable.

13. Open the file named "Walker.txt" in Notepad and attempt to add data.

14. Record what occurred.

_____

15. Close Notepad.

16. Create another folder in the root directory of the floppy that has your name in it.

17. Create another folder in the root directory of the floppy named "Windows 2000 Project 4."

18. Show your instructor.

19. Close all windows.

Lab Project 5

The goal of this project is to work with Windows 2000 taskbar options.

1. Turn on the computer and verify that the operating system loads.

2. Log on to Windows 2000 using the userid and password provided by the instructor or lab assistant.

3. Locate the taskbar on the bottom of the screen. If it is not showing, move the mouse to the bottom of the screen and the taskbar pops up.

4. To modify or view the taskbar settings, right-click on a **blank area** of the taskbar. A menu appears. Note: You can also use the **Start** button, point to the **Settings** option, and click on the **Taskbar and Start Menu** option.

5. Click on the **Properties** option. The Taskbar and Start Menu Properties window appears.

6. Click on the **General** tab.

7. The five options available on this screen relate to how things are shown on the taskbar. The items with a check in the check box to the left are active. The **Always on top** option puts the taskbar visible on the screen at all times (even if a window is full size or maximized). The **Auto hide** option hides the taskbar during normal operation. Press **Ctrl + Esc** or the **Start** button on the keyboard (the one with the Windows emblem on it) to make the taskbar reappear. If both the Always on top and Auto hide options are checked, then the taskbar appears when you are in a window that is full size (maximized). The **Show small icons in Start menu** option reduces the size of the Start menu words. The **Show clock** option displays the clock icon in the right corner of the taskbar. The **Use Personalized Menus** option allows you to hide menu items that you rarely use. Make sure the **Always on top** and **Show clock** options are the only ones with check marks in the check boxes. To remove a check mark, click in the **check box** that already contains a check in it. To put a check mark in a box, click once in an **empty box.**

8. Click on the **Apply** button.

9. Click on the **OK** button.

10. Right-click on an **empty space** on the taskbar.

11. Point to the **Toolbars** option. A submenu appears.

12. Ensure that there is a check mark beside the **Quick Launch** option. This setting allows the Quick Launch icons to appear on the desktop by the Start button.

13. Using the skills you just learned, access the **Taskbar Settings** window.

    List the steps you performed to do step 13.

14. Click on the **Properties** option. Click on the **Question mark** icon in the upper right corner of the window. The question mark is an interactive help system. The pointer on the screen changes to an arrow with a question mark attached. Click on the **Always on top** option. A description of the option appears with the on-line help active.

     What text does the help balloon display?

     _____

15. Click on the **Advanced** tab. The Advanced tab is used to customize the Start button and to delete files that list under the Start button's Documents option or previously accessed web sites.

16. Click on the **Cancel** button.

Lab Project 6

The goal of this project is to work with Windows 2000 REGEDIT editor.

REGEDIT is a utility used for editing the Windows registry. With REGEDIT, you can view existing registry settings, modify registry settings values, or create new registry entries to change or enhance the way Windows operates. In this lab, you will use REGEDIT to view the System BIOS and Video BIOS information on your computer.

CAUTION: Editing the registry can cause your computer to run erratically, or not run at all! When performing any registry editing, follow ALL directions carefully, including spelling, syntax use, etc. Failure to do so may cause your computer to fail!

1. From the Start menu, choose **Run**, type **REGEDIT**, and click **OK**. The REGEDIT utility opens.

2. In the left window, expand **HKEY_LOCAL_MACHINE**, **HARDWARE**, and **DESCRIPTION**, and then select **System**.

3. In the right window, the System BIOS and Video BIOS information display.

     What is the System BIOS date?

     _____

     Who is the manufacturer of your System BIOS?

     _____

     When was your Video BIOS manufactured?

     _____

4. When finished viewing the System and Video BIOS information, close the **REGEDIT** utility.

Lab Project 7

The goal of this project is to work with Windows 2000 REGEDT32 editor.

REGEDT32 is a utility used for editing the Windows registry. With it, you can find and view existing registry settings, modify registry settings values, or create new registry entries to change or enhance the way Windows operates. In this lab, you will use REGEDT32 to create and configure a new registry setting to control how many document entries appear in Documents on the Start menu.

CAUTION: Editing the registry can cause your computer to run erratically, or not run at all! When performing any registry editing, follow ALL directions carefully, including spelling, syntax use, etc. Failure to do so may cause your computer to fail!

1. From the Start menu, click on the **Run** option, type **REGEDT32** in the textbox, and click **OK**. The REGEDT32 utility opens.

2. From HKEY_CURRENT_USER on Local Machine window, expand the following options: **Software, Microsoft, Windows, CurrentVersion,** and **Policies,** and then select the **Explorer** option.

3. From the Edit menu option, choose **Add Value**. The Add Value window opens. Editor opens.

4. The default number of documents that appear in the Start menu's Document folder is 15. From the **DWORD Editor** window, select the **Decimal** radio button, enter the value of **20** in the Data textbox, and click **OK**. The new DATA value displays in Hexadecimal format.

   What is the Hexadecimal value of the new DATA value?

   _____

5. Close the **REGEDT32** utility and reboot the computer for the new registry setting to take effect.

   After the computer reboots, how many documents can be displayed in the Start menu's Documents folder?

   _____

Lab Project 8

The goal of this project is to use the Windows 2000 Backup Utility to back up the registry.

1. Turn on the computer and verify that Windows 2000 Professional loads.

2. Log on to Windows 2000 using the userid and password provided by the instructor or lab assistant. Make sure that the userid is an Administrator userid or a userid that is a member of the Backup Operators group.

3. Click on the **Start** button.

4. Point to the **Programs** option.

5. Point to the **Accessories** option.

6. Point to the **System Tools** option.

7. Click on the **Backup** option. The Backup Utility window opens.

8. Click on the **Backup** tab.

9. Click once in the **System State** check box to enable this option. A check mark appears in the check box.

10. At the bottom of the window, click on the **Browse** button to select a hard drive or any other type of media. Contact your instructor for the location to put the backed up files.

11. Type a **name** for the backup and click on the **Open** button.

    What name did you assign for the backup file?

    _____

12. Click on the **Start Backup** button to begin the backup procedure. The Backup Job Information dialog box appears.

13. Click on the **Advanced** option.

14. Click on the **Automatically backup system protected files with the System State** option.

15. Click on the **OK** button.

16. Click on the **Start Backup** button. A Backup Progress window appears.

    How many estimated files will be backed up?

    _____

17. Click on the **Close** button.

18. Close the **Backup** window.

Lab Project 9

The goal of this project is to install an updated driver under the Windows 2000 operating system.

Note: The installation process outlined below may differ due to computer differences. If the process is different, follow the directions on the screen. Contact your instructor or a lab assistant if you are unsure of the steps.

1. Turn the computer on and verify that the operating system loads. Log in to Windows 2000 using the userid and password provided by the instructor or lab assistant.

2. Click on the **Start** button, point to **Settings**, and click on the **Control Panel** menu option.

3. When the Control Panel window opens, scroll down to the **System** icon and double-click on it.

4. Click on the **Hardware** tab.

5. Click on the **Device Manager** button.

6. Click on the **+ (plus sign)** beside the class of device you want to upgrade.

7. Right-click on the **specific device** you want to upgrade.

8. Click on the **Properties** tab.

9. Click on the **Driver** tab.

10. Click on the **Update Driver** button. The Upgrade Device Driver wizard appears.

11. Click on the **Next** button.

12. Click on the **Display a list of known drivers for this device so that I can choose a specific driver** radio button. Click on the **Next** button.

13. Click on the **Have Disk** button.

14. Insert the media that contains your updated driver into the floppy drive, Zip drive, CD-ROM, or DVD drive.

15. In the Copy manufacturer's files from box, type in the drive letter for the device that contains the updated driver followed by a colon and click on the **OK** button. An example of a driver device is A:. The Browse button can be used to locate a device driver as well.

16. Click on the **Next** button and follow the instructions on the screen to complete the upgrade.

Lab Project 10

The goal of this project is install a non-plug and play device into a computer running Windows 2000 and load the proper driver for it. Additionally you will need the following:

• Non-plug and play device
• Latest Windows 2000 driver for the hardware device

Note: The installation process outlined below may differ due to computer differences. If the process is different, follow the directions on the screen. Contact your instructor or a lab assistant if you are unsure of the steps.

1. Turn the computer off and install the piece of hardware according to the manufacturer's directions.

2. Turn the computer on and verify that the operating system loads. Log in to Windows 2000 using the userid and password provided by the instructor or lab assistant. The Found New Hardware window should not appear if the device is truly non-plug and play.

3. Start the Add/Remove Hardware wizard by clicking on the **Start** button.

4. Point to the **Settings** option and then click on the **Control Panel** menu option.

5. When the Control Panel window appears, double-click on the **Add/Remove Hardware** icon. The Add/Remove Hardware wizard initializes. Click on the **Next** button to continue.

6. Ensure the Add/Troubleshoot a device radio button is selected, and then click on the **Next** button. A list of installed devices appears.

7. Click on the **Add a new device** option from the window and click on the **Next** button.

8. Click on the **No, I want to select the hardware from a list** option and then click on the **Next** button.

9. Select the type of hardware you want to install and then click on the **Next** button.

10. Insert your device driver into the floppy drive, Zip drive, CD-ROM, or DVD drive.

11. Click on the **Have disk** button. In the Copy manufacturer's files from box, type in the **drive letter** for the device that contains the updated driver followed by a colon. An example of this would be A:. Click on the **Browse** button if you don't know where the file is located. Click on the **OK** button. You may be required to select your specific device or model from a list on the screen. If so, select the device and click the **Next** button. Contact an instructor or lab assistant if you don't know what to select.

12. Click on the **Finish** button.

Lab Project 11

The goal of this project is to install a printer attached to the computer running Windows 2000 and load the proper driver for it. Additionally you will need the following:

• Printer (with appropriate cabling if necessary)
• Latest Windows 2000 driver for the printer

Note: The installation process outlined below may differ due to printer differences. If the process is different, follow the directions on the screen. Contact your instructor or a lab assistant if you are unsure of the steps.

1. Turn the computer off and install the printer according to the manufacturer's directions. If you do not have the correct driver, go on the printer manufacturer's web site and download the driver or obtain it from the instructor or lab assistant.

2. Turn the computer on and verify that the operating system loads. Log on to Windows 2000 using the userid and password provided by the instructor or lab assistant. The Found New Hardware wizard should appear. Follow the instructions on the screen to install the printer. When prompted for the printer driver, insert the CD or disk that contains the printer driver, and use the Browse button to locate the driver. Continue through the Add Printer wizard until the printer is installed.

3. If the Found New Hardware wizard does not appear, have the Windows 2000 printer driver ready, click on the **Start** button, point to the **Settings** selection, and double-click on the **Printers** option.

4. When the Printers window opens, click on the **Add Printer** icon.

5. Click on the **Next** button and select the **Local Printer** selection.

6. When prompted for the printer driver, insert the CD or disk that contains the printer driver, and use the Browse button to locate the driver. Continue through the Add Printer wizard until the printer is installed.

7. Print a test page to test the printer. If the page did not print, troubleshoot the printer.

Lab Project 12

The goal of this project is to use the Windows 2000 System Information tool to troubleshoot device conflicts.

1. Turn on the computer and verify that Windows 2000 Professional loads.

2. Log on to Windows 2000 using the userid and password provided by the instructor or lab assistant.

3. Click on the **Start** button and point to the **Settings** selection.

4. Point to the **Control Panel** option.

5. Double-click on the **Administrative Tools** control panel icon.

6. Double-click on the **System Information** icon.

7. Click on the + **(plus sign)** next to the **Components** folder.

8. Click on the **Problem Devices** folder. Any devices with hardware conflicts list in the right window. Keep in mind that you use the Device Manager program to correct any problems with resource allocation.

   Do any devices have conflicts listed in the Problem Devices window? If so, write them below.

   _____

9. Another good check for hardware conflicts is through the Hardware Resources option. Click on the + (**plus sign**) next to the Hardware Resources folder.

10. Double-click on the **Conflicts/Sharing** folder. Any device that lists under this folder has a resource conflict or is sharing a system resource. Do not forget that PCI devices can legitimately share system resources.

    List two devices that are sharing system resources in the space below.

    _____

11. Click on the **IRQs** folder located under the Hardware Resources folder in the left window.

    List any IRQ that is available for use by a new device.

    _____

    Are any IRQs shared by two devices? If so, write them in the space below.

    _____

12. Click on the **Memory** folder located under the Hardware Resources folder in the left window.

13. Close the System Information window by clicking on the **Close** box (the one with the **X**) in the upper right corner of the window.

Lab Project 13

The goal of this project is to create an Emergency Repair Disk (ERD) that can be used when the computer does not boot properly.

1. Turn on the computer and verify that Windows 2000 Professional loads.

2. Log on to Windows 2000 using the userid and password provided by the instructor or lab assistant.

3. Click on the **Start** button.

4. Point to the **Programs** option.

5. Point to the **Accessories** option.

6. Point to the **System Tools** option.

7. Click on the **Backup** option. The Backup window appears.

8. Click on the **Tools** menu option.

9. Click on the **Create an Emergency Repair Disk** option.

   Give one situation where an Emergency Repair Disk is useful.

   _____

10. Insert a blank formatted floppy disk when prompted.

11. Click on the **OK** button.

12. When the ERD creation is complete, close the **Backup** window.

Lab Project 14

The goal of this project is to use the Performance utility to track individual computer components.

1. Turn on the computer and verify that Windows 2000 Professional loads.

2. Log on to Windows 2000 using the userid and password provided by the instructor or lab assistant.

3. Click on the **Start** button, point to the **Settings** option, and click on the **Control Panel** selection.

4. Double-click on the **Administrative Tools** control panel icon.

5. Double-click on the **Performance** icon. The Performance utility allows you to track individual computer component's performance. This is done through individual counters.

6. In the left window, click on the **System Monitor** item.

7. Click on the **Add button** (the button that has a plus sign on it) or right-click in the right window and click on the **Add Counters** option. The Add Counters dialog box opens.

8. Click on the **Performance object** down arrow. A list of system components appears such as Processor, physical disk, paging file memory, etc. Select the **Memory** performance object.

9. Once a system component has been selected, individual counters for that component can be selected and monitored. In the Select counters from list window, click on the **Available Bytes** counter. Click on the **Add** button.

10. Click on the **Performance** object down arrow. Select the **Paging File** performance object.

11. In the Select counters from list window, click on the **%Usage** counter. Click on the **Add** button.

    Using the Explain button, find out for what the %Usage counter is used. Write the explanation in the space below.

    _____

12. Using the method outlined in steps 7 through 9, select two more counters to be monitored.

    What two counters did you add?

    _____

13. Click on the **Close** button. The right window in the Performance window displays a graph of the various counters. You may need to start some applications, do some cutting and pasting, or surf the Internet to see some of the counter activity. When finished, close the **Performance** window.

Lab Project 15

The goal of this project is to help you understand how to encrypt and decrypt files at the command prompt. You must complete this lab project on an NTFS partition. This lab project is a bit more advanced because you are expected to determine many of the steps in the lab. If you have trouble with the command prompt mode, refer to Chapter 3 on DOS and the DOS command prompt.

1. Go to a command prompt.

2. At the command prompt, create a directory named C:\Project15.

3. Record the command used.

   _____

4. Create two files named File15A.txt and File15B.txt in Project15.

5. Record the command used.

   _____

6. To make sure you are at the root directory of the C: drive, type **CD C:** and press Enter.

7. To encrypt a directory, type **CIPHER /E Project15** and press Enter. The directory and new files will be encrypted.

8. To verify the encryption status, type **CIPHER** and press Enter. The status of the directories and files are displayed. If an "E" is beside the directory or file name, then the item is encrypted. If there is a "U," the item is not encrypted.

9. Change directory locations to Project15.

10. Create a new file named File15E.txt.

11. To verify the encryption status, type **CIPHER** and press Enter.

12. Record the output.

    _____

    _____

    _____

13. Record the reason for the output.

    _____

    _____

    _____

14. To make sure you are at the root directory of the C: drive, type **CD C:** and press Enter.

15. To decrypt a directory, type **CIPHER /D Project15** and press Enter. The directory and new files will not be encrypted.

16. Create a new file in Project15.

17. Check the encryption/decryption status now.

18. Record the output and reason for the output.

_____

_____

_____

19. Close the command prompt window.

Lab Project 16

The goal of this project is to help you understand how to compress and uncompress files at the command prompt. You must complete this lab project on an NTFS partition. This lab project is a bit more advanced because you are expected to determine many of the steps in the lab. If you have trouble with the command prompt mode, refer to Chapter 3 on DOS and the DOS command prompt.

1. Go to a command prompt.

2. At the command prompt, create a directory named C:\Project16.

3. Record the command used.

   _____

4. Create two files named File16A.txt and File16B.txt in Project16.

5. Record the command used.

   _____

6. To make sure you are at the root directory of the C: drive, type **CD C:** and press Enter.

7. To encrypt a directory, type **COMPACT /C Project16** and press Enter. The directory and new files will be compressed.

8. To verify the compression status, type **COMPACT** and press Enter. The status of the directories and files are displayed. If a "C" is beside the directory or file name, then the item is compressed. Otherwise, the item is not compressed.

9. Change directory locations to Project16.

10. Create a new file named File16C.txt.

11. To verify the compression status, type **COMPACT** and press Enter.

12. Record the output.

    _____

    _____

    _____

13. Record the reason for the output.

    _____

    _____

    _____

14. Close the command prompt window.

15. To make sure you are at the root directory of the C: drive, type **CD C:** and press Enter.

16. To uncompress a directory, type **COMPACT /U Project16** and press Enter. The directory and new files will not be compressed.

17. Create a new file in Project16.

18. Check the compression/uncompression status now.

19. Record the output and reason for the output.

_____

_____

_____

20. Close the command prompt window.

Lab Project 17 Challenge

The goal of this project is to upgrade an existing operating system to Windows 2000 Professional. Other requirements are :

• Computer with Windows 95 or 98 installed
• Windows 2000 Professional installation CD

Note: The installation process outlined below may differ due to computer differences. If the process is different, follow the directions on the screen. Contact your instructor or a lab assistant if you are unsure of the steps.

1. Turn the computer on and verify that the operating system loads.

2. Insert the Windows 2000 Professional CD into the CD-ROM or DVD drive.

3. The CD may automatically start. If it does, click on the **Setup** icon. If the CD does not automatically start, click on the **Start** button. Click on the **Run** option. Type in **x:setup** where x: is the drive letter of the device containing the Windows 2000 Professional CD.

4. A message appears on the screen that a newer operating system is being installed and asks if you want to continue. Click the **Yes** button.

5. The Welcome to Windows 2000 Setup wizard appears. Make sure the **Upgrade to Windows 2000** radio button is selected. Click on the **Next** button.

6. The licensing agreement appears on the screen. Read the agreement and click on the **I accept this agreement** radio button and then click on the **Next** button if you agree with the terms.

7. The Your Product Key screen appears. Enter the **product key** provided by the instructor or located on the CD case. Click the **Next** button to continue.

8. The Preparing to Upgrade to Windows 2000 screen appears. Click the **Next** button.

9. The Provide Upgrade Packs screen appears. Click on the **No, I don't have any upgrade packs** radio button. Click the **Next** button to continue.

10. If you have a file system other than NTFS, you will be prompted to upgrade to NTFS. Ask your instructor whether or not to upgrade to NTFS.

    If an upgrade to NTFS is desired, click on the **Yes, upgrade my drive** radio button and then click on the **Next** button.

11. An Upgrade Report screen appears. If there was any incompatible hardware or software found by Windows 2000, it displays on the screen. Go back to the pre-installation steps listed at the beginning of the chapter for more assistance. Print the report or save it if the instructor tells you to. Click the **Next** button to continue.

12. After some files are copied, the computer restarts several times.

13. After the final restart, you must type in the **username and password** entered during the setup process.

Lab Project 18 Challenge

The goal of this project is to install Windows 2000 Professional on a computer without an operating system or on a computer on which the old operating system will be replaced (removed). Other requirements are:

• Computer appropriate hardware
• Four Windows 2000 Professional installation disks
• Windows 2000 Professional installation CD

Note: The installation process outlined below may differ due to computer differences. If the process is different, follow the directions on the screen. Contact your instructor or a lab assistant if you are unsure of the steps.

1. Turn the computer on and verify that the BIOS is set to boot from the floppy drive first, and the hard drive second (A,C sequence). Contact an instructor or lab assistant for assistance on entering the BIOS SETUP program.

2. Insert the Windows 2000 Professional Disk 1 into the floppy drive.

3. Restart the computer. If the computer does not boot from the floppy disk, go back to step 1.

4. Hardware detection starts and then you are prompted to insert Disk 2. Insert the Windows 2000 Professional floppy Disk 2 into the drive and press **Enter**. You will later be prompted for Disks 3 and 4. Follow the directions on the screen.

5. After all files from the floppies have been copied, the Welcome to Windows 2000 Professional Setup screen appears. Press **Enter** to continue with the installation process.

6. The licensing agreement appears on the screen. Read the agreement and press the **F8** key if you agree to the terms.

7. The partitioning options appear next. Contact an instructor or lab assistant to find out if you are to create a partition, use an existing partition, or delete a partition.

8. Follow the directions on the screen and partition the hard drive.

   What type of partitioning will you be doing on this computer? Write the instructions in the space below:

   _____

9. After the hard drive partition is created, more files are copied and the computer restarts.

10. After the computer restarts, a more graphical Setup wizard displays. Click the **Next** button to continue.

11. The regional options such as language and time zone appear. Select the **appropriate option** for your area of the world. Click on either **Customize** button that appears on the screen to change the regional settings. Click on the **Next** button to continue.

12. When the Personalize Your Software screen appears, contact an instructor or lab assistant for the Name and Organization to type.

Name _____

Organization _____

13. The Product Key screen appears. Type in the **product key** that is located on the CD case or type in a **key** provided by your instructor. Click the **Next** button to continue.

14. The Computer Name and Password screen appears. Contact an instructor or lab assistant for the Name and Administrator password.

Name _____

Password _____

15. The Date and Time screen appears. Enter the **correct date and time** and click on the **Next** button.

16. The Network Settings screen appears. Click on the **Typical Settings** button.

17. The Workgroup or Computer Domain screen appears. Click on the **No, this computer is not on a network or is on a network without a domain** radio button. Click on the **Next** button to continue.

18. Setup continues to copy more files. When instructed to, remove the installation CD and click on the **Finish** button.

Lab Project 19 Challenge

The goal of this project is to verify any errors that occurred during the Windows 2000 installation.

1. Turn on the computer and log in to Windows 2000 using the userid and password provided by the instructor or lab assistant.

2. Right-click on the **Start** button. Click on the **Explore** option.

3. In the left window, locate the drive letter on which Windows 2000 was loaded. (Normally this is the C: drive.) Contact the instructor or a student assistant if you are unsure. Click on the **+ (plus sign)** beside the drive letter. If there is no plus sign by the drive letter, but instead it is a minus sign, skip this step.

4. In the left window, locate the folder in which Windows 2000 was loaded. (Normally this is WINNT.) Contact the instructor or a student assistant if you are unsure. Double-click on this folder.

5. In the left window, locate the folder called WINDIR and click on the **+ (plus sign)** beside this folder.

6. Click on the **Tools** menu option.

7. Click on the **Folder options** selection.

8. Click on the **View** tab.

9. Locate the Hidden files and folders option and ensure the radio button beside the Show hidden files and folders option is selected. If it is not, click in the **radio button** to enable it. Click on the **OK** button.

10. In the right window, if files and folders do not appear, click on the **Show files** option. Locate the **SETUPACT.LOG** file and double-click on it. Notepad opens with this file. This file contains a listing of all installation actions performed.

    What was the first listing shown in the SETUPACT.LOG of what was done during the installation process?

    _____

11. Close the SETUPACT.LOG file by clicking on the **Close** button (**X**) located in the upper right corner of the window.

12. In the right Explorer window, locate the **SETUPERR.LOG** file and double-click on it. Notepad opens with this file. The SETUPERR.LOG file contains a listing of any errors that occurred during the Windows 2000 installation process.

    Were any errors logged during the Windows 2000 installation? If so, write one of the errors in the space below.

    _____

13. Close the SETUPERR.LOG file by clicking on the **Close** button (**X**).

Internet Discovery

Internet Discovery Lab 1

1. What is the URL for the Microsoft Windows 2000 Professional on-line help?

2. When Windows 2000 Professional is in Standby mode, a stop 0x9F error appears. What should you do? Write the URL and the answer in the space below.

3. Frank Condron has a web site for Windows 2000. On this web site, Frank describes what to do if you get a blue screen of death with the stop error of DATA_BUS_ERROR. What is Frank's solution and what is the URL where you found the solution?

4. ZD, Inc. has a web site for Windows 2000. Write the URL for this site.
   Locate one book on the Internet that deals with troubleshooting Windows 2000 Professional. Write the name of the book, the author, and the URL in the space provided.

5. How many Service Packs are currently available for Windows 2000 Professional? Write the number and URL where you found the answer in the space below.

Soft Skills: Help Desk Support

1. A customer named Marsha Weng calls you and says that her Windows 2000 Professional computer system is running very slowly. It had been working fine prior to today. Identify at least three problems and resolutions.
2. You attempt to install Windows 2000 Professional on a computer that is a Pentium running at 300 MHz with 24 MB of RAM and 10 GB of hard disk space free. What potential issues, if any, do you see with this configuration before you install the operating system?
3. A customer named Matilda Ayers in Sydney, Australia calls you with a problem. She needs to load third party disk drivers on a computer during the installation process. What do you tell her?
4. You work as a consultant for a large international firm and you are currently assigned to a customer in Madrid, Spain. You are working on the Windows 2000 Professional computer for the Vice President of Production. When booting the computer, you receive a message that NTLDR is missing or corrupt. What should you do?

Critical Thinking

1. What is the difference between *%systemroot% and %systemdrive%*?

2. What do you do if you forget the password for the Administrator account?

3. If you can only boot one operating system at a time, what is the purpose of dual-booting and multi-booting?

4. How could you fix a corrupt Master Boot Record in Windows 2000?

5. Explain how encryption works.

6. Discuss the pros and cons of compression.

Study Skills

The "See" in "Hear, See, Do, Say"

Remember the key to learning are these four points: hear, see, do, say. This section focuses on the visual part or the "see" study skill.

- One of ways we learn and get input into our brain is through our senses. Consider your eyes as an "input device," allowing sensory data to enter the "processor" or brain.
- Your instructor may use visual aids on the overhead such as a PowerPoint presentation or demonstrating a lab activity. You learn a great deal in the computer industry by using your eyes.
- Make sure you can "see" the classroom chalkboard or white board. You may see this again on a test.
- Make sure you can "see" the classroom overhead screen. You could see this material on a test as well.
- Make sure you can "see" the computer screen in front of you.
- Pay attention to the output in front of you. The computer is telling you something whether it is good or bad.
- If you have to move closer to the board, do so. Ask your instructor for assistance.
- If you cannot see the screen in front of you, you may need to get your eyes checked.
- "Seeing" does <u>not</u> involve using another person's answers (known as **cheating**).

Self-Study Question(s)
1. Can you clearly see the necessary screens or boards in the classroom?
2. Identify at least two "see" Study Skills you did this week.

7

Chapter 7
Introduction to
Microsoft
Windows XP

OBJECTIVES

The goal of this chapter is twofold:
- To introduce you to Windows XP.
- To help you prepare and pass the following section of the A+ Operating System Technologies Exam:

A+ **Operating System Technologies Exam Objectives**
covered in this chapter (and corresponding page numbers)

1.1 Identify the major desktop components and interfaces, and their functions. Differentiate the characteristics of Windows 9x/Me, Windows NT 4.0 Workstation, Windows 2000 Professional, and Windows XP.

1.2 Identify the names, locations, purposes, and contents of major system files.

1.4 Identify basic concepts and procedures for creating, viewing, and managing disks, directories and files. This includes procedures for changing file attributes and the ramifications of those changes (for example, security issues).

1.5 Identify the major operating system utilities, their purpose, location, and available switches.

Domain 2 Installation, Configuration and Upgrading

2.1 Identify the procedure for installing Windows 9.x/Me, Windows NT 4.0 Workstation, Windows 2000 Professional, and Windows XP and bringing the operating system to a basic operational level.

2.3 Identify the basic system boot sequences and boot methods, including the steps to create an emergency boot disk with utilities installed fro Windows 9.x/Me, Windows NT 4.0 Workstation, Windows 2000 Professional, and Windows XP.

2.4 Identify procedures for installing/adding a device, including loading, adding, and configuration device drivers, and required software.

Domain 3 Diagnosing and Troubleshooting

In this chapter, you will complete the following sections:
- 7.1 Understanding Microsoft Windows XP
- 7.2 Understanding the Pre-Installation Steps of Windows XP
- 7.3 Understanding How to Install and Upgrade Windows XP
- 7.4 Troubleshooting the Windows XP Installation
- 7.5 Understanding How to Dual-Boot Windows XP
- 7.6 Backing Up and Restoring the Windows XP Registry
- 7.7 Configuring Windows XP
- 7.8 Understanding the Boot Process
- 7.9 Understanding Task Manager, Dr. Watson, and Event Viewer
- 7.10 Monitoring System Performance

7.1 Understanding Microsoft Windows XP

Windows XP is a 32-bit operating system but a 64-bit version is also available. Microsoft created Windows XP in two different versions: Home and Professional. Windows XP Home is for personal computers and Windows XP Professional is for a business environment. The following list shows you the features available in both versions.

- Supports disk quotas to limit users on storage space.
- Core reliability and stability like Windows 2000.
- Enhanced support for movies, pictures, and music.
- Contains Windows Messenger that is a collaboration tool for instant messaging and video conferencing.
- Supports **Internet Connection Firewall** (**ICS**), which can help prevent attacks to your computer from hackers when connected to the Internet.
- Improved boot and power resume performance.
- Can automatically configure 802.1x wireless networks—to be discussed in Chapter 9, *Introduction to Networking*.
- Allows you to view output on multiple monitors—this is called **DualView.**
- CDs can be made using drag-and-drop or the CD Writing Wizard.
- Supports **WIA** (**Windows Image Acquisition**), which allows communication between software application and image-capturing devices. The Scanner and Camera wizard is used to retrieve images from any WIA-enabled device.
- Can switch between users without rebooting.

There are quite a few differences between Windows XP Professional and Windows XP Home. Here is a list of specific features available on the Professional edition:

- Supports two CPUs.
- Supports roaming profiles.
- Supports Remote Desktop and Remote Assistance.
- Has Computer Management and Performance Monitor.
- Supports joining a Windows-based server domain.
- Allows Group Policies with users, which allow you to customize user and computer accounts.
- Provides Encrypting Files System (EFS) as discussed in Chapter 6, *Introduction to Windows 2000.*
- Supports dynamic disk support allowing you to increase the size of a partition.
- Supports **IPSec** (**IP Security**), which allows you to secure and encrypt TCP/IP data.
- Upgradeable from Windows 98, Windows NT, Windows ME, or Windows 2000 Professional.

Here is a list of specific features available on the Home edition:

- Supports only one CPU.
- Supports only Remote Assistant.
- Upgradeable from Windows 98 and Windows ME.

Windows XP Professional supports 32-bit Windows applications, 16-bit Windows applications, and some DOS applications (only those that do not access hardware directly). Every 32-bit application runs in its own 2 GB memory space. All 16-bit Windows applications run in a single virtual machine. A virtual machine simulates a single computer with its own memory, hardware devices, and software configuration.

DOS and 16-bit Windows applications run in a single 2 GB memory space. The Windows XP **VMM (Virtual Memory Manager)** handles allocating memory to applications. A single block called a **page** is 4 KB in size. A page is used to store files and may also retrieve a file located on a disk. This file is called a **paging file**.

Windows XP supports **WFP (Windows File Protection)**. WFP protects system files (files critical to the operating system). WFP runs in the background. When WFP detects that a file has been altered or deleted, it copies a replacement file from the WINDOWS \SYSTEM32\DLLCACHE folder, from the Windows XP CD, or from a network share (a shared folder that contains a copy of the XP CD).

Windows XP uses the same file system hierarchy concept to store files and folders as Windows 2000. So, a section covering file and folder management will not be covered here. For review, refer to Section **6.2 entitled Managing Files and Folders on Windows 2000** in Chapter 6, *Introduction to Windows 2000*.

With Windows XP, the first screen to appear when you boot the computer is the Welcome screen. The Welcome screen allows users to log in. A user can log in to the local computer, to a workgroup (peer-to-peer network), or to a domain (a network with a server). See Chapter 9, *Introduction to Networking*, for more information on network types. Note that the log-in screen cannot be bypassed as it could in Windows 9x.

To log in, click on a user account icon and enter a password (if necessary) or press **Ctrl + Alt + Del** to go to the Log On to Windows box.

Note that the local Administrator account (the master account that is allowed to change everything on the local machine) is not a user icon. You must press **Ctrl + Alt + Del** *twice* and enter the correct userid and password to use the Administrator account.

7.2 Understanding the Pre-Installation Steps of Windows XP

Windows XP can be installed from a central location or locally. The pre-installation of any operating system is more important than the installation process. Technicians that grab a disk or CD and load a new operating system without going through a logical process are just asking for trouble. There are two major portions of any new operating system installation—hardware and software. The hardware and software already installed in the system must be compatible with the operating system and researched before any installation steps are taken. The steps to be taken before installing Windows XP are outlined next.

- Decide whether the installation will be an upgrade or clean install and which version of XP is to be loaded.
- Determine the file system(s) to be used.
- Decide whether the computer will have more than one operating system installed.
- Scan for viruses.
- Determine if the hardware is compatible.
- Obtain any drivers, upgrades, or hardware replacements.
- Decide if the software applications are compatible.
- Obtain any patches, upgrades, or software replacements.
- Delete any unwanted files and uninstall any unwanted applications.
- Back up any data files necessary.
- Remove any power management or disk management tools.

The first decision to make when planning to install Windows XP is whether to upgrade from another operating system or perform a clean install. A **clean install** puts an operating system on a computer without one, or the computer's existing operating system is removed (the hard drive is formatted). Three reasons exist to perform a clean install:

- The computer does not have an operating system already installed.
- The computer's current operating system is not upgradeable to Windows XP.
- The computer's current operating system is upgradeable to Windows XP, but the existing files and applications are going to be reloaded.

If the decision is made to upgrade, then determine what operating system is already installed. Windows XP Professional supports upgrading from Windows 98, Windows ME, NT Workstation 4, 2000 Professional, and XP Home Edition. Windows XP Home edition only supports upgrades from Windows 98 and Windows ME. When Windows XP is installed as an upgrade, the user's applications and data are preserved if the operating system is installed in the same folder (directory) as the original operating system. If Windows XP is installed in a different folder, then all applications must be reloaded.

If any of the following situations exist, Windows XP Professional should be used.

- The computer contains multiple processors. Windows XP Professional supports two processors. Multiple processors are normally used on a database or web server. If the computer runs an application that does intense computations such as financial or scientific applications or graphic rendering such as a game uses, then Windows XP Professional is the best choice.
- The computer is in a networked environment and remote access to the computer as if you were there is desired.
- File system encryption is needed.
- The computer is in a networked office environment.

Another decision you must make if upgrading to Windows XP is whether or not to convert the hard drive partition to NTFS. Once a partition is converted to NTFS, the partition cannot be changed. If you are unsure whether or not to convert the partition, leave it unchanged and later use the CONVERT.EXE program to upgrade. Most people want to convert the partition to NTFS for the following reasons:

- Security (individual files can be protected with NTFS).
- More efficient use of cluster space (the cluster size can be defined based on the user's needs when using NTFS).
- NTFS supports file compression.
- NTFS supports larger hard drive partition sizes.
  The information in Table 7.1 helps when making the file system decision.

Table 7.1: FAT32 and NTFS Comparison

FAT32	NTFS
Hard drive is < 32 GB	Hard drive is > 32 GB and only one OS is installed
XP is dual-booted with Windows 95, 98, or 2000 and the existing OS uses FAT32 and files need to be shared between the two operating systems	File security is needed
	Disk compression is needed
	Upgrading from Windows NT or 2000 and the partition is already formatted as NTFS
	Disk quotas are needed
	File encryption is needed

If a FAT32 partition is desired but the FAT32 partition option is not available, the partition is most likely greater than 32 GB. Make the partition less than 32 GB or use NTFS as a file system.

If upgrading from a prior version of NTFS, the drive is automatically configured for XP's NTFS. If any older NTFS volumes are not powered during the installation process, the volume is automatically upgraded when the drive is mounted. If you want to dual-boot between Windows NT 4 and Windows XP (have both operating systems loaded), make sure that NT Workstation Service Pack 4 or higher is installed because some of the features in XP's NTFS change the data structure on disks. The CONVERT command can be used to change a FAT16 or FAT32 partition to NTFS5. The format of the command is CONVERT *x:* /fs:ntfs (where *x:* is the drive to be converted to NTFS).

In order to take advantage of Windows XP's reliability, enhancements, and security features, sometimes a clean installation is the best choice especially if the current operating system is Windows 98 or Windows ME. Because a clean installation involves formatting the hard drive, the user's data must be backed up and all applications reinstalled once the Windows XP installation is complete. Also, all user-defined settings are lost. Another important point to remember is that not all Windows 3.x, 95, and 98 applications are compatible with Windows XP. You can contact the company that developed your application and see if the application is compatible with Windows XP.

If the computer already has NT Workstation 4 or Windows 2000 Professional, then a Windows XP upgrade is recommended. However, if there are hardware drivers for such devices as a DVD player, power management software, or network utilities loaded on the computer, a clean installation may be a better choice. Whichever the case, the user's data and applications should be backed up and restored once the Windows XP installation is complete.

The third decision that must be made is whether or not Windows XP will be installed with one or more other operating systems. This situation is often called a dual-boot or multi-boot scenario. **Dual-boot** means that the computer can boot from two operating systems. **Multi-boot** means the computer can boot from two or more operating systems. Windows XP can be dual-booted with DOS, Windows 95 (SR2), Windows 98, NT Workstation, or Windows 2000 Professional. If this is desired, a separate hard disk partition should be created and used for each operating system. When doing a dual or multi-boot configuration, make sure that Windows XP is installed _after_ the other operating systems. The fourth step in planning for a Windows XP installation is to scan the system for viruses. Viruses can cause havoc on a new upgrade. Viruses are covered in section A4.4 Understanding the Pre-Installation Steps of Windows 98 of Chapter 4 entitled _Introduction to Microsoft Windows 98_.

The fifth thing to do when installing Windows XP is to determine what computer hardware is installed. Table 7.2 lists the minimum and preferred hardware requirements for installing Windows XP Professional.

Table 7.2: Windows XP Professional Requirements

Component	Minimum	Preferred
CPU	Intel Pentium (or compatible) 233 MHz	Intel Pentium II (or compatible) 300 MHz or higher
RAM	64 MB	128 MB
Free hard drive space	1.5 MB	> 1.5 MB
Input Device	Keyboard, mouse, or other pointing device	Keyboard, mouse, or other pointing device
Multimedia Drive	CD-ROM or DVD	CD-ROM or DVD 12x or faster

Microsoft has a tool called the Upgrade Advisor that checks your system for hardware and software compatibility issues. This tool can be downloaded from Microsoft's web site at http://www.microsoft.com/windowsxp/pro/howtobuy/upgrading/advisor.asp. Be aware that the tool might not be able to detect all hardware devices or software applications.

Do *not* download a BIOS update unless you are sure it is compatible with your computer. Installing an invalid update can damage your computer system and cause it not to operate.

Once you have verified all of your hardware, you may have to get Windows XP device drivers from the hardware device manufacturer or their web site. This is the sixth step. You may also need to upgrade the hardware device which usually means replacing the device. This is sometimes the cost of going to a more powerful operating system. You may also decide at this point not to upgrade but to buy a computer with Windows XP already installed.

The seventh determination you must make before installing Windows XP is whether or not any existing software applications are compatible. The preparation for installing a new operating system is usually more work than the actual installation process, but any omitted step will cost you more time in the long run. Use the Upgrade Advisor or contact the developer of each software application to determine if it is compatible. You may also go to the software developer's web site and the information may be posted there. Microsoft also has a list of compatible software on their web site.

Windows XP has a help function that can check for software compatibility after XP is loaded. To access this tool, click on **Start** button, select **Help and Support**, and in the search box type **Program Compatibility wizard**.

Once you have determined whether the software is compatible with Windows XP, you may have to obtain software patches, upgrades, or buy a new version. This is best done before you install Windows XP. Be proactive, not reactive—solve any problems you can *before* upgrading or installing any operating system.

An installation option is the **Check Upgrade Only Mode**, which does not install Windows XP but checks whether your hardware and software are compatible. At the end of the check there is a report generated that lists potential issues.

As with any upgrade, hardware change, or software change, data needs to be backed up. It is really funny that the worst people in the world for backing up data are technicians, but a user's data is very valuable to them. Backing up data is an essential step. Whether you do a clean install or an upgrade, if the user has data on the computer, it must be backed up before starting the installation process. Also, before backing up data, remove any unwanted files and/or applications that are no longer needed in order to free up hard drive space.

The last step in the pre-installation checklist is to remove any power or disk management tools that are loaded. Computer manufacturers for older operating systems frequently provide these types of tools. Power or disk management tools can interfere with the new tools provided with Windows XP. Sometimes software such as this can prevent an operating system from installing whether it is an upgrade or a clean installation.

7.3 Understanding How to Install and Upgrade Windows XP

Once all the pre-installation checklist steps are completed, you are ready to install Windows XP. The installation process is easy if you performed the pre-installation steps. The number one piece of advice to give you when installing any operating system is to do your homework first. The number of possible problems will be greatly reduced.

During the installation process, the computer must be restarted three times. During the first phase, a selection must be made whether to upgrade or perform a clean installation, the product key must be entered, and a basic hardware check including available disk space is accomplished. The computer restarts. After the restart, the second phase begins and setup runs in text mode. During this process, a partition to install XP can be chosen and setup files are copied to the partition. The computer restarts and the third phase begins. During this portion, devices are installed, the Administrator password is entered, and the operating system is created. The system restarts a final time and the log-on screen is presented.

Microsoft requires activation of the Windows XP operating system within 30 days. No name or personal information is required, but activation must occur. You can activate XP over the phone or the Internet. Most new computers that have XP preloaded do not require activation.

7.4 Troubleshooting the Windows XP Installation

Installation problems can be caused by a number of factors. The following list shows the most common causes and their associated solution during the installation process.

- **Incompatible BIOS**—Obtain compatible BIOS, replace the motherboard with one that has compatible BIOS, or do not upgrade/install Windows XP.
- **BIOS needs to be upgraded**—Upgrade the BIOS.
- **Incompatible hardware**—Replace the hardware or do not upgrade/install Windows XP.
- **Incompatible hardware drivers**—Obtain Windows XP drivers from the hardware manufacturer.
- **Incompatible TSRs**—Remove TSRs or obtain updated ones from the software manufacturer.
- **Incompatible applications**—Obtain upgrades from software manufacturer.

- **Minimum hardware requirements have not been met**—Upgrade the hardware. The most likely things to check are the CPU (233 MHz minimum) and RAM (64 MB minimum).
- **A virus is on the hard drive**—Run an anti-virus program and remove the virus.
- **Pre-installation steps have not been completed**—Go back through the list!
- **The installation floppy disks or CD is corrupted** (not as likely as the other causes)— Try the disk in another machine and see if you can see the contents. For the CD, check to see if any scratches or dirt are on the surface. Clean the CD as necessary.
- **Incorrect CD Key**—Type in the correct CD key to complete the installation. The key is located on the CD case.
- If a **STOP message occurs** when installing a dual boot system, boot from the Windows XP installation CD rather than the other operating system.
- If the **installation halts**, try removing any nonessential hardware such as network cards, modems, and USB devices and start the installation again. Reinstall the hardware once XP is loaded.
- If the **computer locks up** during setup and shows a blue screen, check the BIOS and hardware compatibility.
- If a message appears during setup that **a device driver was unable to load**, obtain the latest device drivers that are XP-compatible and restart the setup program.
- When upgrading from Windows 98 or ME to XP and **setup displays an error that states it has disabled the upgrade option**, clean boot the computer and try to run setup again. If that does not work, copy the I386 folder from the Windows XP CD and run setup manually by locating the folder and double-clicking on the **WINNT32.EXE** file.
- If **setup hangs during the file copy phase**, the SMARTDRV command in the AUTOEXEC.BAT file has switches that interfered with the installation. Modify the AUTOEXEC.BAT file to remove SmartDrive switches.
- After the file copying has been completed, if **setup displays the message that it cannot set the required XP configuration information**, a hardware conflict is normally the cause.
- If a **STOP: 0x0000001E (0x800000003, 0xBFC0304, 0X0000000, 0x0000001) error occurs**, there is either not enough disk space to load XP, an incompatible or outdated driver is installed, or the motherboard BIOS needs updating.

Several text files located in whatever folder Windows XP was loaded can be helpful in determining the installation problem—SETUPLOG.TXT and SETUPAPI.LOG. These two files can be opened with any word processor including Notepad.

7.5 Understanding How to Dual-Boot Windows XP

Sometimes users like to try a new operating system and keep the old operating system loaded with the new system installed too.

Any time a dual-boot situation is desired, the oldest operating system should be installed first. The operating systems need to be in separate hard disk partitions.

If Windows XP is installed on an NTFS partition, only Windows 2000 and NT Workstation (with Service Pack 4 or higher) can access the XP partition. In the situation where Windows XP is loaded on a FAT32 partition, only Windows 95 (SR2), Windows 98, and Windows ME can access the partition. One solution to this scenario is to create three partitions—one for XP using NTFS; one for Windows 95 (SR2), Windows 98, or Windows ME using FAT32; and a third partition for shared data that is FAT32 (a partition type that both operating systems can access).

After installing all files and rebooting, a menu appears with the Microsoft Professional option, NT Workstation option, and NT Workstation (VGA Mode) option. To select which option is the default operating system, right-click on the **My Computer** desktop icon and click on the **Properties** option from the drop-down menu. Click on the **Advanced** tab and select the Startup and Recovery **Settings** button. In the System Startup section, select the **Default Operating System** drop-down menu.

7.6 Backing Up and Restoring the Windows XP Registry

The registry is a database that contains information about the Windows XP environment including installed hardware, software, and users. The registry should be backed up whenever the computer is fully functional and when any software or hardware changes are made.

The registry should be backed up and restored on a working computer *before* disaster hits. The time to learn how to restore the registry is *not* when the computer is down.

The registry can be backed up and restored several different ways. The two most common methods used are the REGEDIT and the Backup tools. The **REGEDIT program** allows you to export the registry to a file that has an extension of .REG. The file can be imported back into the computer if the computer fails. The REGEDIT program and the **Backup utility** both back up the entire registry. Refer to Figure 7.1 for a sample screenshot.

Figure 7.1: The Windows XP Registry

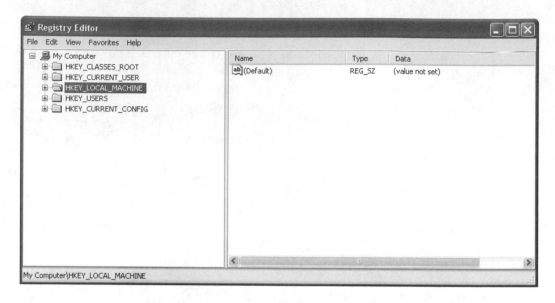

Backing Up the System State

One option available in the Backup utility is the System State. The **System State** is a group of important Windows XP files including the registry, the system files, the boot files, and the COM+ Class Registration database. With the Backup utility, you cannot back up or restore these items individually. They are all needed because they depend on one another to operate properly. The registry files are located in a folder normally labeled *%systemroot%*\Repair\Regbackup, which is normally C:\WINNT (or WINDOWS, depending on the type of installation) \REPAIR\BACKUP. The registry can be restored without having to restore the other System State files.

In order to use the Backup program, you must be an Administrator or a member of the Backup Operators group.

Refer to Figure 7.2 for a sample screenshot of the Backup and Restore Wizard program.

Figure 7.2: The Backup and Restore Wizard

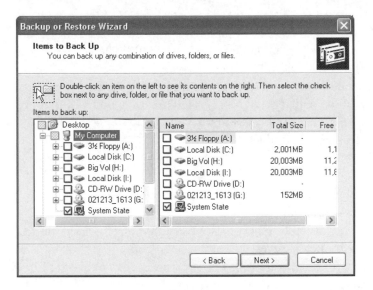

Restoring the System State After a Failure

In order to correct a problem with the system files, registry, or Windows XP boot failure, you must restore the registry. You may also have to restore the System State files (which includes the registry) to make the system operational again. To start the restoration process, install Windows XP to the same folder that it was installed in originally. When you are prompted to format the hard drive volume or leave it, select the **Leave the current file system intact** option. Use the Backup utility to restore the System State and/or the registry (All Programs, Accessories, System Tools, Backup). Click on the **Restore** tab and select the device that holds the backed up files.

Let's perform an exercise to help you understand the registry. Make sure you ask your instructor before performing this exercise. The goal of this exercise is to familiarize you with the REGEDIT registry editing utility. You will need a computer with Windows XP installed and a formatted 3.5″ floppy disk.

REGEDIT is a utility used for editing the Windows registry. With REGEDIT, you can view existing registry settings, modify registry settings values, or create new registry entries to change or enhance the way Windows operates.

In this lab, you will use REGEDIT to view the System BIOS and Video BIOS information on your computer.

CAUTION!: Editing the registry can cause your computer to run erratically or not run at all! When performing any registry editing, follow ALL directions carefully including spelling, syntax use, etc. Failure to do so may cause your computer to fail!

1. From the **Start** menu, choose **Run**, type **REGEDIT**, and click **OK**. The REGEDIT utility opens.
2. In the left window, expand **HKEY_LOCAL_MACHINE**, **HARDWARE**, and **DESCRIPTION** by clicking on the + (plus) symbol located to the left of the name. Click on the **System** option located under DESCRIPTION. The system BIOS and video BIOS information display in the right window.
3. REGEDIT can be used to back up and restore part or all of the registry. To illustrate this point, a portion of the registry will be exported to disk and then imported into the registry. Ensure the following option is still selected in the Registry window: HEKY_LOCAL_MACHINE\Hardware\Description\System
4. Click on the **File** menu option and select **Export**. The Export Registry File window opens.
5. Insert a blank formatted disk into the A: drive. Click on the **Save in** down arrow and select the **3_ Floppy (A:)** option.
6. In the File name text box, type **Registry System Section** and click on the **Save** button. The specific registry key is saved to disk.
7. To restore the registry (or a portion of it as in this exercise), click on the **File** menu option and select **Import**. The screen should list the file located on the A: drive, but if it does not, select the 3_ Floppy (A:) option from the Look in drop-down menu.
8. Click on the **Registry System Section** file name and click on the **Open** button. A message appears when the section is successfully inserted into the registry. Show this message to the instructor or lab assistant.
9. Close the REGEDIT utility and properly power off the system.

7.7 Configuring Windows XP

One of the most common windows used by technicians is the control panel window. A **control panel** is a method for configuring various Windows components. Each control panel icon represents a Windows utility that allows you to customize a particular part of the Windows environment. The number of control panels displayed depends on the type of computer and the components contained within the computer. Windows XP has two control panel views—classic and category. Figure 7.3 shows the Windows XP control panel category view.

Figure 7.3: Windows XP Control Panels

The following exercises will help familiarize you with several control panel categories. The goal of this exercise is to be able to use the appropriate control panels to configure a mouse, keyboard, and enable disabilities options. A computer with Windows XP installed is required.

1. Turn on the computer and verify that the operating system loads.
2. Log on to Windows XP using the userid and password provided by the instructor or lab assistant.
3. Click on the **Start** button and select the **Control Panel** option. **Pick a Category** should display in the right window. If it does not, click on the **Switch to Category View** option in the left pane.

Keyboard Configuration

4. Click on the **Printers and Other Hardware** control panel category. Click on the **Keyboard** icon. The Keyboard Properties window appears.
5. Click on the **Speed** tab. The Keyboard Properties window contains two tabs—Speed and Hardware. The Speed tab has three settings: repeat delay, repeat rate, and cursor blink rate.
6. The **Repeat delay** option configures the duration of wait time before a key starts repeating. This is especially important for people who do not type well or who have to

use a device such as a pencil to press keys down. The **Repeat rate** is an adjustment for how fast characters repeat across the screen. The **Cursor blink rate** controls how many times the cursor blinks per second. Adjust each of these settings and test them using the **Click here and hold down a key to test repeat rate** area.

7. Configure the keyboard settings back to their original configuration.
8. Click on the **Hardware** tab. The Hardware tab is used to access the keyboard trouble-shooting wizard and the keyboard driver. Click on the **Properties** button.
9. Click on the **Driver** tab. The **Update driver** button is used to load a new keyboard driver. Click on the **Cancel** button twice and return to the Printers and Other Hardware control category window.

Mouse Configuration

10. Click on the **Mouse** icon. The Mouse Properties window appears. The options available depend on the mouse manufacturer, but some of the settings are standard.
11. On the Buttons tab, there are three standard options—Button configuration, Double-click speed, and ClickLock. The **Button Configuration** section is where the mouse buttons can be reversed for left-handed people.
12. Adjust the **Double-click speed** and test it using test folder located in the right window of this section.
13. Reset the **Double-click speed** setting to its original configuration. Refer to the answer in the previous question.
14. The **ClickLock** setting is so you can select an option and drag the mouse without holding the left mouse button down. Once a click is made for more than a second, the button locks and the icon can be dragged. When a second click is made, the mouse unlocks.
15. The mouse troubleshooter and driver is accessed through the Hardware tab. Click on the **Hardware** tab. Click on the **Troubleshoot** button. The Mouse Troubleshooter window appears. Close the Mouse Troubleshooter.
16. Access the Mouse control panel's Hardware tab. Click on the **Properties** button.
17. Click on the **Driver** tab. Just like with the keyboard, the **Update Driver** button is used to load a new mouse driver.
18. Click on the **Cancel** button twice to return to the Printers and Other Hardware control panel category. Click on the **Back** button to return to the control panel categories.

Accessibility Options

19. Accessibility options are not just for people with disabilities. The settings can be applicable to any computer user to make their computer environment more comfortable. Click on the **Accessibility Options** category. Click on the first task, **Adjust the contrast for text and colors on your screen**. The Accessibility Options window opens with the Display tab active.

20. The two configuration sections are High Contrast and Cursor Options. Click on the **Use High Contrast** check box to enable it and click on the **Apply** button. A "Please wait" message appears and then the screen changes.

21. Click on the **Use High Contrast** check box to disable it and click on **Apply**. The screen returns to normal. Click on the **Cancel** button and the Accessibility Options control panel category window reappears.

22. Click on the second task, **Configure Windows to work for your hearing, vision, and mobility needs**. The Accessibility Wizard appears. This wizard steps through visual, auditory, and motor skills settings. Click on the **Next** button. The Text Size window appears.

23. Click on the **Next** button and the Display Settings window appears.

24. Click on the **Next** button and the Set Wizard Options window appears.

25. The option that is probably the most vague is **Administrative options**. This option is used to turn certain accessibility features off if the computer sits idle and make the accessibility features available to one user or all users. Click on the **Cancel** button. A Save Changes message box appears. Click on the **No** button so that all configuration changes are not kept.

Controlling Sound

26. Access the **Sounds and Audio Devices** control panel category. Select the **Adjust the system volume** task.

27. The Volume tab is used to control the volume for the entire computer system and speaker configuration. The **Mute** check box is used to mute all of the computer's sound. The **Place volume icon in the taskbar** is used to add a volume control icon in the taskbar in the notification area. The Device Volume slide bar sets the computers value settings. Click on the **Advanced** button located in the Device volume section. The Volume Control window opens.

28. Click on the **Mute all** check box to enable it. If it is already enabled, leave the setting turned on (enabled). Close the Volume Control window. Return to the Sounds and Audio Devices Properties window. You will have to reaccess the control panel category.

29. Return all Volume Control settings back to their original settings and return to the Sounds and Audio Devices Properties window.

30. Click on the **Speaker Volume** button. The Speaker volume screen has a left and right speaker volume. This setting does not affect speakers that simply plug into the Line out connection on the sound adapter or built into the motherboard. Click on the **Cancel** button.

31. The **Advanced** button in the Speakers settings section is used to configure speakers for such things as headphone usage and surround sound. Click on the **Advanced** button in the Speakers settings section and click on the **Speakers** tab. The Speakers setup list is used to specify external speakers. Computers such as ones in a business environment or a lab can be configured for no speakers.

32. To disable speakers, click on the **Speaker setup down arrow** and select the **No Speakers** option. Click on the **Cancel** button twice and close the control panel window.

33. Power off the computer properly.

Operating system specialists must frequently add new hardware and software using the operating system. Windows XP has specific tools for these functions. Using the correct procedure is essential for success. The following sections highlight many of the tasks a technician must perform:

• Adding Plug and Play Devices
• Adding Non-Plug and Play Devices
• Adding a Printer
• Removing Hardware Devices
• Installing and Removing Software

Hardware devices are physical components that connect to the computer. Hardware devices can be either plug and play or non-plug and play. A device driver is a piece of software that allows hardware to work with a specific operating system. Some device drivers are automatically included with Windows XP. A technician must be aware of what hardware is installed into a system so that the latest XP-compatible drivers can be downloaded and installed.

Adding Plug and Play Devices

Plug and play devices are hardware and software designed to automatically be recognized by the operating system. In order for Windows XP to fully support plug and play devices, the computer should have a BIOS that supports **ACPI** or **Advanced Configuration and Power Interface**. ACPI allows the computer's motherboard and operating system to control power needs and modes of operation of various devices. The key to a successful plug and play device installation includes the following:

• Possessing the most up-to-date device driver
• Following the directions provided by the device manufacturer

Install the device according to the device manufacturer's instructions. Once installed, power on the computer. The Windows XP **Found New Hardware** wizard appears. Windows XP attempts to find a driver for the new device. Plug and play devices make use of a special .CAB (cabinet) file called DRIVER.CAB located in *%winroot%*\Driver Cache\i386 folder (where *%winroot%* is normally C:\WINNT or C:\WINDOWS). This file is almost 75 MB and contains more than 2,500 compressed files. If Windows XP detects new hardware, it will automatically search DRIVER.CAB for a driver. If a driver cannot be found, a dialog box appears. The best policy with any operating system is to use the latest driver even if the operating system detects the device. An exercise at the end of the chapter outlines how to install a new hardware driver.

In XP Professional, remember if the operating system cannot configure a plug and play device and prompts for a device driver, you must have Administrator rights to install the driver.

Adding Non-Plug and Play Devices

Devices, known as legacy devices, are also called non-plug and play devices. For devices that are not plug and play, Windows XP has a control panel called **Add Hardware**. If using control panel categories, select the **Printers and Other Hardware** category and then select **Add Hardware** from the left pane. The Add Hardware wizard allows hardware configuration and is used for hardware that is not automatically recognized by Windows XP. It is also used for plug and play devices that don't install properly with Windows XP's automatic detection. You must have Administrator privileges in order to load device drivers for new hardware. An exercise at the end of the chapter explains how to do this.

With both plug and play and non-plug and play devices, the Device Manager tool is used to view installed hardware devices, enable or disable devices, troubleshoot a device, view and/or change system resources such as IRQs and I/O addresses, update drivers, and access the driver roll back option. The **driver roll back** option is a new feature in Windows XP. It allows an older driver to be reinstalled when the new driver causes problems.

You must have Administrator privileges to access or use the driver roll back option.

If the device driver has not been updated, driver roll back will not be possible and a message screen displays stating this fact. The troubleshooting tool should be used instead to troubleshoot the device. An exercise follows that details how to use this feature.

Place older adapters closest to the power supply because XP checks expansion slots in order starting with the closest to the power supply. By putting the older adapters in these slots, XP will allocate system resources to the older adapters first and there is less chance of system resource conflicts.

Sometimes XP can install the wrong driver for an older device or adapter. From Device Manager, right-click on the device and you can uninstall the device driver or disable it. Sometimes the computer must reboot and XP will reinstall the wrong driver (again). The solution to this is to disable the device and then manually install it. The following exercise illustrates how to disable a device on a computer running Windows XP. To manually install new hardware, use the following steps:

1. Log on to the computer.

2. Click on the **Start** button and select the **Control Panel** option.

3. If in Category View, click on **Printers and Other Hardware** and select **Add Hardware** from the left pane. If in Classic View, double-click on the **Add Hardware** icon. The Add Hardware wizard starts.

4. Make sure the new hardware is physically connected and select the **Yes, I have already connected the hardware** option and click on the **Next** button.

5. Use the Installed hardware scroll bar to find and select the **Add a new hardware device** checkbox and click on **Next**.

6. Select the manual option and select the type of hardware being installed. Scroll through the manufacturer list or have an XP-compatible driver ready and click on the **Have Disk** button to install the appropriate driver. Click on the **Next** button to finish the device driver installation.

Adding a Printer

Printers can be connected to a computer through the printer's parallel port, USB port, or through a local area network. Only parallel printers (printers connected to the computer's parallel port) and USB printers will be covered in this chapter. Networked printers will be in Chapter 9, *Introduction to Networking*.

Windows XP can automatically detect printers. If Windows XP detects a printer, the operating system automatically installs drivers, updates the registry, and allocates system resources to the printer.

To install a printer, connect the printer to the appropriate computer port with the appropriate cable. First, power the computer on. The Windows XP wizard normally detects and leads you through the installation process. However, if XP does not detect the printer and it is a printer attached to the parallel port, have the Windows XP-compatible printer driver ready and perform the following steps on a computer running Windows XP:

1. Log on to the computer.

2. Click on the **Start** button and select **Control Panel**.

3. If in Category View, click on **Printers and Other Hardware**. If in Classic View, double-click on the **Printers and Faxes** icon.

4. Access the **Add a printer** icon or selection. The **Add Printer** wizard begins. Click on the **Next** button.

5. Ensure the **Local printer attached to this computer** radio button is enabled and click on **Next**. Windows XP tries to detect a plug and play printer. If it cannot detect it, you are advised of this fact and that a manual installation is required. Click on the **Next** button.

6. Ensure the **Use the following port** radio button is enabled, select the correct port (usually LPT1), and click **Next**.

7. In the Install Printer Software window, click on the **Have Disk** option, insert the XP-compatible driver, and click on **Next**. Select the appropriate model and click on **Next**.
8. Name the printer and select whether the printer will be the default printer.
9. Click **Next**.
10. When prompted if the printer is to be shared, select the appropriate response. If shared, the printer must have a share name and optionally list the location and enter comments. Click **Next**.
11. Select the option to print a test page and click **Next**.
12. Click on the **Finish** button.

If a printer has a TCP/IP connection, manually install the printer using the steps above except instead of selecting the Local printer option, select the **Standard TCP/IP Port** option and click **Next** twice. Enter the printer name or IP address, the port name, and click **Next**. The Install Printer Software window appears and the same steps are used as a local printer.

To configure a printer as a default printer (the printer that applications normally use), locate the printer in the Printers folder. Access the Printers folder by clicking on the **Start** button and clicking on the **Printers and Faxes** selection. Once you locate the appropriate printer icon, right-click on the icon. Click on the **Set as Default Printer** option. In the Printers folder, the default printer has a check mark next to the icon.

Removing Hardware Devices

Windows XP normally detects when hardware has been removed and the operating system automatically removes the device's driver(s). If Windows XP does not automatically detect the device removal, you must manually remove the drivers. An exercise follows that describes how to remove the driver.

If you are removing a printer from the system, use the **Printers** control panel. Access this control panel by clicking on the **Start** button. Point to the **Settings** option and click on the **Printers** option. Right-click on the printer you want to delete and choose the **Delete** option.

Let's perform an exercise.

The goal of this exercise is to install an updated driver under the Windows XP operating system. You will need a computer with Windows XP installed and Internet access. In this lab a new driver is loaded, but then the old driver is reinstalled with the driver roll back feature. The student must be logged in as a user with local Administrator rights to perform this lab.

1. Turn the computer on and verify that the operating system loads. Log in to Windows XP using the userid and password provided by the instructor or lab assistant.
2. Pick an installed hardware device and locate an updated driver using the Internet and download the driver to the hard drive. Note that some drivers may come in a compressed file and must be uncompressed before continuing the procedure.
3. Click on **Start** button, point to **Settings**, and click on the **Control Panel** menu option.

4. When the Control Panel window opens, click on the **Switch to Category View** option if necessary. Select the **Printers and Other Hardware** category. In the left *See also* pane, select the **System** option. The System Properties window opens.

5. Click on the **Hardware** tab and select the **Device Manager** button.

6. Click the **+** (plus sign) beside the hardware category that contains the device being upgraded.

7. Right-click on the device name and click on the **Properties** selection.

8. Click on the **Update Driver** button. The Update Hardware wizard screen appears.

9. Select the **Install from a list or specific location (Advanced)** radio button and click **Next**.

10. Click on the **Don't search. I will select the driver to install** radio button and click **Next**.

11. Click on the **Have Disk** button, use the **Browse** button to locate the downloaded file, and click on **OK**. A list of models might appear. If so, select the correct model and click on **Next**. Finish the driver update.

12. Use Device Manager and right-click on the device name again and select **Properties**.

13. Click on the **Driver** tab and click on the **Roll Back Driver** button. Click on the **Yes** button to roll back the driver. If the device driver has not been updated, driver roll back will not be possible and a message screen displays this fact.

14. Close all windows and power off the computer properly.

Installing and Removing Software

No computer is fully functional without software. One thing you should know about Windows XP is that it does not support some of the older 16-bit software. Most software today is 32-bit and comes on CD and includes an autorun feature. If the CD has the autorun feature, an installation wizard steps you through installing the software when the CD is inserted into the drive. If there is not an autorun feature on the CD or if the software is on a disk, then the Add or Remove Programs control panel is used to install or remove the software.

To access the Add or Remove Programs control panel, click on the **Start** button and click on the **Control Panel** option and then double-click on the **Add or Remove Programs** control panel. Click on the **Add New Programs** icon in the window's left panel. Click on the **CD or Floppy** button and ensure the software disk or CD is inserted in the appropriate drive. If a SETUP.EXE file cannot be found on the designated disk, the system prompts with a dialog box. Use the **Browse** button to locate the installation file. Click on the **Finish** button to complete the process.

To remove a software application, use the same **Add or Remove Programs** control panel; instead of clicking on Add New Programs, click on the **Change or Remove Programs** icon in the left panel. A list of installed applications appears. Locate the software to be removed and click on its name. Click on the **Remove** button. When asked if you are sure you want to remove this software, click on the **OK** button and close the control panel window. Some applications have their own uninstall program. Refer to the application's help file or look in the application's folder where the application was installed for an uninstall icon. The Add or Remove Programs control panel can also be used to update operating system components. An exercise follows that illustrates this concept.

Once an application is installed, launch the application by clicking on the **Start** button and pointing to the **All Programs** option. Locate the application name and click on it. If the application does not appear on the list, do not panic. The most frequently used program names appear in the left Start button panel. Let's perform an exercise to help you understand how to install software.

The goal of this exercise is to be able to install Administrative Tools to Windows XP. You will need a computer with Windows XP installed, the Administrator password and a Windows XP CD.

In this lab, if Administrative Tools is already loaded, it will be removed and reinstalled.

1. Turn the computer on and verify that the operating system loads. Log in to Windows XP using the userid and password provided by the instructor or lab assistant. Ensure the userid is one that has Administrator rights.

2. Click on the **Start** button, point to **All Programs** and look for an Administrative Tools item. Does the Administrative Tools item appear in the All Programs list? If so, proceed with the exercise. If it does not appear, skip to step 5 below.

3. Right-click on the **Start** button and select the **Properties** option. Click on the **Start Menu** tab.

4. Click on the **Customize** button. The Customize Start Menu window opens. In the Start Menu items section, locate the **System Administrative tools** section and click in the **Don't display this item** radio button. Click on the **OK** button. Click on the **OK** button again. Verify that the Administrative Tools no longer displays in the All Programs list.

5. Right-click on the **Start** button and select the **Properties** option. Click on the **Start Menu** tab.

6. Click on the **Customize** button. The Customize Start Menu window opens. In the Start Menu items section, locate the **System Administrative tools** section and click in the **Display on the All Programs menu** radio button. Click on the **OK** button. Click on the **OK** button again. Verify that the Administrative Tools displays in the All Programs list.

7.8 Understanding the Boot Process

With NT Workstation, Windows 2000, and Windows XP, there are two types of partitions that are important during the boot process—the system partition and the boot partition. The **system partition** is the active drive partition that has the files needed to load the operating system. The system partition is normally the C: drive (the active partition). The **boot partition** is the partition or logical drive where the operating system files are located. One thing that people sometimes forget is that the system partition and the boot partition can be on the same partition. These partitions are where certain boot files are located.

Every operating system needs specific files that allow the computer to boot. These files are known as **system files** or startup files. The system files and their specific location on the hard drive are listed in Table 7.3.

Table 7.3: Windows XP Boot Files

Startup File Name	File Location
BOOT.INI	Root directory of system partition.
BOOTSECT.DOS (needed if dual- or multi-boot system)	Root directory of system partition.
CDLDR	Root directory of system partition.
HAL.DLL	%systemroot%\SYSTEM32 (%systemroot% is a variable representing the folder where Windows 2000 is installed. This is normally C:\WINNT.)
HYBERFIL.SYS	%systemdrive% (%systemroot% is the root directory on the drive where Windows 2000 boot files are located which is normally C:\.)
NTBOOTDD.SYS (used with SCSI drives that have the SCSI BIOS disabled)	Root directory of system partition.
NTDETECT.COM	Root directory of system partition.
NTLDR	Root directory of system partition.
NTOSKRNL.EXE	%systemroot%\SYSTEM32
System Key	%systemroot%\SYSTEM32\CONFIG

Reading about Windows XP files can be confusing because the file locations frequently have the entries *%systemroot%* and *%systemdrive%*. This is because computers can be partitioned differently. If you install Windows XP onto a drive letter (a partition or logical drive) other than the active partition (normally C:), the startup files can be on two different drive letters. Also, you do not have to take the default folder name of WINNT or WINDOWS (depending on the type of installation) to install XP. To account for these different scenarios, Microsoft uses the *%systemroot%* to represent the boot partition, the partition and folder that contains the majority of the Windows XP files. *%systemdrive%* represents the root directory and on a computer with a single operating system, this would be C:\.

If Windows XP is installed onto the C: drive and the C: drive is the active partition, then the BOOT.INI, BOOTSECT.DOS, HYBERFIL.SYS, NTBOOTDD.SYS, NTDE-

TECT.COM, and NTLDR files would all be in the root directory of C:. The HAL.DLL and NTOSKRNL.EXE files would be located in the SYSTEM32 folder (located in either the WINNT or WINDOWS folder) on the C: drive.

Another example would be if you installed Windows XP onto the D: drive, but the C: drive is the active partition. The BOOT.INI, BOOTSECT.DOS, HYBERFIL.SYS, NTBOOTDD.SYS, NTDETECT.COM, and NTLDR files would all be in the root directory of C:. The HAL.DLL and NTOSKRNL.EXE files would be located in the SYSTEM32 folder (that is located under the WINNT or WINDOWS folder) on the D: drive.

The boot process is actually quite involved, but the major steps are as follows:

- Power on the computer.
- POST executes.
- BIOS searches CMOS for the boot device order and checks for a boot sector. If the boot device is a hard drive, BIOS reads the Master Boot Record (MBR), and locates and loads the information into sector 0 of the system partition. The contents of sector 0 define the type of file system, the location of the bootstrap loader file, and start the bootstrap loader. With Windows XP, this file is NTLDR.
- NTLDR starts in real mode so that 8- and 16-bit software can be loaded. Then XP is switched to 32-bit mode and the file system begins to load.
- NTLDR reads the BOOT.INI file and displays the various operating system choices contained within the BOOT.INI file. If something other than Windows XP is chosen, the BOOTSECT.DOS file takes over. If Windows XP is chosen, the NTDETECT.COM file executes.
- NTDETECT.COM detects the computer's hardware and ACPI tables are read so that XP can detect power management features.
- NTLDR passes the hardware information to the NTOSKRNL.EXE file.
- The operating system kernel, NTOSKRNL.EXE, executes and the HAL.DLL file loads. **HAL** (**Hardware Abstraction Layer**) is a layer between the operating system and the hardware devices. The HAL allows Windows XP to run with different hardware configurations and components without affecting (or crashing) the operating system.
- The registry key HKEY_LOCAL_MACHINE\System loads. This registry key is located in the %*systemroot*%\System32\Config\System file. This key has information found during the hardware detection process and is used to determine which device drivers to load.
- The operating system kernel initializes and NTLDR passes control to it. The Starting Up process bar displays. During this time, a hardware key is created, device drivers load, and services start.
- The WINLOGON.EXE file executes and the log-on screen appears. While the log-on process is occurring, XP detects plug and play devices.

Troubleshooting the Boot Process

Quite a few things can cause XP to not boot properly. XP has a wealth of tools and start modes that can be used to troubleshoot the system. If XP boots, but still has a problem, try to solve the problem without booting into one of these special modes. For example, if one piece of hardware is not working properly and the system boots properly, use Device Manager and the troubleshooting wizards to troubleshoot the problem. Another problem can be caused by an application that loads during startup.

To disable startup programs, hold the **Shift** key down during the log-on process and keep it held down until the desktop icons appear. For a permanent change to an application starting automatically, move or delete the startup shortcuts from the one of the following places:

%systemdrive%\Documents and Settings*Username*\Start Menu\Programs\Startup
%systemdrive%\Documents and Settings\All Users\Start Menu\Programs\Startup
%windir%\Profiles*Username*\Start Menu\Programs\Startup
%windir%\Profiles\All Users\Start Menu\Programs\Startup

Four major tools can be used to troubleshoot Windows XP boot problems: Last Known Good Configuration, Safe Mode, Recovery Console, and Automated System Recovery wizard.

The Last Known Good Configuration

Last Known Good Configuration is used when XP has been just changed by adding hardware or software that is incompatible with XP, has configuration settings that conflict with XP or other hardware/software, or an important service such as the one used to initialize SCSI (Small System Computer Interface) hard drives has been accidentally disabled. It reverses the most recent changes that have been made since the last successful XP boot. Whenever XP boots successfully (without any problems), a copy of the Clone control set is made to be used if the Last Known Good Configuration options is chosen. A control set is a registry subkey located under HKEY_LOCAL_MACHINE\System and contains information about devices and services. To access Last Known Good Configuration, press the **F8** key during the boot process and select the **LastKnownGood Configuration** option from the Windows Advanced Options menu.

Whenever the **Last Known Good Configuration** option is used, all configuration changes made since the last successful boot are lost! However, since the changes are the most likely cause of XP not booting correctly, **Last Known Good Configuration** is a useful tool when installing new devices and drivers that do not work properly. If Last Known Good Configuration does not work properly, boot the computer into Safe Mode, which is covered in the next section.

If Windows XP works, but a hardware device does not work and a new driver has been recently loaded, use the driver roll back option for the device.

Safe Mode

Safe Mode is used when the computer stalls, slows down, does not work right, improper video, intermittent errors, or new hardware/software installation causes problems. Safe Mode is used to start XP with minimum device drivers and services. Software that automatically loads during startup is disabled in Safe Mode and user profiles are not loaded. When the computer boots in Safe Mode, the mouse, keyboard, CD-ROM, and VGA video device drivers are all that are loaded. Safe Mode allows you to access configuration files and make necessary changes, troubleshoot installed software and hardware, disable software and services, and adjust hardware and software settings that may be causing XP from starting correctly. The bottom line is that Safe Mode puts the computer in a "bare bones" mode so you can troubleshoot problems.

Another option similar to this is **Safe Mode with Networking** that operates the same as Safe Mode except that it allows network drivers and services to load. **Safe Mode with Command Prompt** is used to start the system with the minimum files and drivers and a command prompt where you must use commands instead of a graphic interface. Access Safe Mode by pressing the **F8** key during the boot process and selecting **Safe Mode** from the Windows Advanced Options menu.

The Windows Advanced Options menu contains other options used to troubleshoot Windows XP which are Enable Boot Logging, Enable VGA Mode, Directory Services Restore Mode, and Debugging Mode. **Enable Boot Logging** is used to create a log file, NTBTLOG.TXT, that records drivers and services loaded during the boot process. This text file is located in the folder that contains the operating system, which is normally C:\WINNT or C:\WINDOWS. The file can be used to identify problems that occurred during the boot process and that relates to a driver or service. Each loaded driver or service file is marked as loaded or not loaded and this information can be used to narrow suspect problems. Boot logging is enabled for every Windows Advanced Options menu selection except for Last Known Good Configuration.

Enable VGA Mode is used to start XP with a standard VGA video driver instead of the one currently installed. This option is used when the current video driver will not load or loads in such a way that it is hard to reconfigure XP. With this option enabled, XP loads and the desktop can be seen and used to load a newer driver or adjust the current video settings. All Windows Advanced Options menu selections except for Last Known Good Configuration load the standard VGA driver.

Directory Services Restore Mode is for server operating systems and is used to restore the SYSVOL directory and Active Directory service on a network domain controller. The **Debugging Mode** option is used by programmers and engineers to send debugging information through a serial cable to another computer. Other options may be available on the Windows Advanced Options menu if you are using or have installed and have used the remote installation service.

Recovery Console

Recovery Console is used when Safe Mode and other startup options do not solve a problem. Recovery Console allows access to hard drive volumes without starting the GUI (Graphical User Interface). In other words, the Recovery Console allows you access to a command prompt from which you use commands to start and stop services, repair and access hard drive volumes, replace corrupt files, and perform a manual recovery.

You must have the administrator password to access the full potential of this option.

Recovery Console is not loaded onto the system by default, but it can be installed from the XP CD and then loaded through the boot menu or executed from the XP CD. Normally technicians run Recovery Console from the CD because Recovery Console is needed when there is a problem and is not preloaded.

To run Recovery Console from the Windows XP CD, use the XP CD to start XP. If the CD is unavailable or the computer does not support booting from CD, use XP setup boot disks (obtained from Microsoft). Press **R** at the Welcome to Setup screen to select the repair the installation option and the Recovery Console window appears. Press the number that corresponds to the partition where XP is loaded. An Administrator password prompt appears. Type the Administrator password and press **Enter**. A command prompt appears.

To install Recovery Console, insert the XP CD into the CD drive. If the welcome screen appears, close it. Click on the **Start** button, click on the **Run** option, and type in the appropriate path to access the **i386** folder and the **WINNT32 /CMDCOMS** command. An example is F:\I386\WINNT32 /CMDCOMS. This starts the Recovery Console installation process. When asked if you want to install Recovery Console, click on the **Yes** button and follow the prompts on the screen. Once Recovery Console is installed, restart the computer and Recovery Console appears as a boot option. Use the arrow keys to select the Recovery Console option and press **Enter**. A command prompt appears.

You must sometimes work from the command prompt when the system is not working properly. That is what the Recovery Console tool is all about. You may want to go through the Chapter 3, *DOS and the DOS Command Line Interface*, to understand the process and procedures needed when working from a command prompt.

If the registry is corrupt or has been deleted, you are not prompted for an Administrator password. Instead, the system boots to a prompt where you can use basic commands like CHKDSK, FIXBOOT, and FIXMBR to repair the system. However, you cannot access any folders on the hard drive.

The drive letters available at the Recovery Console command prompt might not be the same ones you used in the GUI environment. Use the MAP command to see the drive letters (and the volumes that do not have drive letters). The syntax for the MAP command is covered later in this chapter.

There are several ways to access a command prompt when the computer is functional. These methods are listed below.

- Click on the **Start** button, click on the **Run** option, and type **cmd** in the dialog box.
- Click on **Start** button, click on the **Run** option, and type **command** in the dialog box. Note that when this option is used, the keyboard arrow keys do not bring up previously used commands as the CMD command does.
- Click on **Start** button, point to the **All Programs** option, point to the **Accessories** option, and click on the **Command Prompt** option.

Some of the most frequently used commands used from the Recovery Console command prompt are outlined in Chapter 6, *Introduction to Windows 2000*. Some commands have different options when used within Recovery Console. To get help from a command prompt running within Windows XP or from within Recovery Console, type **Help** to see a list of commands; type *Help command-name* (where *command_name* is the command itself); or type *command_name /?*.

If using FAT, the Recovery Console is installed, and the partition is converted to NTFS, Recovery Console will have to be reinstalled.

The copy command can be used through Recovery Console to restore the two important files, system and software, that are used to build two important registry keys, HKEY_LOCAL_MACHINE \System and HKEY_LOCAL_MACHINE \Software. If the unresolved problem relates to hardware, try replacing the System file first. If the problem relates to software, replace the software file first. Do not replace both files at the same time because the System or Software files may not be current, which means that drivers or service packs may have to be reinstalled after replacement.

Recovery Console has four default limitations of which a technician should be aware. They are as follows:

- No text editor is available in Recovery Console by default.
- Files cannot be copied to removable media such as floppy disks while in Recovery Console. Write access is disabled.
- The Administrator password cannot be changed from Recovery Console.
- Some folders such as Program Files and Documents and Settings are inaccessible from the Recovery Console prompt.

Automated System Recovery

Automated System Recovery replaces the Emergency Repair Disk used by NT and 2000 Professional and it uses the Backup tool to backup up important system files used to start Windows XP. Automated System Recovery does not backup data files (although the Backup program can be used to back data up too).

To create an Automated System Recovery disk, you will need a 1.44 MB floppy disk and media such as a CD (if the machine has a CD-RW drive) or tape (for a tape drive). The floppy is used to boot the system and then you can restore the files if the hard drive crashes or operating system is inoperable.

To access the Backup tool, which is used to back up the system files, click on the **Start** button, point to the **All Programs** option, point to **Accessories**, point to **System Tools**, and click on the **Backup** option. The Backup wizard begins. On the initial screen, click on the words **Advanced Mode**. The words are underlined in the window. Click on the **Tools** menu option and select **ASR Wizard**. The Automated System Recovery wizard starts.

To use the disk and media created with Automated System Recovery, you will need the floppy disk created when the system was backed up, the backup media written to when the system was backed up, and the original Windows XP CD. Start the computer using the Windows XP CD. During the Setup process, press the **F2** key. A prompt appears to insert the Automated System Recovery floppy disk into the floppy drive. Insert the disk and follow the screen directions to restore the system.

System Configuration Utility

The **System Configuration utility** is used to disable startup programs and services selectively one at a time or several at once. This graphical utility reduces the chances of possible typing errors, deleting files, and other misfortunes that occur when technicians work from a command prompt. Only an Administrator or a member of the Administrators group can use System Configuration utility.

To start System Configuration utility, click on the **Start** button, click on the **Run** option, type in **msconfig** and press **Enter**. Figure 7.4 shows the System Configuration utility's General tab.

Figure 7.4: System Configuration Utility General Tab

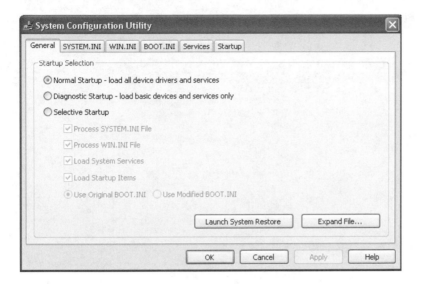

The **General** tab has three radio buttons: Normal Startup, Diagnostic Startup, and Selective Startup. **Normal Startup** is the default option and all device drivers and services load normally when this radio button is selected. The **Diagnostic Startup** radio button is selected when you want to create a clean environment for troubleshooting.

When **Diagnostic Startup** is chosen and Windows XP restarts, the system boots to Safe Mode and only the most basic device drivers and services are active. The Selective Startup radio button is the most common troubleshooting tab on the General tab. When **Selective Startup** is chosen, you can selectively pick which startup options load.

You can narrow down the startup file that is causing boot problems. Start with the first check box, **Process SYSTEM.INI File**, and deselect the check box. Click on the **OK** button and restart the computer. Once you determine which file is causing the problem (the problem reappears), click on the **System Configuration Utility** tab that corresponds to the problem file and deselect files until the exact problem file is located.

The System Configuration Utility Services and Startup tabs are also quite useful when troubleshooting boot problems. Certain applications, such as an anti-virus program, run as services and many services are started during the boot process. The Services tab can be used to selectively disable and enable these boot services. Enabling the **Hide All Microsoft Services** option allows you to view and manipulate third-party (non-Microsoft) services. The Startup tab allows you to enable and disable Windows-based startup programs. Figure 7.5 shows a sample Startup tab screen.

Figure 7.5: System Configuration Utility Startup Tab

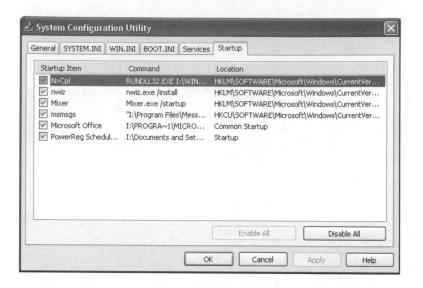

 Windows XP does not support creating an Emergency Repair Disk (ERD) in the same way that NT Workstation and Windows 2000 Professional do. A set of installation disks can be obtained from Microsoft and would only be needed if the computer does not support booting from CD.

Let's perform an exercise to help you understand the System Configuration Utility.

The goal of this exercise is to be able to use the System Configuration utility to troubleshoot boot problems. You will need a computer with Windows XP installed and you will need Administrator rights.

1. Turn the computer on and verify that the operating system loads. Log in to Windows XP using the userid and password provided by the instructor or lab assistant. Ensure the userid is one that has Administrator rights.

2. Right-click on the **Start** button and click on the **Explore** option.

3. Locate the **Documents and Settings** folder and expand it if necessary. Locate the **All Users** folder (located under Documents and Settings) and expand it if necessary. Locate the **Start Menu** subfolder (located under All Users) and expand it if necessary. Locate the **Programs** subfolder (located under the Start Menu folder) and expand it if necessary. Click on the **Startup** folder located under the Programs folder.

4. Use the **Search** Start button option to locate the original Notepad application (notepad.exe). Create a shortcut to the Notepad application and place it in the Startup

folder located under the Programs folder (see step 3). A previous exercise explains how to create a shortcut.

Have a classmate verify your shortcut (especially that it is a shortcut and not a copy of the application or the application itself). Is the icon in the STARTUP folder a shortcut icon?

Classmate's printed name: _____

Classmate's signature: _____

5. Restart the computer and verify that the Notepad program starts automatically when the computer boots. If it does not, re-do step 4.
6. Click on the **Start** button, click on the **Run** option, type in **msconfig** and press Enter. The System Configuration utility window opens.
7. Click on the **Diagnostic Startup—load basic devices and services only** radio button. Click on the **Apply** button and then click on the **Close** button. A System Configuration message box appears. Click on the **Restart** button. When the computer restarts, log in with the same userid used previously.

What is different about the way Windows XP loads?

Did the Notepad application automatically start?

8. Click on the **OK** button. Click on the **Selective Startup** radio button found on the General tab. Check boxes are now available that you can select the startup files that are to be loaded the next time the computer boots. Click in the **Load Startup Items** check box. Click on the **Apply** button and then click on **Close**. Click on the **Restart** button and the system restarts. Log in using the same userid and password.

Did the Notepad application automatically start? Why or why not?

9. Click on the **OK** button. Click on the **Normal Startup—load all device drivers and services** option located on the General tab.
10. Click on the **Startup** tab. Click on the **Shortcut to notepad** check box to disable it.
11. Click on the **Apply** button and then click on **Close**. Click on the **Restart** button. When the computer restarts, log in using the same userid and click on **OK**.

Did the Notepad application automatically start? Why or why not?

What is different about the System Configuration Utility's General tab?

12. Click on the **General** tab and select the **Normal Startup** radio button. Click on the **Apply** button and then **Close**. Click on the **Restart** button. Log in using the same userid.
13. Once the computer reboots, remove the shortcut to the Notepad application from the Startup folder.

7.9 Understanding Task Manager, Dr. Watson, and Event Viewer

Task Manager is a Windows-based utility that displays applications currently loaded into memory, processes that are currently running, microprocessor usage information, and memory usage data. To activate the Task Manager utility, press the **Ctrl + Alt + Del** keys. Another way of accessing Task Manager is to right-click on the taskbar and then click on the Task Manager option.

One of the common uses of Task Manager is to exit from an application that is "hung up" or not responding. Task Manager can help with exiting the program. Once inside the Task Manager window, click on the **Applications** tab. Locate the name of the troublesome application and click on it. Normally, if an application is causing a problem, the status shows the application as "not responding." Select the problem application and click on the **End Task** button. Figure 7.6 shows a sample Task Manager's Applications tab.

Figure 7.6: A Screenshot of Task Manager

Let's perform a few exercises to help you understand Windows XP Task Manager and Event Viewer.

The goal of this exercise is to use Task manager to halt an application. You need a computer with Windows XP installed.

1. Turn the computer on and verify that the operating system loads. Log in to Windows XP using the userid and password provided by the instructor or lab assistant. Ensure the userid is one that has Administrator rights.

2. From the **Start** menu, choose **All Programs, Accessories,** and then select **Notepad**. The Notepad utility opens.
3. To access Task Manager, simultaneously press **Control, Alt**, and **Delete**. The Task Manager window opens.
4. Select the **Applications** tab.
5. Click on the **Untitled—Notepad** option and click on the **End Task** button. Notepad closes.
6. Close the Task Manager window.

Dr. Watson is a utility that automatically loads when an application starts. Dr. Watson can detect and display troubleshooting information as well as create a text log file (DRWTSN32.LOG) when a system or application error occurs. A technician might need this information when communicating with Microsoft or the application developer's technical support. Make notes of any messages that appear on the screen when any type of problem occurs.

To start Dr. Watson in Windows XP, click on the **Start** button, click on the **Run** option, and type **drwtsn32** and press Enter. Click on the application error and click on the View button. The default location for the log file is C:\Documents and Settings\All Users \Application Data\Microsoft\DrWatson. When an error occurs, Dr. Watson appends information to the end of this log file.

Event Viewer is a Windows tool used to monitor various events in your computer such as when a driver or service does not start properly. The EventLog service starts automatically every time a computer boots to Windows XP. This service is what allows the events to be logged and then Event Viewer is used to see the log.

Access the Event Viewer by clicking on the **Start** button, then click on the **Control Panel** option, and if in Category view, click on the **Performance and Maintenance** category, click on **Administrative Tools** icon, and double-click on the **Event Viewer** icon. If in Classic Control Panel view, double-click on the **Administrative Tools** control panel icon. Then, double-click on the **Event Viewer** icon. The left window contains the type of Event Viewer logs such as the application log, the security log, and the system log. The application log displays events associated with a specific program. The programmers who design software decide which events to display in the Event Viewer's application log. The security log displays events such as when different users log in to the computer (both valid and invalid log ins). A technician can pick which events are displayed in the security log. All users can view the system log and the application log, but only a member of Administrators can enable security log information.

The most commonly used log is the system log. The system log displays events that deal with various system components such as a driver or service that loads during startup. The type of system log events cannot be changed, added, or deleted. Click on the system log option in the left panel. The system log events displays in the right window. Figure 7.7 shows an example of Event Viewer's system log.

Figure 7.7: A Screenshot of Event Viewer

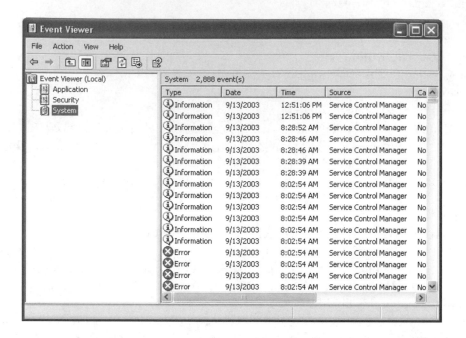

 Double-click on an Event Viewer event to see more information about it.

Event viewer logs can be saved as files and viewed later. This is especially useful with intermittent problems. Use the **Action** menu item to save and retrieve saved event viewer log files. Let's perform a few exercises to help you understand Windows XP Task Manager and Event Viewer.

The goal of this exercise is to use Event Viewer. You need a computer with Windows XP installed.

1. Turn the computer on and verify that the operating system loads. Log in to Windows XP using the userid and password provided by the instructor or lab assistant. Ensure the userid is one that has Administrator rights.

2. Event Viewer is used to monitor various events such as when drivers and services load (or fail to load and have problems). Click on the **Start** button, click on the **Control Panel** option. If in Control Panel Category view, click on **Performance and Maintenance** category, and click on **Administrative Tools**. If in Classic view, double-click on **Administrative Tools** control panel icon, and then double-click on the **Event Viewer** icon. The Event Viewer window opens.

3. Click on the **Application** log located in the left pane. Application events list in the right pane.

4. Double-click on any application event.
5. Close the Event Properties window. Click on the **System** log located in the left plane. System events list in the right pane.
6. Double-click on any of the individual events. Click on the button that looks like two pieces of paper. It is the button directly under the up and down arrow buttons.
7. Click on the **Start** button. Click on the **Run** option. Type **clipbrd** and press Enter. The event is copied to the Clipboard and the **clipbrd** command opens the Clipboard Viewer.
8. Open Notepad by clicking on the **Start** button, pointing to **All Programs**, pointing to **Accessories**, and clicking on the **Notepad** option.
9. Click on the **Edit** menu option and select **Paste**.
10. The event information can be saved as a text file and referenced later especially when there is a problem. Close Notepad without saving the document.
11. Close Event Viewer.

Shutdown Problems

Windows XP should be shut down properly when all work is finished. Before Windows XP can shut down, the operating system sends a message to all devices, services, and applications. Each device that is running sends a message back saying it is okay to shut down now. Any active application saves data that has not been previously saved and sends a message back to the operating system. Active system services also respond that it is okay to shut down. If the system has trouble shutting down, it is due to one of the following three things. The most common problem is an application that is not responding back. When this happens, press **Ctrl + Alt + Del** to access Task Manager. Manually stop any applications that show a status of not responding. You can also click on any applications and stop them to see if they are causing the problem. Sometimes a program will not show a status of not responding until you try to manually stop the application from within Task Manager. If a single application continually prevents Windows XP from shutting down, contact the software manufacturer to see if there is a fix.

For services problems, boot the computer into Safe Mode and then shut the computer down. Take note as to whether or not the computer had any problems shutting down. If the process works, access the BOOTLOG.TXT file that is located in the root directory of the drive that contains Windows XP. Once inside the file, take note of each service that is disabled from booting into Safe Mode. Boot the computer normally. Stop each service one at a time to see which service is causing the problem.

To troubleshoot devices not responding, eliminate services and applications first because a device frequently does not cause a shutdown problem. Then, while working on the computer, take notice of what devices you are using. Common ones are video, hard drive, CD-ROM, keyboard, and mouse. Verify that all of your devices have the most up-to-date drive loaded and that the driver is compatible with Windows XP.

If you cannot stop the problem application or determine if the problem is a service or hardware, try restarting the computer instead of shutting down. Once the computer restarts, try shutting down again. As a last resort, use the computer's power button to power the computer off.

7.10 Monitoring System Performance

It is important for a technician to understand how a computer is performing and be able to analyze why a computer might be running slow. In order to do that, a technician must know what type of applications are being run on the computer and the effect of these applications on the computer resources. A technician must also be able to monitor the computer's resource usage when problems occur, change the configuration as needed, and observe the results of the configuration change.

Three utilities are commonly used to monitor system performance: Task Manager, Performance tool's System Monitor, and Performance tool's Performance Logs and Alerts. **Task Manager** is used to monitor your current system's performance. **System Monitor** is used to monitor real-time data about specific computer components. **Performance Logs and Alerts** allows you to create logs about the computer's performance and create alerts that notify you when a specific instance being monitored reaches a threshold that you define. It includes a summary graph of processor and memory usage.

Task Manager has been discussed in a previous section. How to use it to monitor your computer's performance is as follows. Access Task Manager and click on the Performance tab. Task Manager immediately starts gathering CPU and memory usage statistics and displays them in graph form in the window. Figure 7.8 shows the Task Manager graphs.

Figure 7.8: A Screenshot of Task Manager Showing the Performance Tab

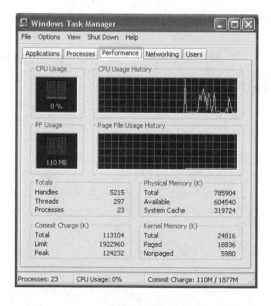

The first window on the left, **CPU Usage**, shows the CPU usage percentage or what percentage of time the processor is working. Actually, it is a percentage of time the processor is running a thread. A thread is a type of Windows object that runs application instructions. The first window on the right, **CPU Usage History**, is a graph of how busy the microprocessor is over a period of time.

The second window on the left (**PF Usage**) shows the amount of virtual memory (the paging file) being used. The amount shown is in megabytes as evidenced by the M after the number. If the display shows that the paging file is near the maximum, you can adjust the page file size. The following steps allow you to set the paging file size:

1. Click on the **Start** button and select **Control Panel**.
2. If in category view, select **Performance and Maintenance** followed by the **System** control panel icon. If in classic control panel view, double-click on the **System** control panel icon.
3. Click on the **Advanced** tab, locate the Performance section, and click on the **Settings** button.
4. Click on the **Advanced** tab, locate the Virtual Memory section, and click on the **Change** button.

The default paging file size is 1.5 times the total amount of physical RAM installed in the computer. Two values are selectable—Initial size and Maximum size. Both of these values should be the same for maximum computer performance. Once you change the values, click on the **Set** button. The Virtual Memory window can also be used to change the amount of space reserved for the registry.

The second window on the right, **Page File Usage History**, is a graph of the virtual memory used over time.

Memory is a frequent bottleneck for computer performance issues. Task Manager can also be used to see the total amount of RAM installed and how much RAM is available. Look in the Physical Memory information section in the Task Manager window to see this.

If you determine that memory is a problem, there are several things you can do including increasing the amount of RAM installed in the system, create multiple paging files when multiple hard drives are installed in the system, manually set the paging file size, run applications that require a lot of memory with all other applications closed, close any unnecessary windows, avoid having too many applications open, upgrade the hard drive or add another hard drive, and run the disk defragmenter program provided with XP.

Task Manager also has the Networking tab that is useful to technicians. The **Networking** tab shows a graph of network performance. The information shown can also be changed by selecting the **View** menu option, clicking on the **Select Columns** option, clicking in the available check boxes, and clicking on the **OK** button. Figure 7.9 shows this window.

Figure 7.9: A Screenshot of Task Manager Showing the Networking Tab

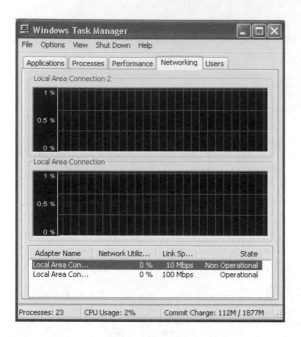

Sometimes a computer can start slowing down. The most common cause of slow-downs are that the computer's resources are insufficient or an application is monopolizing a particular resource such as memory. Other causes of slowdowns include a resource that is not functioning properly or is outdated such as a hard drive, the resource is not configured for maximum performance and needs to be adjusted, or resources such as hard drive space and memory are not sharing workloads properly. They need to be adjusted.

A baseline report is needed before the slowdown occurs. A **baseline** is a snapshot of your computer's performance during normal operations (before it has problems). Task Manager can be used to get an idea of what normal performance is, but the System Monitor and Performance Logs and Alerts tools are better suited to capturing and analyzing specific computer resource data.

To access the Performance tool (which contains System Monitor and Performance Logs and Alerts), perform the following steps:

1. Click on the **Start** button and access **Control Panel**.
2. If in category view, click on **Performance and Maintenance**, click on the **Administrative Tools** control panel icon, and double-click on the **Performance** icon. If in classic control panel view, double-click on the **Administrative Tools** control panel icon, and double-click on the **Performance** icon.

The Performance tool can also be accessed from a command prompt by clicking on the **Start** button, clicking on **Run**, typing **perfmon.msc** and pressing Enter.

The Performance window opens and System Monitor and Performance Logs and Alerts list in the left pane. Click on the **System Monitor** option and the tool starts collecting and displaying real-time data about the local computer or, if configured, from remote computers. A previously captured log file can also be loaded. Data can be displayed in graph, histogram, and report views. Figure 7.10 shows the System Monitor default screen (graph view).

Figure 7.10: A Screenshot of System Manager Showing the Networking Tab

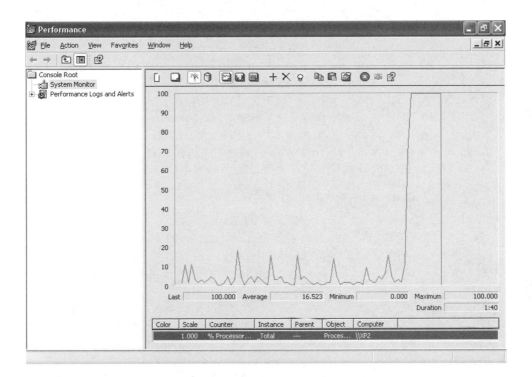

Notice in Figure 7.10 at the bottom of the window is the legend for interpreting the graph including what color is used for each of the performance measures and what counter is being used. A counter is a specific measurement for an object. Common objects include cache, memory, paging file, physical disk, processor, system, and thread.

Running System Monitor also affects your computer's performance especially when you are using the graph view and are sampling large amounts of data. The following helps when running System Monitor or Performance Logs and Alerts:

- Turn off any screen saver.
- Use report view instead of graph view to save on resources.
- Keep the number of counters being monitored to a minimum.
- Sample at longer intervals such as 10 to 15 minutes rather than just a few seconds or minutes apart.

Chapter Summary

- Microsoft Windows XP is a 32-bit operating system that comes in two variations: Home and Professional.
- Before you install Windows XP, you need to run through a pre-installation checklist that includes the file system used, determining hardware compatibility, and virus checking.
- The installation of Windows XP is performed in three phases. During the first phase, you determine whether you are upgrading or performing a clean install. You also enter the Product Key during this phase. During the second phase, you partition the hard drive and enter the Administrator password. The operating system is loaded during this phase. In the final phase, the log-on screen is displayed to you.
- You may be required to troubleshoot your Windows XP installation. Some type of problems you may encounter are: incompatible hardware, hardware drivers and TSRs, incorrect Product Key, a virus, or the BIOS needs to be upgraded.
- Dual-booting allows other operating system to be installed on the same hard drive. Of course, you can only boot one operating system at a time.
- Because the registry is the database containing system information, you need to back it up whenever changes occur to the computer. You use the REDEDIT tool to backup and restore the registry. To backup the registry you would use the backup utility to back up the System State data.
- Configuring Windows XP involves using Control Panel to change the look and feel of your Windows XP system. Examples include: changing mouse, keyboard, and sound settings. You can also add and remove devices.
- You need to understand the boot process of Windows XP so you know how to troubleshoot when problems occur.
- Windows Task Manager allows you view the applications that are running on your system. You get to Task Manager by pressing Ctrl + Alt + Del. Dr. Watson is an application debugging tool and Event Viewer contains a log of events that have occurred on your system.
- You can use System Monitor to monitor the performance of your Windows XP system through the use of objects and counters.

Review Questions

1. Which of the following is not a reason to perform a clean install of Windows XP?
   a) The computer does not have an operating system already installed.
   b) The computer's current operating system is not upgradeable to Windows XP.
   c) The computer's current operating system is upgradeable to Windows XP, but the existing files and applications are going to be reloaded.
   d) You have a new printer and need to add a printer driver.

2. Which product helps protect your Windows XP computer from Internet hacker attacks?
   a) DualView
   b) WIA
   c) ICS
   d) Windows Messenger

3. Which Windows XP software provides for instant messaging and video conferencing?
   a) DualView
   b) WIA
   c) ICS
   d) Windows Messenger

4. Which Windows XP software allows you to view output on multiple monitors?
   a) DualView
   b) WIA
   c) ICS
   d) Windows Messenger

5. Which Windows XP software allows communication between software application and image-capturing devices?
   a) DualView
   b) WIA
   c) ICS
   d) Windows Messenger

6. Which Windows XP software allows you view the applications that are running on your system?
   a) Task Manager
   b) DualView
   c) WIA
   d) NTFS

7. You get to Task Manager by pressing _____.
   a) F8
   b) F2
   c) Alt + Del
   d) Ctrl + Alt + Del

8. _____ is an application debugging tool.
   a) Task Manager
   b) WIA
   c) Dr. Watson
   d) DualView

9. _____ contains a log of events that have occurred on your system.
   a) DualView
   b) Event Viewer
   c) Dr. Watson
   d) WIA

10. _____ is used to monitor the performance of your Windows XP system through the use of objects and counters.
    a) WIA
    b) DualView
    c) Performance Monitor
    d) System Monitor

11. During the boot sequence, _____ detects the computer's hardware.
    a) NTLDR
    b) NTDETECT.COM
    c) NTOSKRNL.EXE
    d) HAL.DLL

12. The name of the Windows XP bootstrap loader is _____.
    a) NTLDR
    b) NTDETECT.COM
    c) NTOSKRNL.EXE
    d) HAL.DLL

13. During the boot sequence, the _____ file is read and is used to display the various operating system choices.
    a) NTLDR
    b) NTDETECT.COM
    c) BOOT.INI
    d) HAL.DLL

14. The operating system kernel is called _____.
    a) NTLDR
    b) NTDETECT.COM
    c) BOOT.INI
    d) HAL.DLL

15. The bootstrap loader file starts in _____ mode so 8-bit and 16-bit software can be loaded.
    a) Protected
    b) Real
    c) WFP
    d) WIA

16. The _____ allows Windows XP to run with different hardware configurations and components without affecting (or crashing) the operating system.
    a) NTFS
    b) Real mode
    c) HAL
    d) WIA

17. The _____ is a group of Windows XP files that includes the registry, system files, and boot files.
    a) HAL
    b) IPSec
    c) ICS
    d) System State

18. The _____ utility is used for editing the Windows registry.
    a) HAL
    b) REGEDIT
    c) DualView
    d) System State

19. Which file system should you use if your hard drive is dual-booted with Windows 95?
    a) FAT32
    b) NTFS
    c) HAL
    d) Explorer

20. The _____ feature allows an older driver to be reinstalled when the new driver causes problems.
    a) ACPI
    b) Disk quota
    c) System State
    d) Driver roll back

Lab Projects

The following projects assume Windows XP is installed and running on a computer unless otherwise noted.

Lab Project 1

The goal of this project is for you to be able to use the Task Manager program to evaluate basic computer performance. You will need a computer with Windows XP installed and you need Administrator rights.

1. Turn the computer on and verify that the operating system loads. Log in to Windows XP using the userid and password provided by the instructor or lab assistant. Ensure the userid is one that has Administrator rights.

2. Press the **Ctrl + Alt + Del** keys to bring up Task Manager. Click on the **Performance** tab. The Performance tab is used to view CPU and page file usage.

3. Open Notepad, access the Internet if possible, open a game if possible, and start other applications.

   What happens to the CPU usage as displayed in Task Manager?

   _____

   What is the page file usage (PF Usage)?

   _____

   What is the total physical memory?

   _____

   How much RAM is available?

   _____

4. Close the Task Manager window.

Lab Project 2

The goal of this project is to use the System Monitor utility to track individual computer components. You need a computer with Windows XP and Administrative Tools loaded.

1. Turn the computer on and verify that the operating system loads. Log in to Windows XP using the userid and password provided by the instructor or lab assistant. Ensure the userid is one that has Administrator rights.

2. Click on the **Start** button, click on the **Control Panel** option. If in Control Panel Category view, click on **Performance and Maintenance** category, and click on **Administrative Tools**. If in Classic view, double-click on **Administrative Tools** control panel icon, and then double-click on the **Performance** icon. The Performance window opens. The Performance utility allows you to track individual computer component's performance. This is done through individual counters.

3. In the left window, click on the **System Monitor** item.

4. Click on the **Add button** (the button that has a plus sign on it) or right-click in the right window and click on the Add Counters option. The Add Counters dialog box opens.

5. Click on the **Performance object** down arrow. A list of system components appears such as Processor, physical disk, paging file memory, etc. Select the **Memory** performance object.

6. Once a system component has been selected, individual counters for that component can be selected and monitored. In the **Select counters from list** window, click on the **Available Bytes** counter. Click on the **Add** button.

7. Click on the Performance object **down arrow**. Select the **Paging File** performance object.

8. In the **Select counters from list** window, click on the **%Usage** counter. Click on the **Add** button.

   Using the Explain button, find out for what the %Usage counter is used. Write the explanation in the space below.

   _____

9. Close the Explain text message box. Using the method outlined in Steps 5 through 8, select two more counters to be monitored.

   What two counters did you add?

   _____

10. Click on the **Close** button. The right window in the Performance window displays a graph of the various counters. You may need to start some applications, do some cutting and pasting, or surf the Internet to see some of the counter activity. When finished, close the Performance window.

Lab Project 3

The goal of this project is to use the System Monitor utility to track individual computer components.

1. Turn the computer on and verify that the operating system loads. Log in to Windows XP using the userid and password provided by the instructor or lab assistant. Ensure the userid is one that has Administrator rights.

2. Click on the **Start** button, click on the **Control Panel** option. If in Control Panel Category view, click on **Performance and Maintenance** category, and click on **Administrative Tools**. If in Classic view, double-click on **Administrative Tools** control panel icon, and then double-click on the **Performance** icon. The Performance window opens. The Performance utility allows you to track individual computer component's performance. This is done through individual counters.

3. Click on the **Performance Logs and Alerts** option in the left pane. Click on the + (plus sign) if necessary to expand the Performance Logs and Alerts category.

   What are the three types of logs tracked by this utility?

   _____

4. Counter logs are used to create a log file using objects and counters you select. Click once on the **Counter logs** option in the left pane.

5. Click on the **Action** menu item and select **New Log Settings**.

6. In the name text box, type **Memory Usage** and click on the **OK** button.

7. Click on the **Add Counters** button. Click on the **Performance object** down arrow and select **Memory**. In the Select counters from list window, click on the **Available bytes** counter, and click on the **Add** button. In the Select counters from list window, click on the **Cache bytes** counter, and click on the **Add** button. Click on the **Close** button. The counters appear in the Counters window.

8. Click on the **Log Files** tab. The Log Files tab is used to select what type of file is created. The default type of file is a binary file, but a text file can be selected. Click on the **Log file type** down arrow and select the **Text File (Comma delimited)** option.

9. The Configure button is used to specify the location of the log file. Click on the **Configure** button.

   What is the default location (folder) for the log file?

   _____

10. Click on the **Cancel** button. Click on the **Schedule** tab. The Schedule tab is used to define the start and stop time for the log file. The default is to start the log and keep going until it is manually stopped. In the Stop log section, click on the **At** radio button. Change the time to two minutes after the current time. Make sure the date is today's date. (The default is one day later.) In other words, you will only be logging for two minutes. Click on the **Apply** button and click on the **OK** button. The Memory Usage log file appears in the right pane.

11. After two minutes, access the **WordPad** accessory. Click on the **File** menu option and select **Open**. Click on the **Files of type** down arrow and select **All Documents**. Use the Look in drop-down box or the icons on the left to locate the Memory Usage file. Reference your answer to the previous question for the name of the folder and drive letter. Click once on the file name and click on the **Open** button. The Memory Usage log file appears. The first set of numbers is the date followed by the time. The next two numbers are the counters that were requested: Available bytes and Cache bytes.

   On the first logged event line, what is the number of available bytes and cache bytes?

   _____

12. Return to the Performance window and click once on the **Memory Usage** counter log that you created earlier. Click on the **Red X** (delete) icon. An alternative method for doing the same thing is to click on the **Action** menu item and click on the **Delete** option.

13. Click once on the **Alerts** log in the left pane. Click on the **Action** menu item and select **New Alert Settings**. In the Name text box, type **Memory Alert** and click on the **OK** button. The Alerts option is used to set a counter that triggers an alert event to be sent to **Event Viewer**.

14. Click on the **Add** button. In the Performance object drop-down menu, select **Memory**. In the Select counters from list window, use the scroll bars to locate the **Available Bytes** counter. Click on the **Explain** button.

   What does the Available Bytes counter log?

   _____

15. Close the Explain Text window. Click on the **Add** button. Click on the **Close** button. On the General tab, type in a **1** in the Limit text box. (Note that this is not a value you would normally pick, but is used for illustration purposes.) Click on the **Action** tab. The Action tab is used to specify what happens when an alert is generated. The default is to send an alert into the application event log.

16. Click on the **Schedule** tab. The Schedule tab is used to define the start and stop time for the log file. The default is to start the log and keep going until it is manually stopped. In the Stop log section, click on the **At** radio button. Change the time to two minutes after the current time. Make sure the date is today's date. In other words, you will only be logging for two minutes. Click on the **Apply** button and click on the **OK** button. The Memory Alert log file appears in the right pane.

17. Open Event Viewer (see previous exercise if necessary) and open the Application event log by clicking on **Application** in the left pane. Look in the right pane. The first few application events should have event code 2031. Double-click on one of these events.

   Write the event description in the space below.

   _____

18. Close Event Viewer and return to the Performance window. Click on the **Alerts** Performance Logs and Alerts category. Click once on the **Memory Alert** log. Click on the **Red X** (delete) icon. An alternative method for doing the same thing is to click on the **Action** menu item and click on the **Delete** option.

Have a classmate verify that the counter log and alert log you created in this exercise are deleted. Have both the Memory Usage counter log and the Memory Alert alert log been deleted?

Classmate's printed name: _____

Classmate's signature: _____

19. Close the Performance window.

Lab Project 4

The goal of this project is for you to be able to back up the Windows XP system state using the Backup utility. You will need a computer with Windows XP installed and a formatted 3.5" floppy disk.

Note: In order to do this exercise, the student must have local Administrator privileges. The system state can be quite large (a common size is 400 MB), so adequate hard drive space must be available.

1. Turn on the computer and verify that the operating system loads.

2. Log on to Windows XP using the userid and password provided by the instructor or lab assistant. Note that the userid must have local Administrator privileges to do this exercise.

3. Click on the **Start** button and point to the **All Programs** selection. Point to the **Accessories** option, point to **System Tools** and click on the **Backup** menu selection. The Backup Wizard starts.

4. Click on the **Advanced Mode** option (which is an underlined option in the words appearing in the window).

5. Click on the **Backup** tab and select the **System State** check box to enable it.

6. Select where the backup is to be stored by clicking on the **Browse** button.

   What hard drive and folder is being used to store the system state backup?

   _____

7. Click on the **Start Backup** button. The Backup Job information dialog box appears. Click on the **Start** button located in the window to start the backup.

8. When the backup finishes, the Backup Progress window shows that the backup is complete. Show this to your instructor.

9. Click on the **Close** button. Close the Backup Utility window.

10. Use Windows Explorer to locate the BACKUP.BKF file and permanently delete it.

    Have a classmate verify the BACKUP.BKF file is *permanently* deleted. Has the BACKUP. BKF file been deleted?

    Classmate's printed name: _____

    Classmate's signature: _____

Lab Project 5

The goal of this project is to disable a driver under the Windows XP operating system.

Note: The student must be logged in as a user with local Administrator rights to perform this lab. In this lab, a driver is disabled and then re-enabled. The purpose of this is to demonstrate disabling a driver because sometimes XP can install the wrong driver. Drivers must sometimes be disabled and then manually reinstalled.

1. Turn the computer on and verify that the operating system loads. Log in to Windows XP using the userid and password provided by the instructor or lab assistant.

2. Using Device Manager, expand the **Network adapters** category.

   What network adapter is installed in the computer?

   _____

3. Right-click on a network adapter and click on the **Disable** selection.

   What message displays on the screen?

   _____

4. Click on the **Yes** button.

   In Device Manager, how is a device that has its driver disabled displayed differently from any other device?

   _____

5. In Device Manager, right-click on the same network adapter and click on the **Enable** option. The device is enabled and appears normally in the window.

6. Close the Device Manager window and all other windows. Log off and power down the computer properly.

Lab Project 6

The goal of this project is to install a new hardware component under the Windows XP operating system. You will need a new device to install and Internet access.

Note: The student must be logged in as a user with local Administrator rights to perform this lab. In this lab, the Internet is used to obtain the device's installation instructions and latest device driver, and then the new hardware device is installed.

1. There are four installation procedures depending on what type of hardware device is being installed. (1) A plug and play external device is normally installed when the computer is powered on. (2) A plug and play internal device is normally installed with the computer powered off and the device is installed according to the manufacturer's instructions. However, with PC Cards, the computer can be usually turned on and they can be inserted. (3) PCI and ISA cards are installed with the computer powered off. When the computer is powered on, Windows XP normally detects the device and starts the installation procedure. **Note:** PCI and ISA are interfaces that cards such as video cards, network cards, and modem cards use to "talk" with the motherboard. (4) Non-plug and play devices are installed with the computer powered off. When the computer powers on, the Add Hardware control panel is used to install the device.

   Of the four types of hardware devices, which one are you installing?

2. Using the Internet, locate the manufacturer's instructions for installing the device.

   Who is the device manufacturer?

   _____

3. Using the Internet, locate the latest device driver that is compatible with XP.

   Does the device have an XP driver?

   What is the device driver version being downloaded?

   _____

4. Connect the device to the computer using the proper installation procedures.

5. Boot the computer. Usually Windows XP automatically detects the new hardware and begins the Found New Hardware wizard. If it does not present this wizard, look to see if the hardware device vendor supplied an installation program. If so, use this program to install the device. If no vendor-supplied installation program is available, use the Add Hardware control panel to install the device. Install the device driver based on the device type and manufacturer's instructions.

   Did the Found New Hardware wizard begin?

   _____

6. Test the device installation either by using the device.

Lab Project 7

The goal of this project is for you to be able to install and remove Windows XP components. In addition to a computer with Windows XP installed, you need the Administrator password and about 18 MB of free hard disk space.

Note: In this lab, Windows XP's Accessories and Utilities component is removed, if already installed, and reinstalled. If the Accessories and Utilities component is not installed, it will be installed, removed, and reinstalled. The Accessories and Utilities component requires about 17.5 MB of hard disk space. The final objective of this lab is to have Accessories and Utilities installed.

1. Turn the computer on and verify that the operating system loads. Log in to Windows XP using the userid and password provided by the instructor or lab assistant. Ensure the userid is one that has Administrator rights.

Verifying if Accessories and Utilities Are Already Loaded

2. Click on the **Start** button and click on the **Control Panel** option. Access the **Add or Remove Programs** control panel by clicking on the category view or double-clicking on the Classic View control panel icon.

3. Click on the **Add/Remove Windows Components** icon located on the left portion of the Add or Remove Programs window. The Windows Components window opens.

   Is the Accessories and Utilities option enabled (checked)?

   _____

   If so, proceed to the **Removing Accessories and Utilities** section. Remove the components and then proceed to the **Installing and Utilities** section to reinstall the components. If the Accessories and Utilities option is not installed (unchecked), proceed to the **Installing and Utilities** section, install the components, then go to the **Removing Accessories and Utilities** section, and uninstall the components, then finally, reinstall the components again. When this lab is complete, the Accessories and Utilities component should be installed.

4. You can double-click on any component to view the subcomponents. Try this procedure on your own. Close all windows and proceed to the appropriate section based on the previous answer.

Removing Accessories and Utilities

5. Click on the **Start** button and click on the **Control Panel** option. Access the **Add or Remove Programs** control panel by clicking on the category view or double-clicking on the Classic View control panel icon.

6. Click on the **Add/Remove Windows Components** icon located on the left portion of the Add or Remove Programs window. The Windows Components window opens.

7. Click on the **Accessories and Utilities** check box to deselect (uncheck) it and click on the **Next** button. The files are deleted.

8. Click on the **Finish** button and verify that Accessories and Utilities are uninstalled using previously described procedures.

   Has the Accessories and Utilities component been removed? Have a classmate verify.

   Classmate's printed name: _____

   Classmate's signature: _____

Installing Accessories and Utilities

9. Click on the **Start** button and click on the **Control Panel** option. Access the **Add or Remove Programs** control panel by clicking on the category view or double-clicking on the classic view control panel icon.

10. Click on the **Add/Remove Windows Components** icon located on the left portion of the Add or Remove Programs window. The Windows Components window opens.

11. Click on the **Accessories and Utilities** check box to select (enable) it. If the box is already checked, go to the **Removing Accessories and Utilities** section. Click on the **Next** button. A prompt appears to insert the Windows XP CD. Insert the CD and the files copy.

12. Click on the **Finish** button, close all Add/Remove Components control panel windows, and verify that Accessories and Utilities are installed using previously described procedures.

    Is the Accessories and Utilities Windows XP component installed? Show this component to your instructor.

    _____

Lab Project 8

The goal of this project is for you to be able to access various Windows XP boot options that are used to troubleshoot startup problems. In addition to a computer with Windows XP installed, you need the Administrator password and about 18 MB of free hard disk space.

Note: In this lab, the students boot without startup programs loaded, boot to Safe Mode, boot to Safe Mode with Command Prompt, boot to Enable Boot Logging and examine the NTBTLOG.TXT file, and boot to Recovery Console and examine commands using the command prompt.

1. Turn the computer on and verify that the operating system loads. Log in to Windows XP using the userid and password provided by the instructor or lab assistant. Ensure the userid is one that has Administrator rights.

Verifying Startup Folder Contents

2. Right-click on the **Start** button and click on the **Explore** option.

3. Locate the **Documents and Settings** folder and expand it if necessary. Locate the **All Users** folder (located under Documents and Settings) and expand it if necessary. Locate the **Start Menu** subfolder (located under All Users) and expand it if necessary. Locate the **Programs** subfolder (located under the Start Menu folder) and expand it if necessary. Click on the **Startup** folder located under the Programs folder.

   Are there any program shortcuts listed in the Startup folder? If so, write at least one of the programs in the space below. If there is no program shortcut, create a shortcut to the Notepad application and place it in the Startup folder. A previous exercise explains how to create a shortcut. Have a classmate verify your shortcut.

   Classmate's printed name: _____

   Classmate's signature: _____

4. Restart the computer and verify that the program listed in the Startup folder starts *automatically* when the computer boots. If it does not, re-do step 3.

Preventing Startup Programs from Loading

5. Restart the computer and while the computer boots and while the log-in process occurs, hold the **Shift key** down until the desktop icons appear. Holding the Shift key down stops startup programs from loading automatically. This technique works when any program that starts automatically is causing problems. If this does not work for you, shut down the computer properly, power off, power back on, log in and hold the Shift key down during the log-in process.

   What indication do you have that holding the Shift key down while booting stopped the application from loading?

   _____

6. Using Windows Explorer, delete the *shortcut* located in the Startup folder.

Have a classmate verify that you only deleted the shortcut and not the application. Has the *shortcut* been deleted?

_____

Classmate's printed name: _____

Classmate's signature: _____

Using Boot Options

7. Restart the computer and press the **F8** key as the computer boots. The Windows Advanced Options Menu appears. If it does not, repeat this procedure until it does. Select the **Safe Mode** option and press Enter.

    What is different about the Windows XP log-in screen?

    _____

    Why do you think the Administrator user icon appears in Safe Mode and not during the regular boot sequence?

    _____

8. Log in as Administrator.

    What indication appears letting you know that the computer is running in Safe Mode?

    _____

9. Click on the **Yes** button.

    Did your program in the startup folder automatically start?

    _____

    Are Administrative Tools available through the Start button's All Programs list?

    _____

    To what Control Panel view does the system default?

    _____

10. Double-click on the **Administrative Tools** control panel. Double-click on the **Computer Management** icon. Access the **Services** folder. Refer to the Computer Management exercise if you forgot how to access it.

    List two automatic services that have a status of *started*.

    _____

11. Notice how there are quite a few services that are automatic services that did not start in Safe Mode. Close the Computer Management screen and the Administrative Tools window.

12. Restart the computer and press the **F8** button to see the Windows Advanced Options menu.

List the boot options available.

_____

Match the following definitions to the appropriate boot option.

_____ Safe Mode

_____ Safe Mode with Command Prompt

_____ Enable Boot Logging

_____ Last Known Good Configuration

   A. Starts the system with minimum files and drivers and only typed commands can be used.

   B. Records the boot process into a text file that can later be viewed and used for troubleshooting.

   C. Starts the system with minimum file and drivers including VGA video drivers.

   D. Used when a newly installed piece of hardware or software causes the system not to boot properly.

13. Select the **Safe Mode with Command Prompt** option and log in as Administrator.

    What is different about the desktop appearance?

    _____

14. Click on the **Minimize** button, which is the left-most button in the upper right corner of the cmd.exe window.

    What does the screen look like now?

    _____

15. The Safe Mode with Command Prompt option is used to start the system with minimum files and drivers and a command prompt where you must type commands instead of working through a graphical interface. Type **exit** at the command prompt.

    What happened to the screen?

    _____

16. Press **Ctrl + Alt + Del** and the Task Manager window appears. Click on the **Shut Down** menu option and select **Restart**. Restart the computer, press the **F8** key to see the Windows Advanced Options Menu. Select the **Enable Boot Logging** option.

    Does the Administrator userid appear as a log-in choice?

    _____

17. Log in to Windows XP.

    How does the desktop appear when using the Enable Boot Logging option?

    _____

18. Using Windows Explorer, locate the file **NTBTLOG.TXT** and double-click on the file icon to open the file.

    List two drivers that loaded properly.

    _____

19. Close the NTBTLOG.TXT window and close all Windows Explorer windows.

Recovery Console

20. Shut the computer down and power off. Insert the Windows XP CD into the drive and power on the computer. The Welcome to Setup screen appears. If the computer does not boot from the Windows XP CD, the BIOS settings probably need to be adjusted. Press **R** at the Welcome to Setup screen. The Recovery screen appears.

21. Press the number that corresponds to the partition that contains XP.

22. Type the Administrator password. Contact a lab assistant or the instructor if the password is unknown. The Recovery Console loads.

    Write down what the prompt looks like.

    _____

23. The Recovery Console is used as a last resort—when other boot options are used and you are not able to solve the problem with these options. At the prompt, type **copy** and press Enter. An error message appears. Command prompt usage must be very precise and exact commands with proper switches must be used.

24. Type **help copy** and press Enter. Help information on the Copy command appears.

25. Type **copy /?** and press Enter. Again, help information appears.

26. Type **help** and press Enter. A list of Recovery Console commands appears. Press the **space bar** to see the rest of the command list.

27. Remove the XP CD and type **exit**. The system boots normally.

Lab Project 9

The goal of this project is to have you create a tree structure on the floppy.

1. Turn the computer on and verify that the operating system loads. Log in to Windows XP using the userid and password provided by the instructor or lab assistant.

2. Create the tree structure shown in Figure 7.11.

Figure 7.11: A Sample Tree Hierarchy

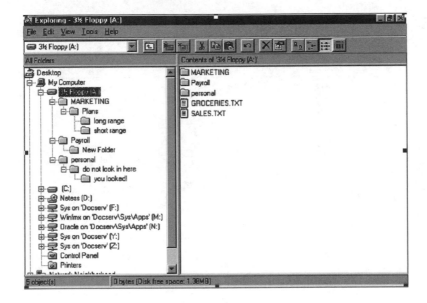

3. Create another folder in the root directory of the floppy that has your name in it.

4. Create another folder in the root directory of the floppy named "Windows XP Project 9."

5. Show your instructor.

Lab Project 10

The goal of this project is to have you create a tree structure on the floppy.

1. Turn the computer on and verify that the operating system loads. Log in to Windows XP using the userid and password provided by the instructor or lab assistant.

2. Create the tree structure shown in Figure 7.12.

Figure 7.12: A Sample Tree Hierarchy

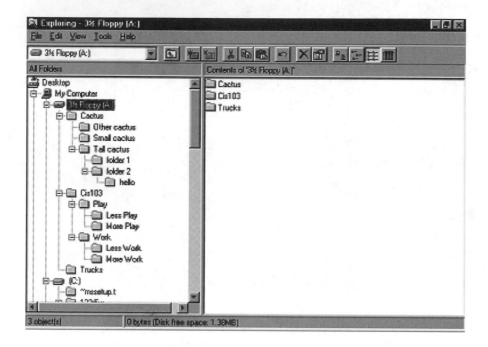

3. Create another folder in the root directory of the floppy that has your name in it.

4. Create another folder in the root directory of the floppy named "Windows XP Project 10."

5. Show your instructor.

Lab Project 11

The goal of this project is to help you understand how to encrypt and decrypt files at the command prompt. You must complete this lab project on an NTFS partition. This lab project is a bit more advanced because you are expected to determine many of the steps in the lab. If you have trouble with the command prompt mode, refer to Chapter 3 on DOS and the DOS command prompt.

1. Go to a command prompt.

2. At the command prompt, create a directory named C:\Project11.

3. Record the command used.

   _____

4. Create two files named File11A.txt and File11B.txt in Project11.

5. Record the command used.

   _____

6. To make sure you are at the root directory of the C: drive, type **CD C:** and press Enter.

7. To encrypt a directory, type **CIPHER /E Project11** and press Enter. The directory and new files will be encrypted.

8. To verify the encryption status, type **CIPHER** and press Enter. The status of the directories and files are displayed. If an "E" is beside the directory or file name, then the item is encrypted. If there is a "U," the item is not encrypted.

9. Change directory locations to Project11.

10. Create a new file named File11E.txt.

11. To verify the encryption status, type **CIPHER** and press Enter.

12. Record the output.

    _____

    _____

    _____

13. Record the reason for the output.

    _____

    _____

    _____

14. To make sure you are at the root directory of the C: drive, type **CD C:** and press Enter.

15. To decrypt a directory, type **CIPHER /D Project11** and press Enter. The directory and new files will not be encrypted.

16. Create a new file in Project11.

17. Check the encryption/decryption status now.

18. Record the output and reason for the output.

_____

_____

_____

19. Close the command prompt window.

Lab Project 12

The goal of this project is to help you understand how to compress and uncompress files at the command prompt. You must complete this lab project on an NTFS partition. This lab project is a bit more advanced because you are expected to determine many of the steps in the lab. If you have trouble with the command prompt mode, refer to Chapter 3 on DOS and the DOS command prompt.

1. Go to a command prompt.

2. At the command prompt, create a directory named C:\Project12.

3. Record the command used.

_____

4. Create two files named File12A.txt and File12B.txt in Project12.

5. Record the command used.

_____

6. To make sure you are at the root directory of the C: drive, type **CD C:** and press Enter.

7. To encrypt a directory, type **COMPACT /C Project12** and press Enter. The directory and new files will be compressed.

8. To verify the compression status, type **COMPACT** and press Enter. The status of the directories and files are displayed. If a "C" is beside the directory or file name, then the item is compressed. Otherwise, the item is not compressed.

9. Change directory locations to Project12.

10. Create a new file named File12C.txt.

11. To verify the compression status, type **COMPACT** and press Enter.

12. Record the output.

_____

_____

_____

13. Record the reason for the output.

    _____

    _____

    _____

14. Close the command prompt window.

15. To make sure you are at the root directory of the C: drive, type **CD C:** and press Enter.

16. To uncompress a directory, type **COMPACT /U Project12** and press Enter. The directory and new files will not be compressed.

17. Create a new file in Project12.

18. Check the compression/uncompression status now.

19. Record the output and reason for the output.

    _____

    _____

    _____

20. Close the command prompt window

Lab Project 13 Challenge

The goal of this exercise is for you to be able to install Windows XP on a hard drive that does not have an operating system installed nor any partitions. You will need a computer with a hard drive without any partitions and you will need the Windows XP CD. If your hard drive has partitions, delete them.

1. Insert the Windows XP CD into the CD-ROM drive and turn on the computer. Some computers require you to press a key to boot from the CD or require special BIOS settings to boot from the CD. Perform the appropriate steps required on the computer to allow the computer to boot from CD. The Setup screen displays. If the CD is an evaluation copy of XP, **Enter** must be pressed. Press **Enter** to display the licensing agreement.

2. Press **F8** to accept the licensing agreement. The hard drive partitioning screen appears.

3. Check with the instructor on how much space is desired for the partition. The partition must be a minimum of 2 GB in size. Record the partition size you chose.

   _____

4. Select a hard drive area that is not partitioned and press **C** to create a partition. A prompt appears asking for the partition size. Use the answer obtained in the previous question and enter this information. Note that **Enter** can be pressed to use all available space for the partition. Press **Enter** to install Windows XP on the partition just created.

5. A screen appears prompting to format the newly create partition. Select **NTFS** as the type of file system used on the formatted partition. The drive is formatted and setup files are copied to the drive.

6. When prompted to restart the computer, **remove the CD from the drive** and press **Enter**.

7. When prompted to insert the CD into the drive, reinsert the Windows XP CD into the drive and click on the **OK** button.

8. Accept the default path for the Windows XP installation by clicking on the **OK** button. After copying more files, you are prompted for regional settings. Set the appropriate language and click **Next**.

9. The Personalize Your Software page appears. This is what applications use for product registration and document identification. Leave this information blank and click on the **Next** button. The Product Key page appears.

10. Enter the product key located on the back of the Windows XP case, on the CD, or provided by the student assistant or instructor. Click on the **Next** button. Check with the lab assistant or instructor for the name that will be given to the computer as well as the Administrator password. Write this information in the space below:

    Computer name: _____

    Administrator password: _____

11. The setup program prompts for the computer name and Administrator password. Type the computer name and Administrator password that were written in the previous step. Click on the **Next** button.

12. If a modem is installed, the modem dialing information is displayed. The correct country, area code, number to access an outside line, etc., are required. Contact the lab assistant or instructor for this information (if displayed), enter it and click on the **Next** button. If a modem is not installed (or after this information is entered), the date and time page appears. Set the **date** and **time** as appropriate and click on the **Next** button.

13. If a network card is installed in the computer, the network settings page displays. If this is the case enter the appropriate networking information as provided by the lab assistant or instructor and click on **Next**. The setup process continues copying files and installing the operating system and then restarts the computer. After the restart, the Welcome to Microsoft Windows screen appears. Does the welcome screen appear? Show this screen to the instructor.

_____

14. Click on the **Next** button and the Internet Connection screen appears. Select to skip this step.

15. When asked if you want to activate Windows, select the **wait until later** option.

16. When prompted, do not select to set up user accounts. This process is covered in another lab.

17. Click on the **Finish** button. Does Windows boot properly after the installation process?

Lab Project 14 Challenge

The goal of this project is for you to be able to install Windows XP on a hard drive that already has an operating system installed. You will need a computer with a hard drive that has Windows 98, Windows ME, NT Workstation 4, or Windows 2000 installed and that has a CD drive installed. You will also need the Windows XP CD.

1. Power on the computer and log on as necessary. Contact the instructor or lab assistant for the userid and password if necessary. Insert the Windows XP CD into the CD-ROM drive. The Welcome to Microsoft Windows XP screen should appear. Click on **Install Windows XP**. The setup process collects information about the computer to ensure it is upgradeable.

2. When asked what type of installation to use, select **Upgrade** and click on the **Next** button.

3. Click on the radio button to accept the licensing agreement and click **Next**.

4. Enter the product key located on the back of the CD case, written on the CD, or provided by the instructor or lab assistant and click on the **Next** button.

5. The Dynamic Update is optional and can only be used if an Internet connection is available. The Dynamic Update updates installation files. Contact the instructor or lab assistant to determine if the Dynamic Update is necessary.

   Was the Dynamic Update performed?

   _____

6. The setup process copies installation files and restarts. When prompted to choose a Windows installation, ***do not select anything***. Windows automatically selects the correct version. The XP logo appears, more files are copied, and the computer reboots again.

7. You may be asked to enter a computer name if the old computer name is not appropriate. If necessary, type an appropriate computer name and click on the **Next** button. The Tour Windows screen appears.

8. When asked to activate the product, select the option to bypass this step. Click the **Finish** button.

9. When asked to setup user accounts, contact the instructor or lab assistant to verify if any are to be created.

   Were any user accounts created? If so, list them below.

   _____

10. Click the **Next** button and the XP desktop appears.

Internet Discovery

Internet Discovery Lab 1

Access the Internet to obtain specific information regarding a computer or its associated parts.

1. Find a web site that offers Windows XP freeware tools. Write the URL in the space below.

2. What is the latest service pack available from Microsoft for Windows XP Professional? Write the answer and the URL where you found the answer in the space below.

3. Find a web site that details how to set up DualView. Write the web address in the space below as well as the steps to configure DualView.

4. Microsoft always has minimum requirements for any of its operating systems. Find a web site that tells you what your system should really have to run Windows XP efficiently. Write the name of the company that posts the recommendation as well as the URL.

5. You are upgrading a computer from Windows 2000 Professional to XP Professional. After the installation, Device Manager is showing that the Windows Sound System Compatible (WDM) drive is having a problem. When you view the device properties, the description shows a code 28 error and states that the drivers are not installed. Find a web site that describes this error and write the cause and URL in the space below.

6. Find one book that prepares you for the Microsoft Windows XP Professional exam. Write the name of the book and the URL in the space below.

Soft Skills: Help Desk Support

1. A customer calls you and wants to install Windows XP. They have a Pentium II running at 233 MHz with 32 MB of RAM and a 2.0 GB hard disk drive. What do you tell them?

2. A user named Tom Smith calls and tells you that he gets some strange data in his spreadsheet. How do you help him? List possible problems.

3. You receive a call from a Support Engineer working for your same company. He wants to be able to pull up specific Administrative Tools easily. How do you help him?

Critical Thinking

1. How large is the System State data?
2. Why is the term "dual-boot" used when you cannot boot but one operating system at a time?
3. What is the purpose of disabling hardware?

Study Skills

The "Say" in "Hear, See, Do, Say"

Remember the key to learning are these four points: hear, see, do, say. This section focuses on the visual part or the "say" study skill.

- If your eyes are considered input and your brain is considered processes, then what would your mouth (used for saying) be considered?
- When you say or repeat something, you are reinforcing your understanding of the topic.
- Your instructor may call on you in class to answer a question. If you have prepared well, you should have no problem with the answer. And, you don't want to be embarrassed because you don't know the answer.
- Your instructor may assign a classroom research paper and presentation. When you present your paper, you are "saying" your work to the class. Who knows, you may teach a classmate something.
- When you have a test, you are "saying" what you know on paper.
- When studying, you can actually make some note cards of key points and definitions and have someone go over them with you. You can have the question on one side and the answer on the other. Have them ask the question and you can "say" the answer. It will really help you to retain the chapter material.
- Go to a place where you cannot disturb others. Read key points and definitions out loud. Again, this will help you remember the material.
- "Saying" does <u>not</u> involve telling someone an answer on a test (known as **cheating**).

Self-Study Question(s)

1. Have you gone over your note cards out loud this week?
2. Identify at least two "say" Study Skills you did this week.

8

Chapter 8
Introduction to
UNIX/Linux

OBJECTIVES

Although the contents of this chapter do not apply directly to the A+ Operating System Technologies Exam, the Linux operating system is becoming increasingly popular and should be included in an Operating Systems Concepts book.

In this chapter, you will complete the following sections:
- 8.1 Understanding the Linux Operating System
- 8.2 Learning about the Linux Hierarchy
- 8.3 Understanding the Types of Commands
- 8.4 Navigating the Tree Structure
- 8.5 Creating and Removing Directories
- 8.6 Managing Files
- 8.7 Learning about Additional Commands
- 8.8 Learning about Pattern Matching (Wildcard Characters)
- 8.9 Learning about Redirection and Filtering
- 8.10 Learning to Write Shell Script Programs

8.1 Understanding the Linux Operating System

AT&T developed the UNIX operating system in the early 1960s. It was based on the C language with some code written in the Assembler language. That version was called AT&T System Release V, or "SR5" for short. At the time, the operating system code was given out to major universities and the employees and students made modifications to it. The University of California at Berkley developed a version named Berkley Source Distribution, or BSD. The UNIX operating system became more commercialized over time. Multiple vendors bought rights to it and made changes to it in hopes of carving a niche in the market. There have been numerous versions of the original UNIX operating system. One such version is Linux.

Linus Torvalds, a student at the University of Helsinki, is credited with being instrumental in the development of Linux. He wrote Linux to be PC-based so anyone could use it at home. It comes with many built-in features with a full array of programming languages, compilers, and system administration tools at a very reasonable cost. Although there are differences between the many versions of UNIX, the term "Linux" is used throughout this chapter. This is because most of the concepts and lab exercises apply to most versions of UNIX and Linux. The screenshots in this chapter were done with Red Hat Linux 8.0.

8.2 Learning about the Linux Hierarchy

Like many operating systems, Linux is hierarchical in nature. This hierarchy is sometimes called a tree structure. Let's look at an analogy. Think of a tree that you buy at a garden store. There is the ball root with branches and leaves. Now, turn it upside down. The root is at the top with branches and leaves flowing from it. A branch can have other branches or leaves. However, a leaf cannot contain a branch or another leaf—it is an endpoint.

The Linux file system hierarchy is identical to the DOS file system hierarchy. However, there are a few differences. As you know from Chapter 3, the root symbol in DOS is \; however, the root symbol in Linux is /. Branching from the root are directories or files. As in DOS, a directory is analogous to a branch and a file is analogous to a leaf. A directory can contain other directories or files. A file cannot contain a directory or another file—like a leaf, it is an endpoint. However, a file does have contents. For example, a file can contain payroll data used for processing paychecks, or a file can contain a picture of someone (like a JPEG file), or it can contain executable code (like a command). Figure 8.1 shows an example of the Linux file system.

Figure 8.1: The Linux Hierarchical File System

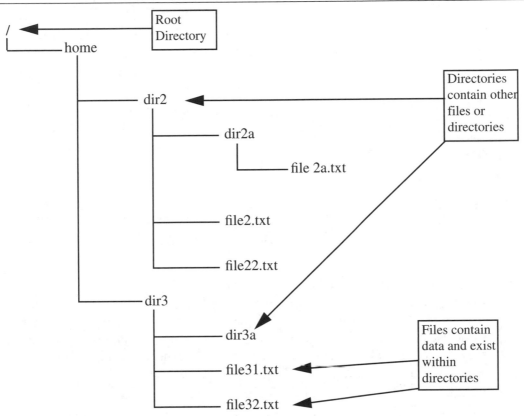

You can see in Figure 8.1 that the root symbol, "/" is at the very top of the hierarchy, or tree. Below the root directory is a directory named home and below that are two directories named "dir2" and "dir3." Each of these directories can contain other directories and files. Figure 8.2 shows a similar tree structure in Linux using the **tree** command.

Figure 8.2: A Screenshot Showing the Tree Structure of Figure 8.1 Using the Linux Tree Command

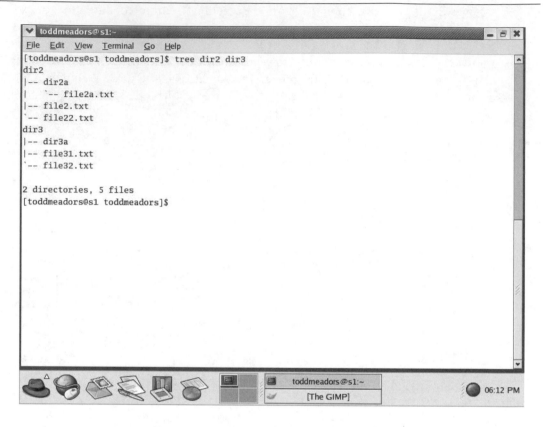

8.3 Understanding the Types of Commands

Commands are classified into two broad categories: 1) directory commands, or 2) file commands. In other words, some commands deal only with directories and some commands deal only with files. There are a few exceptions. For example, the **ls** command displays both files and directories. Most commands use the following generic syntax:

command options

Note *command* is the name of the command and *options* are either file and directory names or actual options that alter the command in some way.

Almost all commands use options to alter the way the command operates. The symbol for using options is the hyphen (-). Refer to Table 8.1 for a list of a few basic commands, a brief description of them, and their DOS/DOS command prompt equivalents. Linux and DOS have commands that operate similarly. Of the commands in the table, the **cd**, **ls**, and the **pwd** commands are the most common.

Table 8.1: A Listing of a Few Basic Linux Commands

Command	Description	DOS and DOS Command Line Equivalent
cd	Allows you to change directory positions, or location, within the tree.	CD
mkdir	Allows you to create a directory.	MD or MKDIR
rmdir	Allows you to remove a directory.	RD or RMDIR
pwd	Allows you to display your current working directory.	CD or look at your prompt
ls	Allows you to see directories and files in a given directory.	DIR

8.4 Navigating the Tree Structure

Before discussing tree structure navigation, let's discuss the concept of command prompts and user accounts. The Linux operating system is a multi-user and multitasking operating system. It acts like a small mainframe in that users can connect to it remotely using the **telnet** command, or they can log in locally as they would on a Microsoft Windows operating system. Because it is a multi-user environment, all users can potentially see the same directories and files.

One specific user account exists that has complete control over the system. That user is the **root** user. This is not the same as the root directory. With the root user account, you actually log in using the user name of "root." If you don't have root user access you can still log in with a standard user account. The root user can create standard user accounts. The standard user account used in this chapter is "toddmeadors." You can do many of the same commands as the root user account. However, the root user account has access to additional system administration commands and tools. If you have root access, it is not a good practice to log in as the root user on a routine basis. You should create a standard user account and log in using that account. You should only use the root user account when you cannot do what you need to do using your standard user account. The author of this chapter worked in UNIX technical support and has seen System Administrators accidentally delete all of the files on their system as the root user using a single command. The command they executed was **rm –r *** in the / directory as the root user. This will remove all of the software on your system. Fortunately, most of us don't have complete access to the root user account and the root directory.

 Be careful using the root user account on a daily basis! Log in as a user and only log in as root when you need to do something that the standard user account cannot accomplish. For this chapter, you will only need a standard user account.

Linux provides a command prompt for users to enter their commands. The root user and standard user accounts have different prompts. The root user's prompt includes the # sign. The standard user account includes the $ sign. If you don't have root access, then you'll see a dollar sign as part of your prompt. In Figure 8.3, you can see the differences. Notice that the user name is included in the command prompt.

Figure 8.3: A Screenshot Displaying the Root User and Standard User Prompts

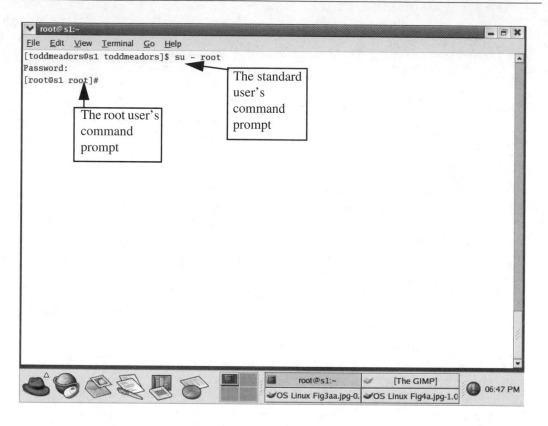

Each user has a home directory represented by the variable **$HOME**. The home directory is the default path, from the root directory, where your files and directories are stored. Generally the home directory begins in /home. Each users home directory ($HOME) is unique. For example, the $HOME variable for the user "toddmeadors" is /home/toddmeadors. The $HOME variable for the user "mickimeadors" is /home/mickimeadors. When you first log in to the Linux system your current directory is your home directory ($HOME). You can **cd** to **$HOME** and that will take you to your own home directory. The home directory will be used for storing your directories and files in this chapter.

Now let's discuss navigating the tree structure. For now, assume the tree structure shown previously in Figure 8.2 has already been created in a user's home directory.

Assume the current directory is root, or /. In order to change directory locations to dir2, you would issue this command: **cd dir2** and press Enter. You can follow up the **cd** command with the **pwd** command. Figure 8.4 shows for a sample screenshot. Notice the prompt changes to reflect the current directory.

Figure 8.4: A Sample Screenshot Displaying the Results of the cd and pwd Command for dir2

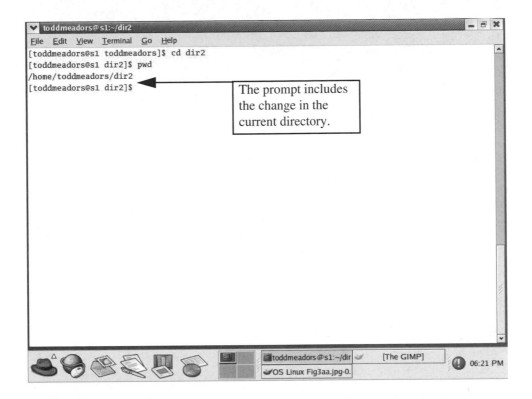

You must watch the case of the letters as you type in a command. Most Linux commands are lowercase, and the operating system is case-sensitive. In other words, entering **CD dir2** would generate an error. This is easy to fix, just toggle the Caps Lock key. The Linux operating system can use uppercase characters for file and directory names. For example, dirTM, is valid and is different from dirtm, which is also valid. The rule is if you use an uppercase when you create a file or directory, you must use an uppercase when using the file or directory. Also, most versions of Linux allow you to use most characters in a file name as long as you place double quotes around the file name. For example, the file name, "Payroll Data for June.dat" would be valid. Notice the use of spaces within the name itself. If double quotes surround the name, it should work. We'll discuss file name characters later.

Now let's navigate to the directory beneath dir2 named dir2a. Any directory beneath another directory is called a **subdirectory**, but they will simply be referred to as directories in this chapter. Figure 8.5 is a screenshot of the change to dir2a. Notice that the command prompt has changed to reflect the change.

Figure 8.5: A Screenshot Displaying the Results of the cd and pwd Command for dir2a

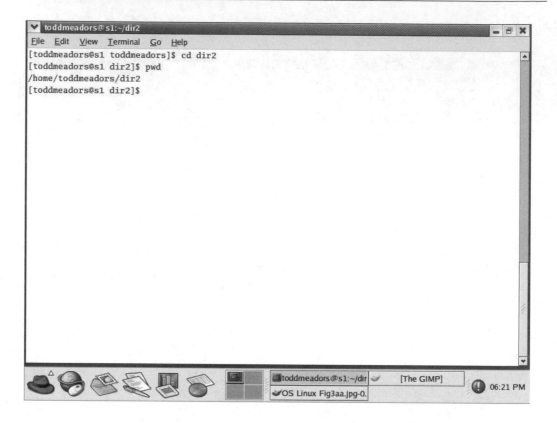

Now let's use the **ls** command to take a directory listing. By entering in just **ls** you will see any directories and files in the current directory. You can also use options with the **ls** command. Options modify the command in some way. For instance, the **ls –l** command performs a long listing. This gives you more information about the file or directory. This information will be discussed later. The **ls –a** command displays all files and directories. This includes files or directories that are classified as hidden. The **ls** command by itself won't show you hidden files or directories. We'll discuss creating hidden files and directories later too. You can combine options on some commands. For example, the **ls –la** command gives you a long listing of all files. Or, interchangeably, **ls –al** gives the same result. Figure 8.6 shows a screenshot of the **ls** command. The **ls, ls –a,** and **ls –al** commands are shown.

Figure 8.6: Results of the **ls** Command and Some of Its Options

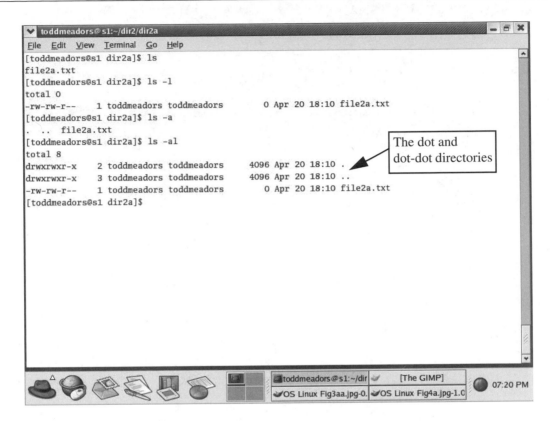

Referring back to Figure 8.1, the directories named dir2 and dir3 are children of root. Or said another way, root is their parent. A **parent directory** is the directory that holds a given file or directory. It is up one level in the hierarchy from the current directory. A **child directory** is a file or directory contained within its parent. It is down one level. The parent directory is represented by two dots, **..**, commonly called dot-dot. The current directory is represented by a single dot. Notice the dot and dot-dot displayed in Figure 8.6.

Let's use this knowledge of the parent and child relationship to change directory locations. If you wanted to change to the root directory, you could type the **cd /** command. If you want to change to your parent directory, you can type **cd ..** command.

Notice in Figure 8.1 that dir2a is a child of dir2 and dir2 is a child of the root directory. The root directory is a grandparent of dir2a. If you wanted to change to root from dir2a, you would issue either one **cd /** command or two **cd ..** commands. You might ask why would you change to the root directory with two commands instead of one? To change to root by name, you would issue the single command: **cd /**. However, to change to any parent directory, you would enter the **cd ..** command. To change to a parent directory going up one level, you must use two dots, or dot-dot. To change to a parent's parent (grandparent), you would enter **cd ..** twice. Or, you can combine levels on one command. For example, you could issue the **cd ../..** command to go back up two levels.

Figure 8.7 shows a screenshot demonstrating both the **cd ..** and **cd /** command variations.

Figure 8.7: Using the cd .. and cd / Commands

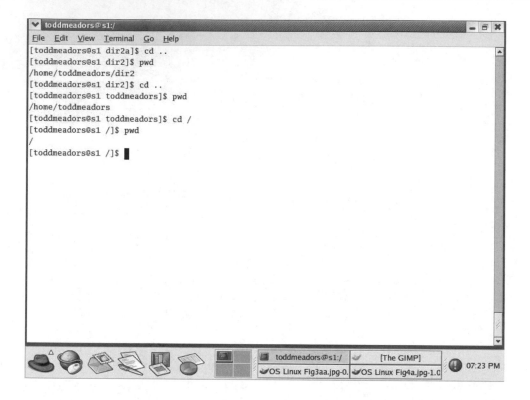

Each directory and file in the tree has a location called its **path**. There are two types of paths. There is a **full path**, sometimes called an absolute path, which identifies the directory or file location beginning from root. A command employing the full path always begins at root. There is a **partial path**, sometimes called the relative path, which identifies the directory or file location relative to your current location. A command employing the partial path depends upon your present location in the tree. Table 8.2 shows a few examples. Notice how each full path has the root symbol as the very first character. Also, each of these paths can be used with most any command.

Table 8.2: Examples of Full and Partial Paths

Path	Path Type
/	Full
/dir2	Full
/dir2/dir2a	Full
/dir3/file31.txt	Full
..	Partial
../..	Partial
dir2	Partial
../dir3	Partial

8.5 Creating and Removing Directories

In order to create a directory, you would use the **mkdir** command. For example, to create a directory named Payroll, you would enter **mkdir Payroll**. Notice that the directory Payroll is a partial path. You really don't know what parent directory it is located in. You can use the mkdir command with either the partial path or full path name of a directory. For example, you could enter **mkdir /Acct/General** to create a directory using a full path. Or, you could use a partial path, as in these examples: **mkdir ../dir5, mkdir dir6/dir6a/dir6b.** or **mkdir dirTM**. You can also create multiple directories using one mkdir command. To create three directories in the current directory, you could enter **mkdir a b c**. They will be siblings.

Next, you need to understand how to remove directories. The command to remove a directory is **rmdir**. In order to remove a directory, it must be empty and your current directory location cannot be in the directory. Refer to Figure 8.2 from earlier in the chapter. You would enter the **rmdir dir3a** command. If you get an error indicating you cannot remove the directory, try doing a **pwd** command. If your current directory is the one you are trying to delete, then go to the parent, with the **cd ..** command. Retry the failed **rmdir** command and it should work.

Let's do an exercise to get you started. You may have telnet access to the computer. If so, get the necessary information from your instructor on how to telnet to the school's Linux computer. Because the login steps are the same for all the exercises and hands-on projects in this chapter (at the end), subsequent exercises and hands-on projects will use to step 1 of this exercise in this chapter. This step will be presented only once—in this exercise.

1. Based on your situation, choose the appropriate method to get to get to a login prompt.
   a) **Remote access using telnet:** Telnet to the computer that has Linux loaded. For example, if your computer's IP address is 160.100.100.1, then enter TELNET

160.100.100.1. If you know the name of the computer, then issue the name instead of the IP address. For example, if the computer's name is xxx.yyy.edu, then enter TELNET xxx.yyy.edu.

b) **Local access with Windows running Red Hat Linux:** On the taskbar located at the bottom of the screen, locate the **Terminal emulation program** by moving your cursor along the icons. A new window will open with a prompt.

3. Log in to your Linux computer and get to the command prompt. In order to make sure you are in your HOME directory, type **cd $HOME** and press Enter.

4. To create a directory, type **mkdir dira** and press Enter. The directory is created and your prompt returns.

5. Now we need to change directory locations to it. Type **cd dira** and press Enter. Your current directory changes to dira.

6. Let's make a directory within the directory named dira. Type **mkdir diraa** and press Enter. The directory is created and your prompt returns.

7. Type **cd diraa** and press Enter. Your current directory changes to dira.

8. In order to remove the directory, type **rmdir diraa** and press Enter. You receive an error message because you cannot remove your current directory.

9. In order to change to the parent directory, type **cd ..** and press Enter. Your prompt changes to your parent directory.

10. In order to display a listing of files, type **ls** and press Enter. The listing displays.

11. Type **rmdir diraa** again and press Enter. This time it works because you are at the parent directory.

12. Type **pwd** and press Enter. Your current directory displays.

13. Type **cd ..** and press Enter. Your prompt changes to your parent directory.

14. Refer to Figure 8.8 for a sample screenshot.

15. Close your window.

Figure 8.8: A Screenshot of the Previous Exercise

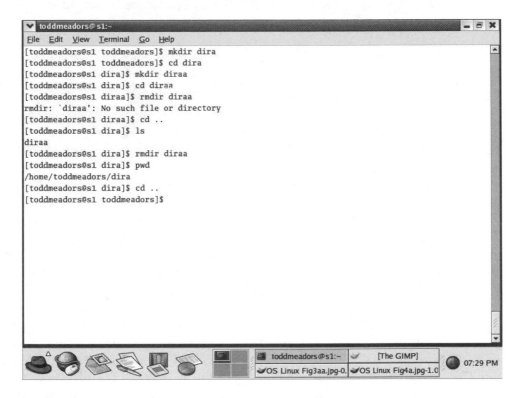

8.6 Managing Files

In this section, we will learn how to create, delete, modify, copy, and move files. There are several ways to create files. We are not concerned about all the methods, nor are we really concerned about the data within the files. We just need to have files created so we can manipulate them within the tree structure. Table 8.3 highlights a few ways to create a file.

Table 8.3: A Few Methods Used to Create a File

Command	Description	Example
echo	Displays text on the screen. Can be used to create a file.	echo "hi" > file1.txt
touch	Used to update the modification time of a file. It will create an empty file if the file does not exist.	touch pay4.dat
cat	Displays the contents of a file to the screen. Can be used to create a file.	cat > sales2.dat
vi	The most common text editor available on most Linux computers.	vi marketing.txt

You need to know that a file name in the Linux operating system is typically comprised of characters making up the name, followed by a dot, and then a three- character extension. The general format is as follows:

> *filename.ext*

Note *filename* is the name of the file, the dot is literally a period, and *ext* is the extension. For example, these file names are valid: **pay5.dat**, **file5.txt**, **Paychecks03.txt** and **Acct.doc**. The extension is not required and it can also be more than three characters. File names are case-sensitive. For example, **Pay5.dat** is not the same as **pay5.dat**. There are several characters that you generally don't use in a file name. They are: / \ > < * . | and space. The general rule of thumb is if you use the symbol elsewhere, don't use it for a character in a file name. If you stick to letters and numbers, you will be fine. Rules about file names also apply to directory names.

Let's explore the methods used to create files in greater detail. In order to create a file with the echo command, you would enter **echo "Text" > filename**. The echo command normally displays text to the screen. However, you can place a greater-than symbol after the text, followed by a file name and the echo command will put the text in the file name instead. We will discuss the use of the > symbol later in this chapter.

The touch command simply updates the modification time for an existing file. You would use the **ls –l** command to verify that the time has changed. However, if there is no file, the touch command creates a new empty file with an updated time. This command is quick for creating files when practicing because you usually don't care if it is empty.

The **cat** command is usually used for displaying the contents of a file. However, it can be used to create a file with some text in it. Let's try an exercise using the **cat** command. Follow these steps:

1. Log in to Linux and access the command prompt.
2. Go to your home directory.
3. In order to create a file using the cat command, type **cat > file2.txt** and press the **Enter** key. The cursor will move to the beginning of the next line. There will be no prompt on that line.
4. Type the following and press **Enter** after each line of text is typed. Note, if you make a mistake on a line and press Enter, you cannot change it using the **cat >** method. You would have to use **vi** to modify a previous line.

Working with Linux is fun!
I like ice cream - chocolate.
My birthday is in February.

5. Press **Ctrl + D** to send an End-of-File (EOF) character to the cat command. This is how you save the file.
6. In order to display the contents of the file using **cat**, type **cat file2.txt** and press Enter.
7. Close your window.
8. Refer to Figure 8.9 for a screenshot.

Figure 8.9: A Screenshot of the Previous Exercise

Most Linux computer systems come with the **vi** (VIsual) Editor. Although there are other editors around, and it is a bit tricky to use, you should become familiar with it because it's widely used. Not all versions of Linux use those other editors. So there's no guarantee that one editor specific to one version will be on another version of the operating system.

There are two modes of **vi**: **command mode** and **text mode**. In command mode you tell the **vi** editor what you want to do by issuing various commands. In text mode you actually type in the text you want to appear in the file. The **Escape** key is used to toggle back and forth between command mode and text mode. You must issue a command before you enter the text. Table 8.4 lists some of the more common **vi** commands.

Table 8.4: Common vi Commands

Command	Description
i	The vi editor will insert text at the current location of the cursor.
o	The vi editor will open a new line below the current line.

a	The vi editor will allow you to append text immediately after the current location of the cursor.
Shift + G	Move the cursor to the end of the file.
N followed by Shift + G	Move the cursor to the Nth line in the file. For example, 3 and Shift + G, takes you to the third line.
d followed by d	Delete the current line.
u	Undo the previous command.
y followed by y	Yank, or copy, one line into the **vi** buffer.
p	Put lines previously yanked from the **vi** buffer. Using yank and put is like copy and paste in MS Windows.
N followed by y and y	Yank, or copy, N lines into the **vi** buffer.
x	Delete one character.
:wq!	Write and quit **vi**.

Now you will use **vi** to create a file.

1. Log in to Linux and access the command prompt.
2. Change directory locations to your home directory.
3. Type in **vi file27.txt** and press **Enter**. The **vi** editor opens the file named file27.txt. The ~ symbol is called the tilde.
4. In order to insert text, press **Escape** and then **i**.
5. Enter the following text, pressing the **Enter** key at the end of each line:

   This is line 1.
   This is line 2.
   This is line 3.
   Last line of the file.

6. In order to save the file, press **Escape** followed by the colon symbol, **:**, and the cursor will move to the bottom of the screen with a colon to its left.
7. In order to write the file, press the **w**, then **q**, then **!,** and then press **Enter**. Your prompt returns.
8. Close your window.
9. Refer to Figure 8.10 for a screenshot.

Figure 8.10: A Screenshot of the Previous Exercise

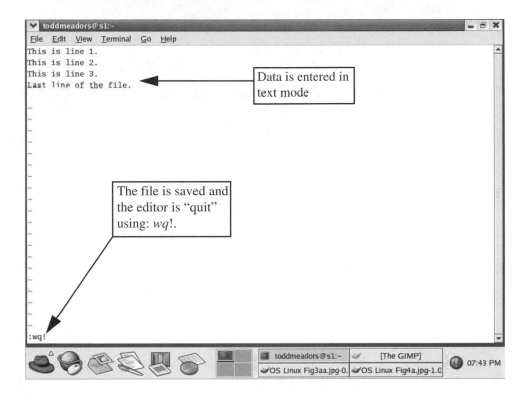

There are more hands-on projects using the **vi** editor in the Lab Projects section near the end of this chapter. Let's look at creating hidden files and directories.

You can create hidden files in the Linux operating system by preceding the file name with a dot. For example, the file named **.resume.txt** is considered a hidden file. In order to see a hidden file using the **ls** command, you must use the **–a** option. This shows all files including those starting with a dot. You can also create a hidden directory by simply placing a dot in front of the name when you create the directory using the **mkdir** command. Let's go through an exercise to create hidden files and directories.

1. Log in to Linux and access the command prompt.
2. Change directory locations to your home directory.
3. In order to create a hidden file, type **touch .salesreport.dat** and press Enter.
4. Type **ls** and press Enter. A listing displays.
5. Type **ls –a** and press Enter. A listing of all files and directories displays.
6. To create a hidden file using the **echo** command, type **echo "hello" > .secret.txt** and press Enter. The file is created and your prompt returns.
7. To create a hidden directory, type **mkdir .secretdir** and press Enter. The directory is created and your prompt returns.

8. Type **ls –a** and press Enter. A listing of all files and directories displays.
9. Close your window.
10. Refer to Figure 8.11 for a screenshot.

Figure 8.11: A Screenshot of the Previous Exercise

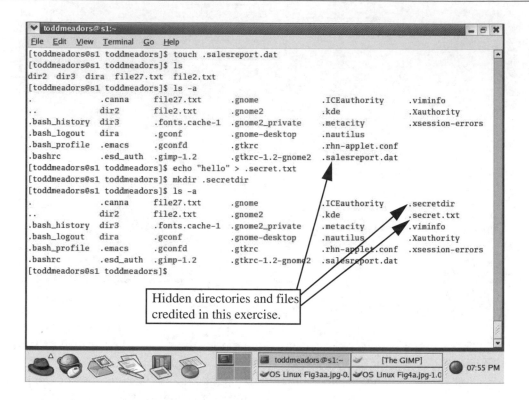

Now you will learn how to copy and move files. The Linux command to copy a file is **cp**, and the Linux command to move a file is **mv**. Use caution because these two commands are very different. The **cp** command will make a duplicate of an existing file. Two files will exist once the **cp** command completes. The move command will actually change to name of an existing file. Only one file will exist once the **mv** command completes. The **cp** command would be used to create a backup file whereas the mv command would be used to move a file from one directory to another or to rename a file.

If you wanted to copy a file named Sales.rpt to SalesBackup.rpt, you would enter **cp Sales.rpt SalesBackup.rpt** at the command prompt. In Figure 8.12, the first **ls** command shows the presence of Sales.rpt, but not SalesBackup.rpt. Then the **cp** command is executed. Finally, the **ls** command is executed again to show you that SalesBackup.rpt was created. Also, you notice that Sales.rpt is still present.

Figure 8.12: Using the cp Command to Make a Copy of a File

If you wanted to move or rename the file named Sales.rpt to MonthlySales.rpt, you would enter **mv Sales.rpt MonthlySales.rpt** at the command prompt. In Figure 8.13, the first **ls** command shows the presence of Sales.rpt, but not MonthlySales.rpt. Next, the **mv** command is executed. Finally, the **ls** command is executed again to show you that MonthlySales.rpt was created, but Sales.rpt is no longer present.

Figure 8.13: Using the mv Command to Move or Rename a File

The **cp** and **mv** commands have two useful options. The –**i** (for interactive) option is used to prompt before you overwrite the destination file. The –**v** (for verbose) option will display a message indicating action taken. So, to copy, verify, and prompt before over-writing fileA.txt to fileB.txt, you would enter **cp –iv fileA.txt fileB.txt**. To move the file, enter **mv –iv fileA.txt fileB.txt**. You can use either option alone. For example, **cp –i fileA.txt fileB.txt** and **cp –v fileA.txt fileB.txt** would work as well.

Let's go through an exercise to copy and move files.

1. Log in to Linux and access the command prompt.
2. Change directory locations to your home directory.
3. Create a file named Pay1.rpt so you can copy and move it. Type either **touch Pay1.rpt** or **echo "data" > Pay1.rpt**. The file is created.
4. To copy the file to Pay2.rpt, type **cp Pay1.rpt Pay2.rpt** and press Enter. The file is copied and your prompt returns.
5. Run the **ls** command. The output displays.
6. To move or rename the file, type **mv Pay2.rpt PayOld.rpt** and press Enter. The file named Pay2.rpt will be renamed to PayOld.rpt and your prompt will return.
7. Run the **ls** command. The output displays.

8. To copy using the verbose and prompt options, type **cp –iv Pay1.rpt PayOld.rpt**. Because PayOld.rpt exists (from the are previous move), you are prompted whether you want to overwrite the file.

9. Enter **y** to overwrite.

10. Next, due to the verbose setting a line displays with an arrow point from Pay1.rpt to PayOld.rpt indicating the file was copied.

11. To move using the verbose and prompt options, type **mv –iv Pay1.rpt PayOld.rpt**. Because PayOld.rpt exists, you are prompted.

12. This time enter **n** so the file will not be overwritten. Your prompt returns and no message appears because you entered **n**.

13. Close your window.

14. Refer to Figure 8.14 for a screenshot.

Figure 8.14: A Screenshot of the Output of the cp and mv Commands

In order to remove a file, you use the **rm** command. For example, to remove the file named resume2.txt, you would enter **rm resume2.txt** and press Enter. On some versions of Linux, the **rm** command will go ahead and remove the file. On others, it will prompt you with an "Are you sure?" type message.

Use caution when executing the **rm** command. There is no way to get a removed file back.

Let's go through an exercise to remove a file.

1. Log in to Linux and access the command prompt.
2. Change directory locations to your home directory.
3. First you need to create a file so you can remove it. Create a file named Accounting2.rpt. Type **touch Accounting2.rpt**. The file is created.
4. Run the **ls** command. The listing shows files and directories including the one just created.
5. Type **rm Accounting2.rpt** and press Enter. The file is removed.
6. Run the **ls** command. It is not in the listing.
7. Create a new file to use. Type **touch Accountin3.rpt**.
8. To use the interactive and verbose options on the **rm** command, type **rm –iv Accounting3.rpt**. You are prompted whether you want to overwrite or not.
9. Enter **y** to remove.
10. A message displays indicating the file has been removed.
11. Close your window.
12. Refer to Figure 8.15 for a screenshot.

Figure 8.15: A Screenshot of the Output of the rm Command

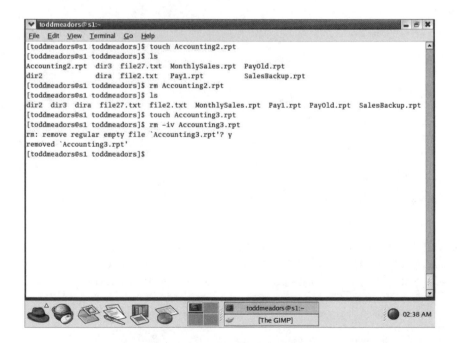

The "-l" option on the **ls** command displays additional information about a file or directory. Look at Figure 8.16 and you'll see additional information using the **ls –l** command.

Figure 8.16: A Screenshot of the Output of the ls –l Command

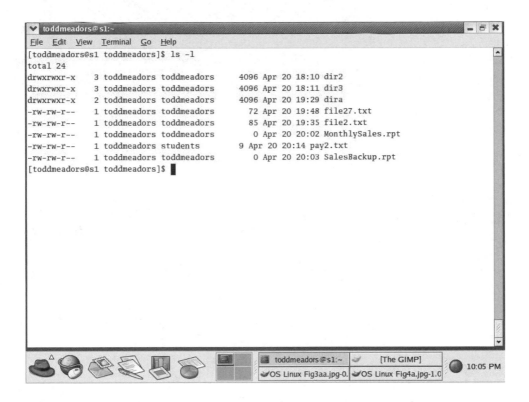

Each row of information deals with either a file or directory. Take a look at the file named pay2.txt. Its name is in the last column. The row for pay2.txt is as follows:

-rw- rw - r - - 1 toddmeadors students 9 Apr 20 20:14 pay2.txt

The first character is a dash. This means that pay2.txt is a file. A "d" in place of the dash would indicate a directory instead of a file. Next, you have "rw- rw- r - -. " These are the file permissions. An "r" means you can read the contents of the file. A "w" means you can write to the file, and an "x" means the file is executable. If there is a dash, then the permission is not allowed. The permissions are taken in sets of three. The first set of three permissions is for the owner; the second set is for members of the group; and the last set of permissions is for everyone else, called the world. So, the owner has read and write but not execute (the first set: r w-), the members of the group have read and write (the middle set: r w -), and the world has read only (the last set: r - -).

The number after the permissions is the number of links, or shortcuts, to the file. In our case, there is only one. The name after the links is the owner of the file. In our case the owner is "toddmeadors." Next, you see the group member name as "students." This means that members of the group "students" have the middle set of permissions on the file. The number to the right of the group name is the file size in bytes. Next, you have the date and time, and finally you have the file name.

You can change permissions with the Linux **chmod** command. You can change permissions of a file you own. When a file is first created, it has certain default permissions. For the root user, the permissions are read and write for the owner and read for the group and others. For standard users, the default file permissions are read and write for the owner and group members, and only read for others. The **chmod** command uses numbers to set permissions. Table 8.5 shows the permission numbers for the different permissions along with their binary equivalent.

Table 8.5: Permission Numbers Used by the chmod Command

Permissions	Permission Number	Binary Equivalent
---	0	000
--x	1	001
-w-	2	010
-wx	3	011
r--	4	100
r-x	5	101
rw-	6	110
rwx	7	111

Here's how you use this table. Look for the permission you want to assign and then look to the right to find the appropriate number. The three permission columns represent a 4, 2, and 1 for read (r), write (w), and execute (x), respectively for the user owner, group, and other users. You add up the numbers for the permissions you want.

For example, if you want read only, you simply need 4 for the permission. If you want read and write, you would add 4 and 2 to get the number 6. If you want read, write, and execute, you add up 4, 2, and 1 to get the number 7. If you want read, write, and execute for the owner; read and execute for the group; and only read for all others for a file named SalesA.dat, you would enter **chmod 754 SalesA.dat**.

Let's go through an exercise to help you understand how this works.

1. Log in to Linux and access the command prompt.
2. Change directory locations to your home directory.
3. Type **echo "secret" > secure.dat**. The file is created.

4. Type **ls –l secure.dat**. A long listing appears.
5. Type **cat secure.dat**. The contents are displayed.
6. To remove all permissions and prevent the file from being changed, type **chmod 000 secure.dat**. The permissions change to no permission.
7. Type **ls –l secure.dat** to prove the permissions changed.
8. Attempt to open the file. Type **cat secure.dat**. A permission denied message appears because the **cat** command cannot read the contents of the file.
9. Type **chmod 740 secure.dat**. The permissions change to read, write, and execute for the owner; read for the group members; and no permissions for everyone else.
10. Type **ls –l secure.dat** to prove the permissions changed.
11. Attempt to open the file. Type **cat secure.dat**. This time you can.
12. Type **chmod 444 secure.dat**. The permissions change to read only for all.
13. Attempt to write to the file. Type **echo "data" > secure.dat**. You receive a permission denied message because you are attempting to write to the file.
14. Type **chmod 664 secure.dat** to set the permissions to the default of read and write for owner and group members, and read for all others.
15. Type **ls –l secure.dat** to prove the permissions changed.
16. Close your window.
17. Refer to Figure 8.17 for a sample screenshot of these steps.

Figure 8.17: A Screenshot of the Previous Exercise Using the chmod Command

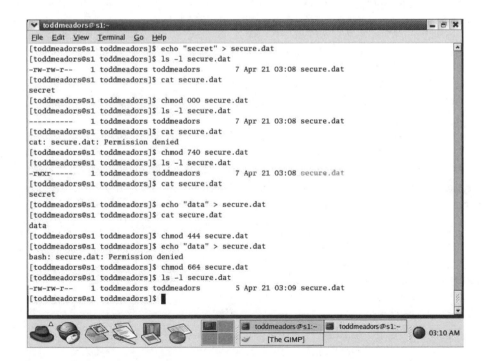

8.7 Learning about Additional Commands

Table 8.6 lists a few other commands with which you should become familiar.

Table 8.6: Additional Commands

Command	Description
awk	The **awk** command is a pattern scanning and processing language.
cal	The **cal** command is used to display a calendar.
cut	The **cut** command allows you to take text out of files. It is useful for extracting columns from a file.
date	The **date** command displays the current date and time.
df	The **df** command displays the amount of free disk space you have available.
diff	The **diff** command displays the differences between two files. This is useful for verifying the integrity of a file against a master file.
find	The **find** command allows you to search for files.
grep	The **grep** command allows you to search for text within files.
man	The **man** command displays help on a command.
more	The **more** command allows you to scroll through a page of text at a time.
passwd	The **passwd** command allows you to change your password.
sort	The **sort** command allows you to sort the contents of a file in both ascending field and descending field order.
who	The **who** command displays the list of current users on the system.

Let's go through an exercise using some of these commands.

1. Log in to Linux and access the command prompt.
2. Change directory locations to your home directory.
3. Create a file named Checks.txt, which holds employee data. There are three fields separated by a colon. The first field is the Employee ID, the second field is the Employee Name, and the last field is the Salary.

   104:Tina Brownlee:55000
   103:Roger Avery:45060
   105:Mary Davis:78000
   101:Sue Smith:45000
   102:Tom Jones:56000

4. In order to view the **man** pages for the **cut** command type **man cut**.

5. Review the **man** pages on **cut**. You can press the Spacebar to move down a page or press Enter to move down one line.

6. Press **q** (for quit) when finished.

7. In order to use the cut command to select the Employee Name column, you would type **cut –d: -f2 Checks.txt** and press Enter. The **–d:** option means to use the colon as the delimiter (separator). The **–f2** option means to cut field 2, or the Employee Name. The command displays all the names of the Employees.

8. In order to view the **man** pages for the **sort** command type **man sort**.

9. Review the **man** pages on **sort**.

10. Quit when finished.

11. In order to sort the data in ascending order, type **sort Checks.txt** and press Enter.

12. In order to sort the data in descending order, type **sort –r Checks.txt** and press Enter. In order to search for the name "Tom" in the file, type **grep Tom Checks.txt** and press Enter.

13. View **man** pages on **grep** if necessary.

14. In order to display users logged on the system, type **who** and press Enter.

15. In order to display the amount of free disk space, type **df** and press Enter.

16. In order to display the current system date and time, type **date** and press Enter.

17. In order to display the calendar for the current month, type **cal** and press Enter.

18. In order to display help about the **diff** command, type **man diff** and press Enter.

19. Review the **man** pages for **diff**.

20. Quit when finished.

21. The **awk** command is really a complete programming language, and fully discussing it is beyond the scope of this book. However, a few of its options will be discussed here. In order to use **awk** to display the fields in the Checks.txt file in different order, type **awk –F: '{print $2, $1, $3}' Checks.txt**. The **–F** option means to use the following character, which is the colon, as a field separator. Look back at the file and you'll see that a colon separates each field. The $1 equates to field 1, which is Employee ID, $2 equals field 2, which is the Employee Name and $3 is the Salary. So, in this case, the Employee Name is displayed first, followed by the Employee ID and then Salary.

22. View the **man** pages on **awk** if necessary.

23. To display just the Employee Name and Salary, type **awk –F: '{print $2, $3}' Checks.txt**. The Employee Name and Salary are displayed for each employee.

24. Close your window.

8.8 Learning about Pattern Matching (Wildcard Characters)

The Linux operating system allows for pattern matching of files. You use certain symbols to match characters following a certain pattern. Suppose you wanted to see all of the files that began with an uppercase S or all of the files that ended in .dat. How would you do this? Or, what if you wanted to see all files that began with either a lowercase t or an uppercase T?

The answer to these questions is pattern matching, sometimes called wildcarding. The operating system provides three pattern matching mechanisms:

- The asterisks symbol (*), which matches all characters.
- The question mark symbol (?), which matches a single character positions.
- The square brackets ([...]), which match one of several characters positions listed between the brackets. The dots in [...] represent characters.

Let's look at the asterisks symbol. You can use the * symbol to match all character positions. For example, if you wanted to display all files that begin with the letter m, then you would enter **ls m*** at the prompt. If you wanted to display all the files that had **dat** after the dot in the file name, then you would enter **ls *.dat**.

Let's take a look at the use of the question mark to wildcard a single character position. Suppose you wanted to display all files that had a 5 in the fourth character position, and you didn't care about the other characters. You would have to use the question mark. You would type the command **ls ???5*** at the command line prompt. You could not use the asterisks prior to the 5 in this case, because it would display all files with a 5 anywhere in the file name—not just in the fourth character position.

Let's look at the use of the square brackets as a wildcard technique. If you wanted to display all files that began with either an uppercase L or a lowercase l, you would need to use the square brackets. You would type **ls [Ll]*** in this case. If you wanted to display all files that began with either an S or an s followed by the letters pa, then you would enter **ls [Ss]pa*** at the command line prompt. An example of some of the files that would be displayed using this command are: Spanish, Spaniel, sparkle, sparkling, and spa.

Figure 8.18 shows a screenshot of the wildcard symbols in use. First the **ls** command displays the files. The second command, **ls t***, displays all files that begin with the letter t. The third command, **ls ?3***, displays all files with a 3 in the second position. The last command, **ls [Tt]***, displays all files that begin with a T or t.

Figure 8.18: A Screenshot of Using Wildcard Characters

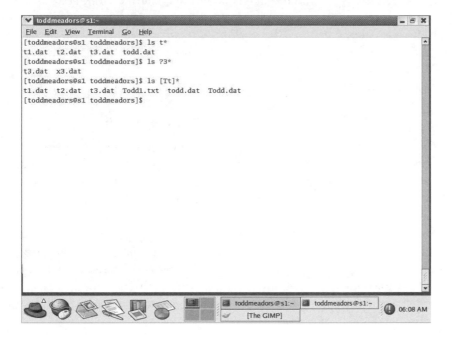

8.9 Learning about Redirection and Filtering

Let's look at the concept of redirection. These are the redirection symbols:

- A single less-than symbol (<) to redirect input.
- A single greater-than symbol (>) to redirect output.
- Two greater-than symbols (>>) to redirect and append output.

With redirection, you change the direction of the normal input and output locations for a file. When you run a command, the output normally displays on the screen. This is called **standard output** (or **stdout**). **Standard input** (or **stdin**) is the normal input that a command will get data from—which is the keyboard. With redirecting, you are indicating that you want to change standard output or standard input to be some other device, usually a file.

The general format of redirecting input is

> *command < filename.ext*

For example, the sort command will accept input from the command line. However, you can redirect input from a file by typing **sort < unsort.dat** at the command line prompt. This command will read input from the file named "unsort.dat" and display the sorted results on the screen.

Let's look at redirecting output. The general format of redirecting output is

> *command > filename.ext*

The standard output for the ls command is the display screen. If you wanted to redirect that output to a file, you would enter **ls > listing.txt** at the command prompt. The output would be redirected to a file named "listing.txt." When you redirect standard output to a file using the > symbol, the ***original contents of the file are deleted***. If you wanted to add standard output from another command to the same file name, you would issue the redirect and append output symbols, >>.

The general format of redirecting and appending output is

command >> filename.ext

Now, if you wanted to display the output of the who command and append it to the same file, you would need to use two greater-than symbols. For example, the command **who >> listing.txt** means to append the list of current users to the file named "listing.txt." If you had issued **who > listing.txt**, you would have written over the listing of the ls command.

Figure 8.19 shows an example of redirecting input and output. The command **cat unsort.dat** displays the data before it is sorted. The next command **sort < unsort.dat** redirects input but notice the output is on the screen following the command. The next command **sort < unsort.dat > sort.dat** redirects both input and output. Notice that no output appears from this command. Finally, the command **cat unsort.dat** demonstrates that the original contents did not change.

Figure 8.19: A Screenshot of Redirecting Input and Output

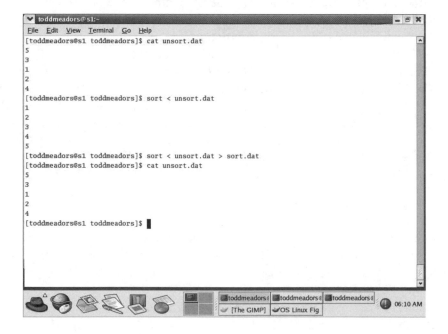

You can use filtering to modify the output of a command in some way. The filter symbol is the pipe symbol (|). It is the broken vertical bar on your keyboard but prints as a solid vertical bar. It is located on the same key as the backslash symbol. The general format of commands using the pipe symbol is

> *command1 | command2*

The command on the left side of the pipe symbol will have its output sent as input to the command on the right side of the pipe symbol. For example, the ls command displays files and directories, and the more command allows you to scroll through a page of data at a time. If the output of the ls command scrolled several pages, how would you be able to see all of the files and directories? The answer is to use the pipe symbol. The command **ls | more** will display a list of file and directories one page at a time.

If you wanted to sort the listing, you could enter **ls | sort**. If this command displayed too many lines, then you could enter **ls | sort | more** command instead. This will allow you to scroll through the sorted listing. You can have several pipes going on one command line. For example, you could have the following:

> *command1 | command2 | command3 | command4*

The way this works is that command1's output is pipe as input to command2. Then, that filtered output is piped to command3 and then the filtered output of command1, command2 and command3 is piped as input into command4.

Refer to Figure 8.20 for a sample screenshot of the pipe symbol in action. The command **ls | sort | more** displays a sorted directory listing and shows you a page at a time. Notice the text **---More---** at the bottom of the screen. If you press the Spacebar or the Enter key, you can scroll forward through the list of files and directories.

Figure 8.20: A Screenshot Displaying the Use of the Filtering (Pipe) Symbol

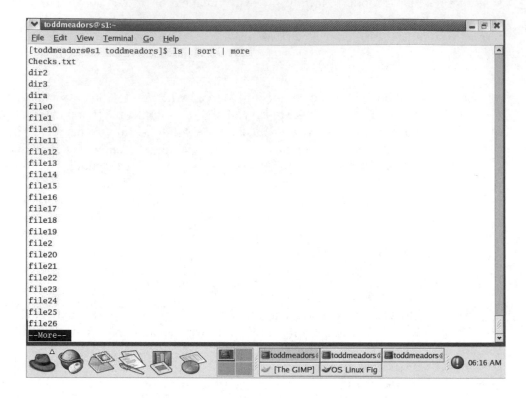

Just remember, redirection occurs to or from a file. Filtering (piping) occurs from one command to another command.

8.10 Learning to Write Shell Script Programs

The shell is the Linux program that interprets the command you enter on the command line. The default shell on Linux is called **bash** for Bourne-Again SHell. The bash program is like the DOS COMMAND.COM program. The bash program gives you your command prompt. A shell script program is similar to a DOS batch program. A **shell script** is a file containing Linux operating system commands and programming structures such as conditional processing and looping. Shell scripts execute their statements sequentially beginning from the first line in the file. We'll give you a general overview of the process here. Learning all aspects of shell script programming is beyond the scope of this book.

Let's learn shell scripts by performing an exercise. In this exercise, you will use operating system commands to create your own shell script.

1. Log in to Linux and access the command prompt.
2. Change directory locations to your home directory.
3. Open a file named **script1** using the **vi** editor.

4. Insert the following text. Note that the # sign is like the DOS REM statement and is used for a comment. Change *Your-Name* to your actual name. Make sure to save the file and quit the editor before executing it.

   # Author: *Your-Name*
   # Script Name: **$HOME/script1**
   echo My first shell script:
   pwd
   who

5. To execute the shell script, type **bash script1** and press Enter. The script executes. With this command, you are giving the bash shell the name of your script called script1 and the script executes.

6. Let's create another shell script. Open a file named **script2** using the **vi** editor.

7. Insert the following:

   # Author: *Your-Name*
   # Script Name: **$HOME/script2**
   echo Shell script to create a small tree structure:
   mkdir scriptsA scriptsB scriptsC
   touch scriptsA/fileA1.dat
   touch scriptsB/fileB1.dat
   touch scriptsC/fileC1.dat
   touch scriptsC/fileC2.dat
   tree –f $HOME/scripts*

8. To execute the script, type **bash script2** and press Enter.

9. For a challenge, create a script on your own that removes the directories and files created in **script2**.

10. Log out.

Chapter Summary

- The Linux file system is hierarchical (tree-like) in nature. The file system uses a directory to store files or other directories. A file cannot store another file or directory. The root directory (/), is the top level directory.
- The two basic categories of commands are file and directory commands.
- The **ls** command displays files and directories. The **cd** command changes directory locations. The **mkdir** command creates a directory. The **rmdir** command removes a directory.
- The **vi** editor is used to create a file. The **cp** command is used to copy or duplicate a file. The **mv** command is used to rename a file.
- You create a hidden file by placing a dot (period) as the first character in the file name.
- You can use wildcard characters to match on certain patterns. The * symbol matches all character positions. The **?** matches a single character position. The use of **[...]** matches one of several character positions.

- You can redirect standard output by using the > symbol. You can redirect and append output by using the >> symbols. You can redirect input by using the < symbol.
- You can filter command output by using the pipe symbol, which is | .
- You can write shell script programs to automate your operating system commands. Shell scripts allow you to combine commands with programming structures such as conditions and loops.

Review Questions

1. Which of the following is not a valid Linux command?
   a) cd
   b) mkdir
   c) md
   d) dir

2. Which of the following are full paths?
   a) ../..
   b) fun2/file2.txt
   c) /
   d) /sales/southern/regionA/sales.dat

3. You are working as a Technical Support Specialist. A customer named Consuela Gomez calls and asks you how to copy a file named /monthly/jan1.dat to her home directory. A pwd shows her current directory is /tmp. What command do you give Consuela to accomplish this goal?
   a) mv /monthly/jan1.dat $HOME/jan1.dat
   b) rm /monthly/jan1.dat
   c) move /monthly/jan1.dat $HOME/jan1.dat
   d) cp /monthly/jan1.dat $HOME/jan1.dat

4. What symbol is used to redirect and append output to a file?
   a) >
   b) >>
   c) |
   d) ?

5. You want to make a duplicate of a file named data3.dat. The new file is to be named data4.dat located in the same directory. What command will do this for you?
   a) copy data3.dat data4.dat
   b) cp data4.dat data3.dat
   c) cp data3.dat data4.dat
   d) mv data3.dat data4.dat

6. Which of the following will wildcard a single character position?
   a) ?
   b) *
   c) [...]
   d) >

7. A customer calls while you are carrying the beeper for the weekend. She is having trouble accessing a file in a certain directory. She cannot recall the exact name of it but she knows it begins with a dot. She runs the **ls** command with no options and she cannot see the file. What command should you suggest to her in order for her to see the file name on the screen?
   a) ls –l
   b) ls –a
   c) ls *
   d) dir

8. Which of the following will redirect input?
   a) >
   b) >>
   c) *
   d) <

9. You want to change the name of a file named data3.dat. The new file is to be named data4.dat located in the same directory. What command will do this for you?
   a) copy data3.dat data4.dat
   b) cp data4.dat data3.dat
   c) cp data3.dat data4.dat
   d) mv data3.dat data4.dat

10. You want to remove the file named data3.dat and no other file. What command will do this for you?
   a) del data3.dat
   b) rm data3.dat
   c) rm data3.*
   d) rm *.*

11. Which of the following will create a directory named /Spooling?
   a) mkdir Spooling
   b) mkdir \Spooling
   c) md /Spooling
   d) mkdir /Spooling

12. Which of the following will write and quit a file in the vi editor?
    a) :W
    b) :Q
    c) :q
    d) :wq!

13. The name of the command that will display your current working directory is
    _____.
    a) mkdir
    b) cd
    c) pwd
    d) ls

14. Which of the following will allow you to display file permissions?
    a) ls
    b) ls –l
    c) ls –a
    d) pwd

15. You are working for a small consulting firm. A customer, George Patel, calls you and
    has a problem with the **ls** command. He wants to review the online help for this com-
    mand. What command do you tell him to run?
    a) ls
    b) man ls
    c) ls *
    d) ls ???

16. Jessie wants to determine the names of the users currently logged onto the system.
    What command will do this for her?
    a) ls
    b) pwd
    c) who
    d) cd

17. Zac wants to display all the files that have a 0 for the fourth position and a 7 for the
    fifth position in their file name regardless of the characters in later positions. What
    command will accomplish this?
    a) ls *05?
    b) ls ?05*
    c) ls 4 and 5
    d) ls ???07*

18. Sue Weng wants to make sure that the permissions of a file are set as follows: owner has read and execute, the group has read, and the world has no access. Choose the correct answer that meets this requirement in the order as they would appear on the **ls –l** command.

   a) -r - - r x - r - -
   b) -r - - r - - r - x
   c) -r - - r - x r - -
   d) -r - x r - - - - -

19. What command will allow you to change to the root directory?

   a) cd \
   b) cd /
   c) cd ..
   d) cd .

20. What of the following would allow you to change to the parent directory?

   a) cd \
   b) cd /
   c) cd ..
   d) cd .

Lab Projects

Lab Project 1

In this lab project you will create the tree structure displayed in Figure 8.1. However, instead of building it starting in the root directory, we will build it starting from your own home directory, $HOME.

1. Log in to Linux and access the command prompt.

2. In order to change directory locations to your home directory, type **cd $HOME** and press Enter. Make sure you specify the HOME directory in uppercase and precede the text with a dollar sign, $. The dollar sign, $, refers to the contents of the variable named HOME. It's just the Linux syntax.

3. Type **mkdir Project1** and press Enter.

4. Record what this command accomplishes.

   _____

5. Type **cd Project1** and press Enter.

6. Record what this command accomplishes.

   _____

7. Type **mkdir dir2** and press Enter.

8. Type **mkdir dir3** and press Enter. **Note:** You could have typed **mkdir dir2 dir3** on the same line to create both directories. This would have replaced the previous two **mkdir** commands.

9. Type **cd dir2** and press Enter. Notice your prompt changes.

10. Type **pwd** and press Enter.

11. Record what the command accomplishes.

    _____

12. Record the output of the **pwd** command.

    _____

13. Type **mkdir dir2a** and press Enter.

14. Type **cd dir2a** and press Enter.

15. Type **pwd** and press Enter.

16. Record the output.

    _____

17. Type **touch file2a.txt** and press Enter.

18. Type **cd ..** and press Enter.

19. Record what this command accomplishes.

_____

20. Type **pwd** and press Enter.
21. Record the output.

_____

22. Type **touch file2.txt** and press Enter.
23. Type **touch file22.txt** and press Enter.
24. Record another way you could have created file22.txt.

_____

25. Type **cd ..** and press Enter. This should take you to $HOME\Project1.
26. Type **pwd** and press Enter.
27. Record the output.

_____

28. Type **echo "hi" > file32.txt** and press Enter. This is another method to create a file.
29. Type **cp file32.txt fil33.txt** and press Enter.
30. Type **cp $HOME/Project1/dir2/dir2a/file2a.txt $HOME/Project1/dir3/ file3a.txt** and press Enter.
31. Record the output.

_____

32. Log out of your account.

Lab Project 2

In this lab project you will review the commands given and then draw the tree based upon the commands.

1. Review the list of commands below. Pay careful attention to the order of the commands listed. Your instructor may allow you to enter the commands on the computer. If so, log in and run them.

   cd $HOME
   mkdir Project2
   cd Project2
   mkdir payroll
   mkdir general
   mkdir fun
   cd payroll
   mkdir paychecks
   mkdir payday
   cd payday
   touch pay1.dat
   cp pay1.dat pay2.dat
   cd ../../general
   echo "Hi" > gen1.dat
   cp gen1.dat gen2.dat
   cp gen2.dat $HOME/Project2/gen3.dat
   cd $HOME/Project2/payroll
   cp payday/pay1.dat paychecks/payC.dat
   cd $HOME

2. Draw the tree structure. You have been given a starting point, $HOME. The rest is up to you.

                    $HOME
                      |
                      (Draw the tree here!)

Lab Project 3

In this lab project you will reinforce your understanding of the tree structure. As with Lab Project 2, you will review the commands given and then draw the tree based upon the commands.

1. Review the list of commands below. Pay careful attention to the order of the commands listed and the case (lower vs. upper) of the files and directories. Your instructor may allow you to enter the commands on the computer. If so, log in and run them.

```
cd $HOME
mkdir Project3
cd Project3
mkdir DirA
mkdir DirB
mkdir DirC
mkdir DirA/DirAA
mkdir DirA/DirAB
mkdir DirA/DirAC
mkdir DirB/DirBB
cd DirA/DirAA
mkdir Fun5
cd Fun5
mkdir MoreFun
cd MoreFun
touch fileT.dat
touch fileM.dat
cp MoreFun/*.dat $HOME/Project3/DirC
touch $HOME/Project3/DirB/DirBB/Games.dat
ls >> $HOME/listing.txt
cd $HOME
```

2. Draw the tree structure. You have been given a starting point, $HOME. The rest is up to you.

$HOME

| (Draw the tree here!)

Lab Project 4

In this lab project you will review a tree structure and create it using Linux commands. Note that directories do not have an extension, but files do have an extension so use the appropriate command. Notice that checks.dat is hidden (initial character is a dot).

1. Log in to Linux and access the command prompt.

2. Type **cd $HOME** and press Enter.

3. Type **mkdir Project4** and press Enter.

4. Type **cd Project4** and press Enter.

5. Review and create the following tree structure.

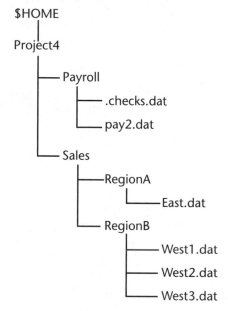

6. List the exact steps you took to create this tree.

Lab Project 5

In this lab project you will review a tree structure and create it using Linux commands. Note that directories do not have an extension, but files do have an extension so use the appropriate command.

1. Log in to Linux and access the command prompt.

2. Type **cd $HOME** and press Enter.

3. Type **mkdir Project5** and press Enter.

4. Type **cd Project5** and press Enter.

5. Review and create the following tree structure.

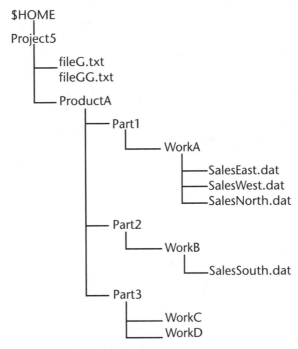

```
$HOME
 |
Project5
     |____ fileG.txt
     |     fileGG.txt
     |____ ProductA
                |____ Part1
                |        |____ WorkA
                |                  |___SalesEast.dat
                |                  |___SalesWest.dat
                |                  |___SalesNorth.dat
                |____ Part2
                |        |____ WorkB
                |                  |___SalesSouth.dat
                |____ Part3
                         |____ WorkC
                         |____ WorkD
```

6. List the exact steps you took to create this tree.

Lab Project 6

The goal of this lab project is for you to be able to distinguish between navigating to a parent, a child, and a sibling directory. A sibling directory is defined as a directory that is on the same hierarchical level as another. You must complete Lab Project 4 before beginning this project.

1. Log in to Linux and access the command prompt.

2. Change directory locations to $HOME. Your prompt will reflect the change.

3. Type **cd Project4** and press Enter. (Yes, it is Project4!)

4. Type **cd Sales** and press Enter. Your prompt will reflect the change.

5. In order to change to a child directory, you must use the child directory's name. Type **cd RegionA** and press Enter. Your prompt will reflect the change.

6. In order to change to a sibling directory, type **cd ../RegionB** and press Enter. The format is cd ../*sibling-name* where *sibling-name* is the name of a sibling directory. In the tree structure in Project 4, RegionA and RegionB are sibling directories.

7. Record at least one other sibling relationship in the tree structure in Project 4.

   _____

8. To copy West1.dat from RegionB to RegionA using the sibling concept, type **cp West1.dat ../RegionA/West1.dat** and press Enter. The file is copied.

9. In order to change back to RegionA, type **cd ../RegionA** and press Enter.

10. Record what this command accomplishes.

    _____

11. To move East.dat located in RegionA to RegionB using the sibling concept, type **mv East.dat ../RegionB** and press Enter.

12. In order to change to the parent directory, type **cd ..** and press Enter.

13. Record what this command accomplishes.

    _____

14. Log out of your account.

Lab Project 7

The goal of this lab project is to be able to utilize the vi editor to append, insert, yank, put, and delete text in a file.

1. Log in to Linux and access the command prompt.

2. Change directory locations to $HOME. Your prompt will reflect the change.

3. Type **mkdir Project7** and press Enter.

4. Type **cd Project7** and press Enter.

5. Type in **vi file27.txt** and press Enter. The tilde (~) symbol appears. The vi editor opens the file named file27.txt.

6. In order to insert text, press **Escape** and then the **i**.

7. Enter the following text and make sure you press the Enter key at the end of each line:

   This is the first line of the vi editor.
   This is the second line of the vi editor.
   Last line of the vi editor.

8. In order to save the file, press **Escape** followed by the colon symbol, **:**. The cursor moves to the bottom of the screen with a colon to its left.

9. In order to write the file, press the **w**, then **q,** then **!,** and then press Enter. Your prompt returns.

10. Do not close your terminal window because we'll use it in the next step.

11. In order to open the file again, type in **vi file27.txt** and press Enter. The vi editor opens file27.txt.

12. Now you will work with the yank and put commands to copy and paste text. In order to copy the first three lines and place them at the end of the file, move your cursor to the first character on the first line.

13. Next, press the number **3,** and then press the letter **y** twice. This will yank, or copy, the three lines and place them in the vi buffer.

14. Move your cursor to the end of the file by pressing **Shift + G**. The cursor moves to the first character on the last line.

15. Now you will paste the three previously yanked lines by pressing the letter **p**. This will place a copy of the three lines at the end of the file.

16. Now, let's delete the first two lines. Move the cursor to the first character on the first line by pressing the number **1** followed by **Shift + G**. Your cursor moves to the first character of the first line in the file.

17. Press the letter **d** twice to delete the first line of text.

18. Repeat the previous step to delete the second line of text. When this step is complete, the file will look like this:

Last line of the vi editor.
This is the first line of the vi editor.
This is the second line of the vi editor.
Last line of the vi editor.

19. Now you will change the word "Last" in the very first line to "First." Press **Escape**, followed by pressing the number **1** and then by **Shift + G**. Your cursor moves to the first character of the first line in the file.

20. To delete each letter in the word "Last," press **x**, to delete a single character, four times (one for each letter). The word is deleted.

21. To insert the word "First," press **i**, for insert. Type **First**. The word "First" appears.

22. To append a line of text at the end of the file, use the **a** command in vi. Press **Escape**, then press **Shift + G**. Your cursor moves to the first character of the last line in the file.

23. Press **Escape**, followed by **o** and then Enter. The cursor drops down to a blank line. Type **The End!** and press Enter. The completed file will appear as follows:

First line of the vi editor.
This is the first line of the vi editor.
This is the second line of the vi editor.
Last line of the vi editor.
The End!

24. Press **Escape**, followed by **:wq!** and then Enter. The file is saved and your prompt is returned.

25. Log out of your account.

Lab Project 8

The goal of this lab project is to create files and directories and identify the columns of the output of the **ls -l** command.

1. Log in to Linux and access the command prompt.

2. Change directory locations to $HOME. Your prompt will reflect the change.

3. Type **touch pay1.dat** and press the **Enter** key.

4. Repeat the previous step five additional times. Change the number within the file name each time. You should end up with six files you created.

5. Create four directories. You decide upon the names.

6. Issue a long listing of the directory. Type **ls –l** and press Enter.

7. On a separate piece of paper, label and identify each column. Refer to the man pages on the **ls** command if necessary.

8. Log out of your account.

Lab Project 9

The goal of this lab project is for you to be able to utilize the **sort** command to sort data in ascending (lower to higher) and descending (higher to lower) order.

1. Log in to Linux and access the command prompt.

2. Change directory locations to $HOME. Your prompt will reflect the change.

3. Type **mkdir Project9** and press Enter.

4. Type **cd Project9** and press Enter.

5. Type **vi inputfile.dat** and press Enter. The **vi** editor opens.

6. In order to insert text, press **Escape** and then the **i**.

7. Enter the following text and make sure you press the Enter key at the end of each line. Each row is considered a Part Record with the colon separating the fields. The first field is the Part Number. The second field is the Part Name, the third field is the Part Amount, and the last field is the Part Quantity.

   102A:Wrench:$12.00:56
   105T:Drill:$129.00:7
   103F:Saw:$55.00:18
   101A:Hammer:$35.00:24

8. In order to save the file, press **Escape** followed by the colon symbol, **:wq!** and press Enter. Your prompt will return.

9. Sort the data in ascending order. Type **sort inputfile.dat** and press Enter.

10. Record the output.

    _____

11. Sort the data in descending order. Type **sort –r inputfile.dat** and press Enter.

12. Record the output.

    _____

13. Sort the data in descending order and redirect the output to another file. Type **sort –r inputfile.dat > sorted_reverse.dat** and press Enter.

14. Sort the data in ascending order and redirect the output to another file.

15. Record the command you used.

    _____

16. Record the output of the command you executed in the previous step.

    _____

17. Log out.

Lab Project 10

The goal of this lab project is for you to be able to utilize the **cut** and **diff** commands to manipulate text in a file.

1. Log in to Linux and access the command prompt.

2. Change directory locations to $HOME. Your prompt will reflect the change.

3. Type **mkdir Project10** and press Enter.

4. Type **cd Project10** and press Enter.

5. Copy **inputfile.dat** from the Project9 directory to the Project10 directory.

6. Record the command you used.

_____

7. To cut the first field, type **cut -d: -f1 inputfile.dat** and press Enter. The cut command does not remove the data from the file it simply displays the requested information on the screen.

8. Record the result.

_____

9. To cut the second field, type **cut -d: -f2 inputfile.dat** and press Enter.

10. Record the result.

_____

11. Cut the last field in the file and record the command you used.

12. Record the result.

_____

13. To cut both the Product Name and Product Price, type **cut -d: -f2, f3 inputfile.dat** and press Enter.

14. Record the result.

_____

15. To cut both the Product Name and Product Price and redirect the output to a file name **outfile.dat**, type **cut -d: -f2, f3 inputfile.dat > outfile.dat** and press Enter.

16. Record the result.

_____

17. Cut the first and last fields of **inputfile.dat** and redirect the output to a new file named **outfile2**.

18. Record the command you used.

_____

19. Record the result.

   _____

20. Now, let's look at the **diff** command. First, you'll need to make a copy of the input file and add some additional records. To make a copy of inputfile.dat, type **cp inputfile.dat inputfile2.dat** and press Enter.

21. Open the file named inputfile2.dat and add three additional records.

22. To display the differences between the two files, type **diff inputfile.dat inputfile2.dat** and press **Enter**. Refer to the man pages on the **diff** command if necessary.

23. Record the output.

   _____

24. Close the window.

25. Log out of your account.

Lab Project 11

The goal of this lab project is for you to be able to utilize the **grep** to search for data within a file and to use the **find** command to search for file names.

1. Log in to Linux and access the command prompt.

2. Change directory locations to $HOME. Your prompt will reflect the change.

3. Type **mkdir Project11** and press Enter.

4. Type **cd Project11** and press Enter.

5. Copy **inputfile.dat** from the Project9 directory to the Project11 directory.

6. Record the command you used.

_____

7. In order to search for the record with the text "Drill" in the file named inputfile.dat, type **grep Drill inputfile.dat** and press Enter.

8. Record the output.

_____

9. Issue the command to search for the record with the text "Saw" in the file named **inputfile.dat**.

10. Record the command.

_____

11. Record the output.

_____

12. If you type the text "Drill" using a lowercase "d," the **grep** command would not find a match. It is case-sensitive. However, you can use an option on the **grep** command to ignore the case. Type **grep –i drill inputfile.dat** and press Enter.

13. Record the output.

_____

14. Issue the command to search for the text "saw" in the file named inputfile.dat.

15. Now let's look at the **find** command. In order to locate a file somewhere in the tree, you can issue the **find** command. To find the file named inputfile.dat, type **find inputfile.dat** and press Enter. The file name should be displayed.

16. The **find** command really comes in handy when you are in one directory and a file is in another part of the tree but you don't know where it is. Let's say you want to find a file, in this case inputfile.dat, that is somewhere in your home directory but you aren't sure where. You would type **find $HOME –name inputfile.dat** and press Enter. The file name should be displayed. The $HOME variable tells the **find** command to start the search in that directory. You could start the search from another directory. The "-name"

option tells the **find** command to look for the name that follows; in our example it would be **inputfile.dat**.

17. Create another file in the Project11 directory.

18. Record the name.

_____

19. Change to the parent directory.

20. Record the command used.

_____

21. Issue the **find** command to search for the location of the file you just created. Instead of using the $HOME variable, use the dot (.) for your current directory.

22. Record the command used.

_____

23. Log out of your account.

Lab Project 12

The goal of this lab project is for you to be able to utilize the redirection and pipe symbols.

1. Log in to Linux and access the command prompt.

2. Change directory locations to $HOME. Your prompt will reflect the change.

3. Type **mkdir Project12** and press Enter.

4. Type **cd Project12** and press Enter.

5. Create 10 files and five directories within the Project12 directory. You decide upon their names.

6. Record the name of the command you used to create one of the files.

   _____

7. Record just the name of the command you used to create one of the directories.

   _____

8. In order to redirect output, type **ls > listing.dat** and press Enter.

9. In order to redirect and append output, type **date >> listing.dat** and press Enter.

10. In order to redirect and append output again, type **who >> listing.dat** and press Enter.

11. Verify the contents of the file named listing.dat.

12. Record the command used.

    _____

13. In order to remove the contents of the file named listing.dat, type **echo > listing.dat** and press Enter.

14. Issue another command to redirect output.

15. Record the command used.

    _____

16. Issue another command to redirect and append output.

17. Record the command used.

    _____

18. To redirect input to a command, type **cat < listing.dat** and press **Enter**. In reality, you could leave off the less-than symbol. You don't redirect input that often, hence there are very few practical examples. You redirect output more often.

19. Issue another command to redirect input.

20. Record the command used.

    _____

21. To use the pipe symbol, type **ls | more**. The listing is displayed one screen at a time.

22. To use the pipe symbol again, type **who** | **more**. The list of current users is displayed one screen at a time.

23. Issue another command to redirect output.

24. Record the command used.

_____

25. Log out of your account.

Lab Project 13

The goal of this lab project is for you to be able to utilize wildcard symbols.

1. Log in to Linux and access the command prompt.

2. Change directory locations to $HOME. Your prompt will reflect the change.

3. Type **mkdir Project13** and press Enter.

4. Type **cd Project13** and press Enter.

5. Create the following 12 files within the Project13 directory using the **touch** command. These represent payroll files with the three-character month followed by the two-digit day of the month.

jan07.dat	**dec04.dat**	**jul04.dat**
jan02.dat	**dec21.txt**	**jun06.dat**
jan03.dat	**jul04.txt**	**jul13.txt**
jan04.txt	**feb07.dat**	**dec07.dat**

6. In order to display all files ending in .dat, type **ls *.dat** and press Enter.

7. Record the output.

   _____

8. Issue a command to display all files ending in .txt.

9. Record the command.

   _____

10. Record the output.

    _____

11. In order to display all files that begin with a j and end with .dat, type **ls j*.dat** and press Enter.

12. Record the output.

    _____

13. Issue a command to display all files that begin with a d and end in .dat.

14. Record the command.

    _____

15. Record the output.

    _____

16. In order to display all files for January, type **ls jan*** and press Enter.

17. Record the output.

    _____

18. Issue a command to display all files for July.

19. Record the command.

_____

20. Record the output.

_____

21. In order to display files for the fourth day of the month (a 0 in the fourth position and a 4 in the fifth position), type **ls ???04*.*** and press Enter.

    Note: ls ???04* would work too.

22. Record the output.

_____

23. Issue a command to display files for the seventh day of each month.

24. Record the command.

_____

25. Record the output.

_____

26. Log out of your account.

Lab Project 14

The goal of this lab project is to help reinforce your knowledge of the tree structure. Additionally, you will list the exact commands you used to create the tree structure.

Optionally, you will write a shell script of the commands you used, delete the existing tree structure and then run the shell script to recreate the tree structure. Because this is optional, ask your instructor before proceeding with this.

1. Log in to Linux and access the command prompt.

2. Change directory locations to $HOME. Your prompt will reflect the change.

3. Create a directory named **Project14** and change locations to it.

4. Create the tree structure shown in Figure 8.21.

5. List the exact steps used to create it.

6. **(Advanced Optional Step—ask your instructor before doing this step!)** Using the **vi** editor, create a shell script with the exact commands you listed in step 5. Make the script executable. Delete the existing tree structure you just created and run your script. The shell script should recreate the tree structure.

7. Log out of your account.

Figure 8.21: A Screenshot of the Tree Structure for Project 14

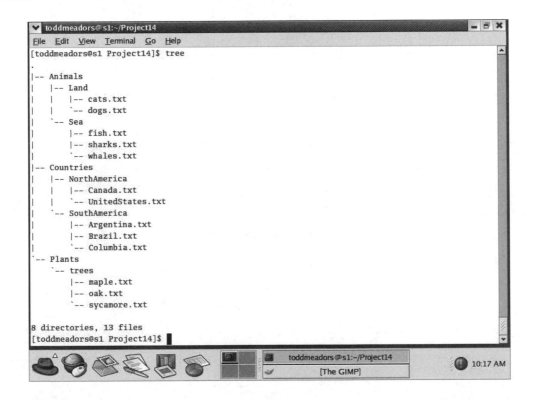

Lab Project 15

The goal of this lab project is to further your understanding of **sort**, **grep** and **awk** commands will be used.

1. Log in to Linux and access the command prompt.

2. Change directory locations to your home directory.

3. Create a directory named Project15 and change locations to it.

4. Create a file named Parts.txt, which holds part data. There are four fields separated by a colon. The first field is the Part ID, the second field is the Part Name, the third field is the Part Quantity on Hand and the last field is the Part Price.

   1009:Hammer:10:15.99
   1006:Saw:4:39.99
   1005:Tool Kit:3:99.99
   1004:Anvil:1:399.99
   1008:Pipe Wrench:50:5.49
   1002:Drill:14:12.49
   1001:Chain Saw:15:100.00
   1003:Box of Nails:40:4.99
   1000:Hack Saw:6:5.99
   1007:Box of Tacks:53:2.00

5. Save the file, quit the editor and return to the command prompt.

6. Sort the file in reverse order.

7. Record the command used.

   _____

8. Sort the file in reverse order and redirect the output to a file named PartsSortReverse.txt.

9. Record the command used.

   _____

10. Using **grep**, search for the Product Name of "Chain Saw" to display its data.

11. Record the command used.

   _____

12. Search for the Product Name of "Hammer" to display its data.

13. Record the command used.

   _____

14. Search for all the Saw records and redirect the output to a file named Saw.txt.

15. Record the command used.

   _____

16. Display just the second and last fields.

17. Record the command used.

_____

18. Display just the Product ID and Product Name.

19. Record the command used.

_____

20. Display just the Product Name, Product Quantity on Hand and Part Price and redirect the output to a file named PartsA.txt

21. Record the command used.

_____

22. Type **grep Anvil parts.txt | awk –F: '{print $2, $4}'** and press Enter.

23. Record what this command does.

_____

_____

_____

24. Type **grep Saw parts.txt | sort | awk –F: '{print $2, $4}'** and press Enter.

25. Record what this command does.

_____

_____

_____

26. Type the following and press Enter when complete:

    grep Saw parts.txt | sort | awk –F: '{print $2, $4}' > SawsSortedNamePrice.txt

27. Record what this command does.

_____

_____

_____

28. Close your window.

29. Logout.

Lab Project 16 Challenge

You have been hired by ZJ Industries to design its tree structure. The company has four departments: Accounting, Sales, Manufacturing, and IT (Information Technology). The Accounting Department has these three work groups: Accounts Receivable, Accounts Payable, and Payroll. The Sales department has regional offices located in the West, East, North, and South. The Manufacturing department runs three shifts. The IT department is currently working on three projects.

1. You need to design the Linux file system hierarchy on paper. Each department, work group, shift, region, and project will become directories. Keep the tree structure on the Linux file system just as you would have an organizational chart for the company. In other words, create a directory named East under Sales, not Accounting. The first part has been done for you to get you started.

$HOME
|

ZJ_Industries
|

2. You need to create this tree structure within the Linux system. Create at least one empty file in each directory. Use the department, work group, shift, region, or project name as the file name.

Lab Project 17 Challenge

1. Create a directory in your home directory named Project17.

2. Change to Project17.

3. Create a shell script that creates the tree you designed for ZJ Industries in Project 16 Challenge.

4. On a separate piece of paper, record your script.

5. Using one command, list all of the files and directories.

6. Log out.

Internet Discovery

Internet Discovery Lab 1

1. Go to a computer with Internet access. Open a web browser.

2. In the **Address** text box, type **http://www.redhat.com** and press the Enter key. The Red Hat web site appears.

3. In the **Search Red Hat:** text box, enter **find** and click **Go**. The screen displays a listing of the links to documents utilizing the **find** command.

4. Click one of the documents and review it.

5. Based on your findings, briefly discuss the **find** command in the space provided.

6. Repeat the search for the **ls** command.

7. In the space provided, discuss something new you've learned about this command by visiting this web site.

8. Browse the web site.

9. In the space provided, discuss something new you've learned by visiting this web site.

Internet Discovery Lab 2

1. Using your browser, search the Internet for any scripts using the **awk** command.

2. Identify at least three new features of the **awk** command.

3. Locate and record an **awk** script.

4. Close your browser.

Soft Skills: Help Desk Support

1. A customer calls you and is very upset. He used the **mv** command to move a file but he thought it would keep the original. The file has been moved but he wants a duplicate of the file instead. What do you tell him?

2. Explain to a customer how you would navigate to a sibling directory.

3. You receive a call in the middle of the night. A customer has deleted a file and wants to get it back. How do you proceed?

Critical Thinking

1. Compare the Linux file system to your family tree. Go back as far as your great-grandparents if you can.

2. Explain why you cannot issue the **cd ..** command at the root directory.

3. Compare and contrast the DOS and Linux file system hierarchy.

Study Skills

The "Hear" in "Hear, See, Do, Say"

Remember the key to learning are these four points: hear, see, do, say. This section focuses on hearing, or listening to your instructor.

- Listen, listen and listen again. Make sure you listen to your instructor at every class meeting to know what you must do for the day and week.
- Good listening habits include focusing on the instructor's lecture. If the instructor gives hand-outs, review them as the instructor reads over them.
- Although you should listen during every class session, listen very carefully the first week of class. The tone of the class is usually set during this first week and you don't want to miss anything.
- Pay attention to detail. If the instructor says do **NOT** begin a lab yet, then do not begin the lab yet—the instructor may have good reason. The lab project could entail formatting a disk or deleting a file that the instructor needed to see before you you're your data.
- Listening does include being attentive and asking questions when you don't understand. However do _not_ waste time over issues the instructor covered but you didn't get because you were not listening!
- Listening does _not_ include searching the Internet while the instructor is lecturing.
- Listening does _not_ include talking to fellow students while the instructor is lecturing.
- Listening does _not_ include asking questions over material you just heard!
- Listening does _not_ include being disruptive! Just because you are in a class that you might pay for, does not give you license to participate in disruptive behavior. This simply wastes time. Remember, you are only hurting yourself but more importantly, you are hurting fellow students. You could be interviewing for a job one day and the interviewer might be a fellow student—how would that job interview turn out?
- Finally, did we mention to listen? Listen and you will learn something!

Self-Study Question(s)
1. Did you hear your instructor's lecture today?
2. Identify at least two listening Study Skills you did this week.

Introduction to Networking

OBJECTIVES

The goal of this chapter is twofold:
- To introduce you to networking concepts.
- To help you prepare and pass the following sections of the A+ Operating System Technologies Exam:

A+ Operating System Technologies Exam Objectives
covered in this chapter (and corresponding page numbers)

Domain 4 Networks

4.1 Identify the networking capabilities of Windows. Given configuration parameters, configure the operating system to connect to a network.

4.2 Identify the basic Internet protocols and terminologies. Identify procedures for establishing Internet connectivity. In a given scenario, configure the operating system to connect to and use Internet resources.

In this chapter, you will complete the following sections:
- 9.1 Understanding Networking
- 9.2 Understanding Network Topologies
- 9.3 Understanding Network Cabling
- 9.4 Understanding Access Methods
- 9.5 Understanding Network Standards
- 9.6 Understanding the OSI Model
- 9.7 Understanding Network Protocols
- 9.8 Configuring Networking
- 9.9 Troubleshooting Networks
- 9.10 Configuring a Networked Printer
- 9.11 Understanding Dial-Up Networking (DUN)
- 9.12 Learning about Internet Software
- 9.13 Understanding Wireless Networks
- 9.14 Installing and Configuring a Wireless Network Card
- 9.15 Understanding Wireless Access Points
- 9.16 Troubleshooting Wireless Networks

9.1 Understanding Networking

A **network** is two or more devices that can communicate with one another and share resources. A network allows computer users to share files; communicate via e-mail; browse the Internet; share a printer, modem, or scanner; and access applications and files. Networks can be divided into two major categories—LANs and WANs. A **LAN** (**Local Area Network**) is a group of devices that can share resources in a single area, such as a room or a building. A **WAN** (**Wide Area Network**) is communication between LANs. The Internet is an example of a WAN as are two networks located in two cities.

Networks are vital to businesses today. They can even be found in many homes. You must have a basic understanding of the devices that make up networks (computers, printers, modems, etc.) and then learn network devices. You cannot bypass computer repair and go straight into networking.

Types of Local Area Networks

There are two basic types of LANs: a server-based network and a peer-to-peer network. With a **server-based network**, computer users log in to a main computer called a server where they are authenticated (authorized to use the network). The server is a more powerful computer than a normal workstation. The server contains information about who is allowed to connect to the network, and to what network resources (files, printer, and applications) the network user is allowed access. A **peer-to-peer network** does not have a central server. Instead, each computer is its own server. The computer user sets up passwords to allow others access to the resources. A user uses the network to access the remote files, printer, applications, and so forth, from their own workstation. Server-based networks are more common in businesses, whereas peer-to-peer networks are more common in homes and very small businesses. A server-based network can consist of 10 or more computers; in contrast, a peer-to-peer network usually has fewer than 10 computers.

A server-based network is more secure than a peer-to-peer network. This is because the server is normally located in a locked network room or wiring closet. Also, the network users and what they are allowed to do (their network rights and permissions) are configured and stored on the network server. Servers have a special operating system loaded on them called a **NOS** (**Network Operating System**). Examples of network operating systems are Novell's NetWare, Microsoft's NT Server, 2000 Server, and 2003 Server. A network operating system has utilities that allow computer user management (who is allowed onto the network), resource management (what network applications, files, printers, etc. a user can use), and security management (what a user is allowed to do with a resource, such as read, write, read and write, etc.). One userid and password is all a remote user needs to access many network resources located throughout the business organization. A network user can sit down at any computer in the organization, log on to the server, and start working with the network resources.

Figure 9.1 shows how a server-based network can be configured. The network has one server in the center, four workstations, and two laser printers labeled LP1 and LP2. The server has a database of users, CSchmidt, RDevoid, and MElkins, and their associated passwords. The server also has three applications loaded—Microsoft Excel, Microsoft Project, and Microsoft Word. These applications and associated documents are stored on the server. Whether or not the users can access these applications and documents and what they can do within each document is also stored on the server. In the Permission column of the table located in Figure 9.1 is either R for Read or R/W for Read/Write. This is an indication of what the user can do in a particular application. For example, user CSchmidt has read and write access to Excel, Project, and Word. User MElkins can only read Excel and Word documents, but she can read and write Microsoft Project documents. User CSchmidt can print to either of the laser printers, but user RDevoid prints only to the LP1 laser printer.

Another benefit of server-based networks is that a user can sit down at any workstation, log in to the server with his or her userid and password, and have access to all of the network resources. For example in Figure 9.1, computer user RDevoid can sit down at any workstation and have access to her Excel and Word documents and print to laser printer LP1.

Figure 9.1: A Server-Based Network

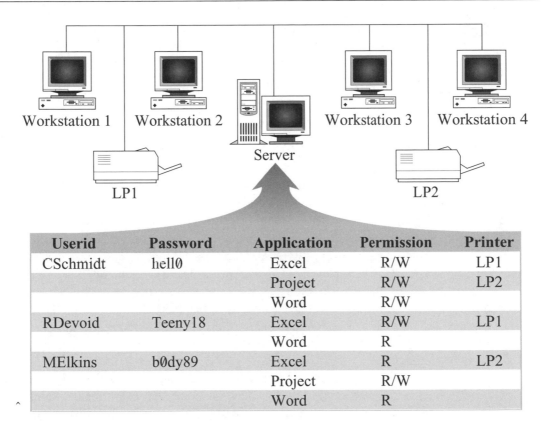

Userid	Password	Application	Permission	Printer
CSchmidt	hell0	Excel	R/W	LP1
		Project	R/W	LP2
		Word	R/W	
RDevoid	Teeny18	Excel	R/W	LP1
		Word	R	
MElkins	b0dy89	Excel	R	LP2
		Project	R/W	
		Word	R	

A peer-to-peer network is not as expensive, nor as secure as a server-based network. A server is more expensive than a regular workstation plus it requires a network operating system. Since peer-to-peer networks do not use a dedicated server, costs are reduced. Instead of a network operating system, each workstation uses a regular operating system such as Windows 95, 98, NT Workstation, 2000 Professional, or XP. A peer-to-peer network is not as secure as a server-based network because each computer must be configured with individual userids and passwords. Figure 9.2 shows how a peer-to-peer network is configured.

Figure 9.2: A Peer-to-Peer Network

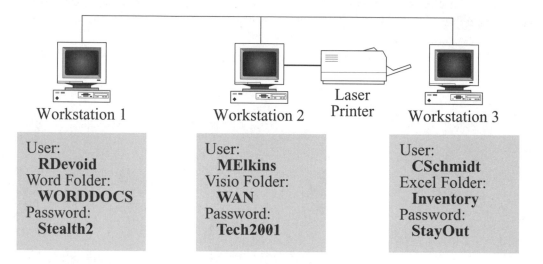

In Figure 9.2, there are three workstations labeled Workstation 1, 2, and 3. Workstation 2 has a shared printer. A shared printer is a printer connected to the computer that has been configured so that other network users can print to it. There are three people in this company, Raina Devoid, Cheryl Schmidt, and Melodie Elkins. Raina Devoid normally works at Workstation 1 and Raina has shared a folder on the hard drive called WORD-DOCS that has a password of Stealth2. Cheryl and Melodie can access the documents located in WORDDOCS from their own workstations as long as they know the password is Stealth2. If Raina (who is sitting at Workstation 1) wants to access Melodie's WAN folder, Raina must know and remember that the password is Tech2001. If Melodie changes the password on the WAN folder, Melodie must remember to tell the new password to anyone who needs access. The password is only used when accessing the WAN folder documents.

A peer-to-peer network password is only effective across the network. The password is not effective if someone sits down at the workstation. For example, if a summer intern, Ken Tinker, sits down at Workstation 3, Ken has full access to the Inventory folder and documents. Even though the folder is passworded for the peer-to-peer network, Ken is not using the network to access the folder so the password is useless.

Management of network resources is much harder to control on a peer-to-peer network than on a server-based network. Each user is required to manage the network resources on one computer and password management can become a nightmare. Remember with peer-to-peer networks, anyone who has the password can access the folder across the network. Server-based networks are normally more secure because (1) passwords are managed centrally at the server and (2) the server is normally locked in a wiring closet.

The problem of having access to a workstation and all its resources simply by sitting down at a computer is not as much of a threat today because of the newer operating systems' features. NT Workstation and 2000 Professional cannot be accessed without a userid and password.

In order to have a network, the following are required: network adapters (NICs), network cabling, and an operating system with network options enabled. The following sections explore these concepts.

9.2 Understanding Network Topologies

Network topology is how network devices connect together. The three major types of network topologies are star, ring, and bus. Keep in mind that a large business may have combinations of these topologies. A topology that combines other topologies is known as a hybrid topology.

The most common network topology used today is the **star topology** because it is used with Ethernet networks. Each network device connects to a central device, normally a hub or a switch. Both the **hub** and the **switch** contain two or more RJ-45 network jacks. The hub is not as intelligent as a switch. The switch takes a look at each data frame as it comes through the frame. The hub is not able to do this. Figure 9.3 illustrates a hub or switch.

Figure 9.3: A Hub/Switch

In a star topology, each network device has a cable that connects between the device and the hub or switch. If one computer or cable fails, all other devices continue to function. However, if the hub or switch fails, the network goes down. The hub or switch is normally located in a central location such as a network wiring closet. Figure 9.4 shows how a star topology is cabled. By looking at how each device connects to a central location, you can easily see why it is called a star.

What is the difference between a hub and a switch? Bottom line: A switch is faster than a hub! A switch is considered a "smart" device because it "learns" what computers are attached to it and forward data packets to the correct port on the switch. (A **port** is a receptacle on a hub or switch where one end of a network cable connects; the other end connects to your computer's network card.)

A hub on the other hand does not "learn" where the computers are. It simply checks each port to determine if the data packet should be sent to the computer attached to it. A switch may sometimes take longer to become operable after it boots because it is learning what ports contain what computers. However, after it is up, it is faster than a hub for network operations. Memory tip: "S" for Switch, "S" for "Smart" and "S" for "Speed."

More cable is used in wiring a star topology than with the bus topology, but the type of cable used is cheap and this is not an issue for today's network managers. Star topologies are easy to troubleshoot. If one network device goes down, the problem is in the device, cable, or port on the hub/switch. If a group of network devices go down, the problem is most likely in the device that connects them together (hub or switch). Look at Figure 9.4. If Workstation 1, Workstation 2, Workstation 3, Workstation 4, and Workstation 5 cannot communicate with one another, the problem is the switch in the middle. If only Workstation 3 cannot communicate with the other network devices, the problem is in Workstation 3, the cable that connects Workstation 3, or in port 13 on the switch.

Figure 9.4: A Star Topology

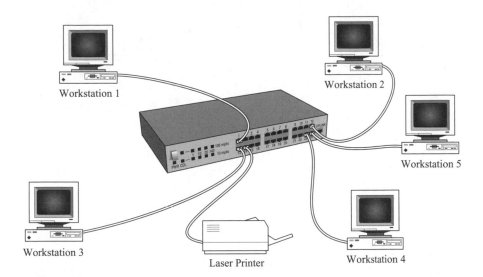

Workstation 1

Workstation 2

Workstation 5

Workstation 3

Laser Printer

Workstation 4

The **ring topology** is physically wired like a star, but operates differently. The ring topology is used in Token Ring networks. A token (a special network packet) passes from one network device to the next in a continuous fashion. Token Ring networks are wired like a star, but they operate like a logical ring. Figure 9.5 shows how the Token Ring network appears to be a ring.

The token passes from one workstation to another in a continuous loop. When the token does not contain data, it is known as a free token. As the free token is passed around the ring, any workstation wishing to transmit data takes the token and adds data. The data is sent around the ring until it reaches its destination. No other workstation can accept the data except for the destination network device. Once the data has been transmitted, a free token is placed on the ring again. No workstation can transmit until the free token comes back around the ring.

Figure 9.5: A Ring Topology

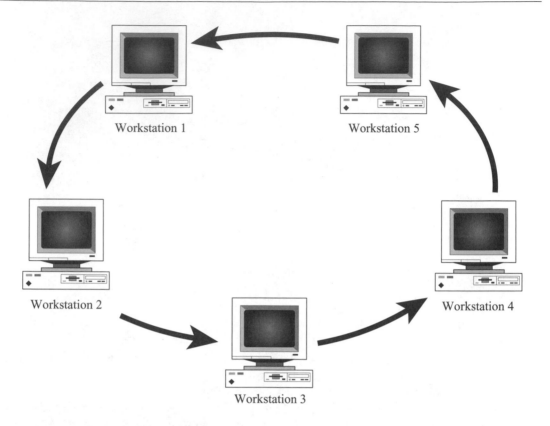

Workstation 1   Workstation 5

Workstation 2   Workstation 4

Workstation 3

Think of a Token Ring topology like an Olympic relay race. In the relay race, several racers pass a baton, or "token," to the next racer down the line. This racer in turn races to the next racer and gives the "token" to that racer. This continues until the race is won. Token ring topologies operate similarly with a token being passed to each computer. The computer wiith the token is the one that can "talk" on the network. IBM developed Token Ring networks.

The **bus topology** is one of the oldest network topologies. All network devices connect to a single cable. If the cable has a break, the entire network is down. Bus topologies are also difficult to troubleshoot when there is a network problem. Figure 9.6 depicts a bus topology.

Figure 9.6: A Bus Topology

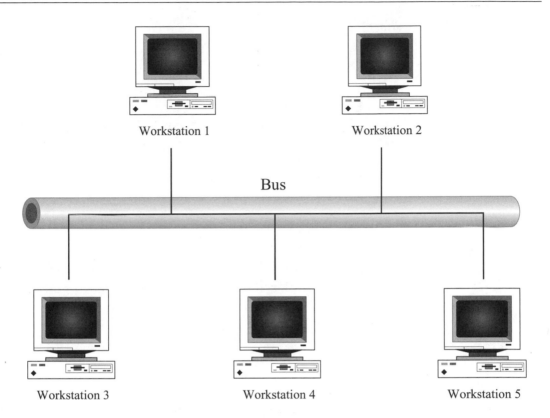

A **mesh topology** is not as common as other topologies, but it is used when all network devices connect to each other. Mesh topology is more likely to be used in a WAN (Wide Area Network) rather than a LAN (Local Area Network). Mesh topologies take a lot of cabling, but if a cable breaks, the network still continues to function.

An example of a mesh topology is a college that has three main campuses—North, South, and West. Each campus has a connection to the other two campuses. For example, the North campus has a connection to the South and the West campuses. Each campus has important servers to which the other campuses need access. If the North campus to South campus connection breaks, the North campus can still reach the South campus by going through the West campus. Whenever a network can still function after a cable break, the network is said to be fault tolerant. A mesh topology provides the most fault tolerance of any network topology. Table 9.1 summarizes network topologies.

Table 9.1: A Comparison of Network Topologies

Topology	Advantage	Disadvantage
Bus	Takes less cable (cheaper)	If a break in the bus, network is down
Mesh	If a break in the cable, network still works (fault tolerant)	Expensive and complex (hard to reconfigure)
Ring	Easy to install	Expensive parts
Star	Easy to install; most common; if a break in workstation cable, network still works (fault tolerant)	More expensive than bus

9.3 Understanding Network Cabling

Networks require some type of medium to transmit data. This medium is normally some type of cable or air (when using wireless networking). The most common types of cable are twisted-pair and fiber-optic, although some very old networks have coax cable.

Twisted-pair cable comes in two types: shielded and unshielded. The acronyms used with this type of cable are **STP** for shielded twisted-pair and **UTP** for unshielded twisted pair. The most common is UTP. With twisted-pair cable, all network devices connect to one central location such as a patch panel, hub, or switch. If one cable breaks, only the one device fails. Most people are familiar with twisted-pair cable because this type of cable is used in homes for telephone wiring. The type used with networking has eight copper wires. The wires are grouped in colored pairs. Each pair is twisted together to prevent crosstalk. **Crosstalk** occurs when a signal on one wire interferes with the signal on an adjacent wire. The wires are wrapped in a vinyl insulator. Figure 9.7 shows unshielded twisted-pair cable.

UTP cabling is measured in gauges. The most common measurements for UTP cabling are 22,- 24-, or 26-gauge unshielded twisted-pair cables. UTP cables come in different specifications called categories. The most common are categories 3, 4, and 5. People usually shorten the name Category 3 to CAT 3, or Category 5 to CAT 5. The categories determine, in part, how fast the network can run. Category 3 was mainly installed for telephone systems in many office buildings. CAT 3 is called a voice grade cable, but it has the ability to run up to 10 Mbps Ethernet or 16 Mbps Token Ring topologies. Networks that run 10 Mbps are known as 10BaseT networks. 100 Mbps Ethernet networks are known as Fast Ethernet, 100BaseT4, and 100BaseT8. The 100BaseT4 networks use two pairs (four wires) of the UTP cable whereas the 100BaseT8 networks use all four pairs (8 wires). The most common type of UTP is CAT 5. Fairly new categories of UTP cable include CAT 5e, which is designed for 100 Mbps on UTP and STP; CAT 6, which is designed for 1000 Mbps on UTP and STP; and CAT 7, which is designed for 1000 Mbps on UTP, STP, and fiber. UTP and STP cable are used in star and ring topologies.

Figure 9.7: UTP Cabling

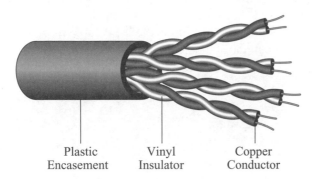

Plastic   Vinyl   Copper
Encasement   Insulator   Conductor

In order to avoid extra troubleshooting time, most businesses install their network cabling according to the ANSI/TIA/EIA-568-A or 568-B standard. This standard specifies how far the cable can extend, how to label it, what type of jack to use, and so forth. Figure 9.8 illustrates the common cabling standards used in industry.

Figure 9.8: UTP Wiring Standards

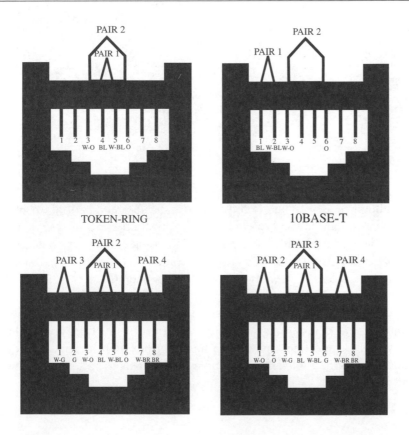

STP (**Shielded Twisted-Pair**) cable has extra foil shielding that provides more shielding. Shielded twisted-pair cable is used in industrial settings where extra shielding is needed to prevent outside interference from interfering with the data on the cable. When installing network cabling, it is important to insert the UTP cable fully into the RJ-45 jack and to insert the colored wires in the standardized order.

One common mistake that you might make when putting an RJ-45 connector on UTP cable is they put the cable into the RJ-45 connector backwards. Figure 9.9 shows the location of pin 1 and pin 8 on an RJ-45 connector. Another common mistake is not pushing the wires to the end of the RJ-45 connector. Before crimping the wires into the connector, look at the end of the RJ-45 connector. You should see each wire jammed against the end of the RJ-45 connector.

Figure 9.9: RJ-45 Pin 1 and Pin 8 Assignments

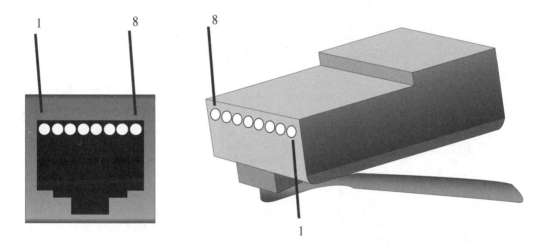

Fiber-optic cable is made of glass or a type of plastic fiber and is used to carry light pulses. Fiber-optic cable can be used to connect a workstation to another device, but in industry, the most common use of fiber-optic cable is to connect networks together forming the network backbone. Copper cable is used to connect workstations together. Then fiber cable is used to interconnect the networks especially when the network is located on multiple floors or multiple buildings.

Fiber-optic cable is the most expensive cable type, but it also handles the most data with the least amount of data loss. The two major classifications of fiber are single-mode and multi-mode. **Single-mode** fiber-optic cable has only one light beam sent down the cable. **Multi-mode** fiber-optic cable allows multiple light signals to be sent along the same cable. Multi-mode fiber is cheaper than single-mode fiber and is good for shorter distance applications. But, single-mode fiber can transmit a signal farther than multi-mode.

Fiber-optic cabling has many advantages including security, long distance transmission, and bandwidth. Fiber-optic cabling is used by many government agencies because of the high security it offers. Light signals that travel down fiber are impossible to detect remotely, unlike signals from other cable media. Also, because light is used instead of electrical signals, fiber-optic cable is not susceptible to interference from EMI or RFI-producing devices.

Each fiber-optic cable can carry signals in one direction, so an installation normally has two strands of fiber-optic cable in separate jackets. Fiber is used in the ring and star topologies. Figure 9.10 shows a fiber-optic cable.

Figure 9.10: Fiber-Optic Cable

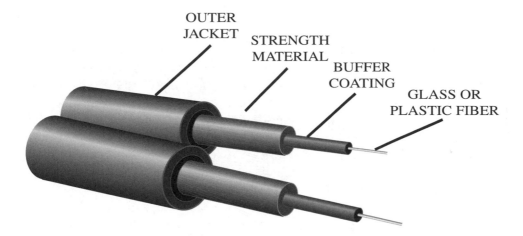

The last type of cable is **coaxial cable** (usually shortened to **coax**). Coax cable is used in older Ethernet 10Base2 and 10Base5 networks as well as mainframe and minicomputer connections. Most people have seen coax cable in their homes. The cable used for cable TV is coax cable, but is a different type than network cabling. Coax cable has a center copper conductor surrounded by insulation. Outside the insulation is a shield of copper braid, a metallic foil, or both, that protects the center conductor from EMI. Figure 9.11 shows a coax cable. Coax is used in star and bus topologies.

Figure 9.11: Coax Cable

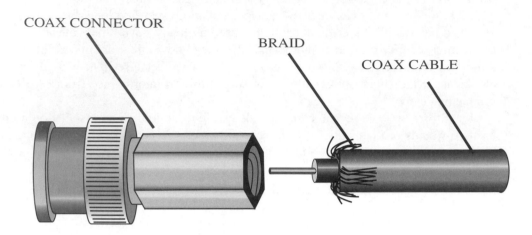

9.4 Understanding Access Methods

Before a computer can communicate on a network it must adhere to a set of communication rules to which all computers on the network comply. This set of communication rules is known as a **common access method**. Ethernet uses a common access method known as **CSMA/CD** (Carrier Sense Multiple Access/Collision Detect), whereas fiber networks and Token Ring use **token passing** as the common access method. Wireless networks and Apple networks use **CSMA/CA** (Carrier Sense Multiple Access/Collision Avoidance). The purpose of the common access method is to ensure that each workstation has an opportunity to communicate with the other workstations.

With CSMA/CD, each workstation can place data onto the network cable at any time, but the network adapter checks the network cable to ensure that no other workstation is already transmitting. In the acronym CSMA/CD, the CS stands for "Carrier Sense," which means that it is checking the network cable for other traffic. "Multiple Access" means that multiple computers can access the network cable simultaneously. "Collision Detect" provides rules for what happens when two computers access the network at the same time. One point to remember is that collisions are common and normal on Ethernet networks.

Take an example of a busy highway. The highway represents the network cable and cars on the highway represent data traveling down the cable. Each intersection that crosses the highway is simply a computer wanting to connect onto the major highway. Using CSMA/CD, the workstation checks that no other traffic is traveling down the highway (cable). If the way is clear, data is allowed to go onto the highway. If two workstations happen to transmit at the same time, a collision occurs. Both workstations have to stop transmitting data for a specified amount of time and then try transmitting again.

A Token Ring adapter uses token passing as the common access method. This method differs from CSMA/CD because there are no collisions in the Token Ring environment. With token passing, a **token** (a small packet of data) is passed from one workstation to another. Only the workstation that possesses the token is allowed to transmit data. The token is passed around the ring from one workstation to another with each workstation receiving a turn. When a workstation wants to transmit, it changes one bit inside the token data frame, adds data, and then places the data frame onto the cable. If a workstation does not want to transmit any data, the token is passed to the next workstation.

CSMA/CA is used with wireless LANs and Apple networks. Network devices listen on the cable for conflicting traffic just like CSMA/CD; however, with CSMA/CA, a workstation that wants to transmit data sends a jam signal onto the cable. The workstation then waits a small amount of time for all other workstations to hear the jam signal and then the workstation begins transmission. If a collision occurs, the workstation does the same thing as CSMA/CD—the workstation stops transmitting, waits a designated amount of time, and then retransmits.

Ethernet Issues and Concepts

Since Ethernet is the most common type of network, more time needs to be spent on some issues that deal directly with Ethernet. Some of these issues are full duplex and half duplex transmissions, network slowdowns, and increasing bandwidth.

Ethernet networks were originally designed to support either half duplex or full duplex data transmissions. **Half duplex** transmission is data transmitted in both directions on a cable, but not at the same time. Only one network device can transmit at a time. One example of half duplex transmission is using a walkie-talkie. **Full duplex** transmission is data transmitted in both directions on a cable simultaneously. This is similar to a phone conversation. Both people can talk at the same time if they want to do so. Ethernet networks were originally designed for half duplex transmission. Ethernet was also designed for a 10 Mbps bus topology and still performs as if it is connected in a bus network. Due to CSMA/CD, each workstation has to listen to the cable to see if any other transmission is occurring. Then, if no other network device is transmitting, the workstation starts transmitting data. In a request for a web page, for example, data would travel back to the workstation from the web server. With half duplex transmission, the workstation transmits and then later the web server transmits. The transmission could not occur simultaneously in both directions. The more workstations on the same network, the more collisions occur and the more the network slows down. In addition, with half duplex Ethernet, less than 50 percent of the 10 Mbps available bandwidth could be used because of collisions and the time it takes for a network frame to transmit across the wire.

Today's Ethernet networks support speeds of 10 Mbps, 100 Mbps, and 1000 Mbps. Most Ethernet NIC cards are 10/100, which means they can run at either 10 Mbps or 100 Mbps. Ethernet networks are also known as 10Base2, 10Base5, 10BaseT, 100BaseT, and 1000BaseT. When considering the term 10Base2, the 10 means that the network runs at 10 Mbps. Base means that the network uses baseband technology. The 2 in 10Base2 means that the maximum coax cable length is 185 meters (which is close to 200 meters). A 10Base2 network has terminators at both ends of the coax cable bus network. The T at the end of 10BaseT means that the computer uses twisted-pair cable. The 100 in 100BaseT means that the network supports 100 Mbps and the 1000 in 1000BaseT means that 1000 Mbps is supported.

Ethernet networks now support full duplex transmissions. With full duplex implemented, collisions are not a problem. This is because full duplex takes advantage of the two pairs of cables, one for receiving and one for transmitting. Full duplex Ethernet creates a direct connection between the transmitting station at one end and the receiving circuits at the other end. Full duplex allows 100 percent of the available bandwidth to be used in each direction. In order to implement full duplex Ethernet, both network cards in the devices must have the ability and be configured for full duplex.

Another way to speed up the network is to use a switch instead of a hub when connecting network devices together. Full duplex Ethernet works great, but replacing hubs with switches also improves network performance. A switch has more intelligence than a hub. When a workstation sends data to a hub, the hub broadcasts the data out all ports except for the port the data came in on. This is inefficient. A switch, on the other hand, keeps a table of addresses. When a switch receives data, the switch forwards the data out the port for which it is destined. A switch looks very similar to a hub and it is sometimes hard to distinguish between the two. Switches are very common devices in today's business network environment.

A classroom setting is much like CSMA/CD. Many times students will simply blurt out questions and answers and interrupt one another (and sometimes the instructor) in a classroom—that is the "multiple access" part. The instructor senses the communication—the "carrier sense" part. And, the instructor decides who will go first—the "collision detect" part.

9.5 Understanding Network Standards

The **IEEE** (**Institute for Electrical and Electronics Engineers**) committee created network standards called the **802 standards**. Each standard is given an 802.x number and represents an area of networking. Standardization is good for the network industry because different manufacturers' network components work with other manufacturers' devices. Table 2 lists the various 802 standards.

For more information about the 802 standards, access the IEEE web site at http://standards.ieee.org/getieee802/index.html.

Table 9.2: IEEE 802 Standards

802 Standard	Purpose
802.1	Bridging and Management
802.2	Logical Link Control
802.3	CSMA/CD Access Method
802.4	Token-Passing Bus Access Method
802.5	Token Ring Access Method
802.6	DQDB (Distributed Queue Dual Bus) Access Method
802.7	Broadband LAN
802.8	Fiber Optic
802.9	Isochronous LANs
802.10	Security
802.11	Wireless
802.12	Demand Priority Access
802.15	WPANs (Wireless Personal Area Networks)
802.16	Broadband Wireless Access
802.17	Resilient Packet Ring

The number 802 comes from the month (02 for February) and the year (80 for 1980) that the standards where first developed.

9.6 Understanding the OSI Model

The **International Standards Organization (ISO)** has developed a model for network communications known as the OSI (Open Systems Interconnect) model. The **OSI model** is a standard for information transfer across the network. The model sets several guidelines including (1) how the different transmission media are arranged and interconnected, (2) how network devices that use different languages communicate with one another, (3) how a network device goes about contacting another network device, (4) how and when data gets transmitted across the network, (5) how data is sent to the correct device, and (6) how it is known if the network data was received properly. All of these tasks must be handled by a set of rules and the OSI model provides a structure into which these rules fit.

Can you imagine a generic model for building a car? This model would state that you need some means of steering, a type of fuel to power the car, a place for the driver to sit, safety standards, and so forth. The model would not say what type of steering wheel to put in the car or what type of fuel the car must use, but is just a blueprint for making the car. In networking, the OSI model is such a model. The OSI model divides networking into different layers so that it is easier to understand (and teach). Dividing up the network into distinct layers also helps manufacturers. If a particular manufacturer wants to make a network device that works on layer 3, the manufacturer only has to be concerned with layer 3. This division makes networking technologies emerge much faster. Having a layered model also helps to teach network concepts. Each layer can be taught as a separate network function.

The layers of the OSI model (starting from the top and working down) are application, presentation, session, transport, network, data link, and physical. Refer to Figure 9.12.

Each layer of the OSI model uses the layer below it (except for the physical layer which is on the bottom). Each layer provides some function to the layer above it. For example, the data link layer cannot be accessed without first going through the physical layer. If communication needs to be performed at the third layer, (the network layer), then the physical and data link layers must be used first.

Figure 9.12: OSI Model Layers

7	Application
6	Presentation
5	Session
4	Transport
3	Network
2	Data Link
1	Physical

Certification exams contain questions about the OSI model and knowing the levels is a good place to start preparing for the exams. A mnemonic to help remember the OSI layers is: **A P**erson **S**eldom **T**akes **N**aps **D**uring **P**arties. Each first letter of the mnemonic phrase is supposed to remind you of the first letter of the OSI model layers. For example, *A* in the phrase is to remind you of the application layer. The *P* in Person is to remind you of the Presentation layer, and so on. Another mnemonic is All People Seem To Need Data Processing. Or, from the bottom to the top: Programmers Do Not Throw Sausage Pizza Away.

Each layer of the OSI model from the top down (except for the physical layer) adds information to the data being sent across the network. Sometimes this information is called a header. Figure 9.13 shows how a header is added as the packet travels down the OSI model. When the receiving computer receives the data, each layer removes the header information. Information at the physical layer is normally called bits. When referring to information at the data link layer, use the term **frame**. When referring to information at the network layer, use the term **packet**.

Each of the seven OSI layers performs a unique function and interacts with the layers surrounding it. The bottom three layers handle the physical delivery of data across the network. The **physical layer** (sometimes called layer 1) defines how bits are transferred and received across the network media without being concerned about the structure of the bits. The physical layer is where connectors, cable, and voltage levels are defined. The **data link layer** (sometimes called layer 2) provides the means for accurately transferring the bits across the network and it groups (encapsulates) the bits into usable sections called frames. The **network layer** (sometimes called layer 3) coordinates data movement between two devices. This layer provides path selection between two networks. Most companies and even some homes have a router that they use to connect to the Internet through their **ISP (Internet Service Provider)**. An ISP is a vendor who provides Internet access.

Figure 9.13: OSI Peer Communication

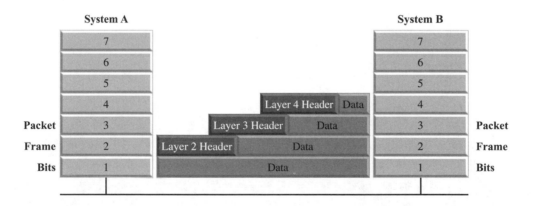

The top four layers handle the ins and outs of providing accurate data delivery between computers and their individual processes, especially in a multi-tasking operating system environment. The **transport layer** (sometimes called layer 4) provides a service to the upper layers so they do not have to worry about the details of how data is sent. The transport layer provides such services as whether the data should be sent "reliably" or not. This is similar to getting a return receipt for a package at the post office.

The **session layer** manages the communication and synchronization between two network devices. The **presentation layer** provides a means of translating the data from the sender into data the receiver understands. This allows all types of computers to communicate with one another even though one computer may be using one language (such as EBCDIC) and another computer using a different language (such as ASCII). Note EBCDIC is a character set developed by IBM for use on mainframe computers. The **application layer** provides network services to any software applications running on the network. The application layer provides network services to a computer. This allows the computer to participate or enter the OSI model (the network). Some of the services the application layer provides include negotiating authentication (what type of authentication will be used in the communication), negotiating who has responsibility for error recovery, and negotiating quality of service across the network.

Certain network devices or components work at a specific OSI layer. For example, cables, connectors, repeaters, hubs, and patch panels all reside at layer 1 of the OSI model, the physical layer parts of the network card reside at layer 1, and part of the OSI model resides at layer 2. A switch also resides at layer 2, the data link layer. A **router**, a network device that determines the best path to send a packet, works at layer 3, the network layer.

The OSI model is very confusing when you are first learning about networking, but it is very important. Understanding the model helps when troubleshooting a network. Knowing where the problem is occurring narrows the field of what the solution may be. For example, if a computer has problems communicating with a computer on the same network, then the problem is most likely a layer 1 or a layer 2 problem because layer 3 takes care of communication between two networks. Check the cabling and NIC settings.

Table 9.3 summarizes the OSI model for you.

Table 9.3: OSI Model

OSI Model Layer	Purpose
Application	Provides network services (file, print, and messaging services) to any software application running on the network.
Presentation	Translates data from one character set to another.
Session	Manages the communication and synchronization between devices.
Transport	Provides the mechanisms for how data is sent, such as reliability and error correction.
Network	Provides path selection between two networks. Routers reside at the network layer. Encapsulated data at this layer is called a packet.
Data Link	Encapsulates bits into frames. Can provide error control. MAC address and switches are at this layer.
Physical	Defines how bits are transferred and received. Defines the network media, connectors, and voltage levels. Data at this level is called bits.

9.7 Understanding Network Protocols

A **network protocol** is a data communication language. There are three primary network protocols used: TCP/IP, NetBEUI, and IPX/SPX. **TCP/IP (Transport Control Protocol/Internet Protocol)** is the most common network protocol and is used when accessing the Internet. Most companies (and homes) use TCP/IP as their standard protocol. **IPX/SPX (Internetwork Packet Exchange/Sequenced Packet Exchange)** is used when connecting to a Novell network, but Novell networks now use TCP/IP as its standard protocol. **NetBEUI (NetBIOS Enhanced User Interface)** is a non-routable network protocol. This means that it can only be used on simple networks, not on multiple networks that are tied together. A common place for NetBEUI is on a peer-to-peer network.

Network Addressing

Network adapters normally have two types of addresses assigned to them—a MAC address and an IP address. The **MAC address** is used when two network devices on the same network communicate with one another. The MAC address is a 48-bit unique number that is burned into a ROM chip located on the NIC and is represented in hexadecimal. A MAC address is unique for every computer on the network. However, the MAC address has no scheme to it except that the first three bytes represent the manufacturer. The MAC address is known as a layer 2 address.

The **IP address** is a much more organized way of addressing a computer than a MAC address and it is sometimes known as a layer 3 address. The IP address is a 32-bit number that is entered into a NIC's configuration parameters. The IP address is used when multiple networks are connected together and when accessing the Internet. The IP address is shown using dotted decimal notation, such as 192.168.10.4. Each number is separated by periods and represents eight bits, and the numbers that can be represented by eight bits are 0 to 255.

IP addresses are grouped into classes. It is easy to tell which type of IP address is being issued by the first number shown in the dotted decimal notation. Class A addresses have any number from 0 to 127 as the first number; Class B addresses have any number from 128 to 191 as the first number; and Class C addresses have numbers 192 through 223. For example, if a computer has an IP address of 12.150.172.39, the IP address is a Class A address because the first number is 12. If a computer has an IP address of 176.10.100.2, it is a Class B IP address because the first number is 176.

An IP address is broken into two major parts—the network number and the host number. The **network number** is the portion of the IP address that represents which network the computer is on. All computers on the same network have the same network number. The **host number** is the portion of the IP address that represents the specific computer on the network. All computers on the same network have unique host numbers or they will not be able to communicate.

The number of bits that are used to represent the network number and the host number depends on which class of IP address is being used. With Class A IP addresses, the first eight bits (the first number) represent the network portion and the remaining 24 bits (the last three numbers) represent the host number. With Class B IP addresses, the first 16 bits (the first two numbers) represent the network portion and the remaining 16 bits (the last two numbers) represent the host number. With Class C IP addresses, the first 24 bits (the first three numbers) represent the network portion and the remaining eight bits (the last number) represent the host number. Figure 9.14 illustrates this point.

Figure 9.14: IP Addressing

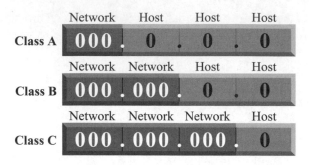

In order to see how IP addressing works, it is best to use an example. A business has two networks connected together with a router. On each network, there are computer workstations and printers. Each of the networks must have a unique network number. For this example, one network has the network number of 193.14.150.0, and the other network has the network number of 193.14.151.0. Notice how these numbers represent a Class C IP address because the first number is 193.

With a Class C IP address, the first three numbers represent the network number. The first network has a network number of 193.14.150 and the second network has a network number of 193.14.151. Remember that each network has to have a different number than any other network in the organization. The last number of the IP address will be used to assign different network devices their IP address. On the first network, each device will have a number that starts with 193.14.150 because that is the network number and it stays the same for all devices on that network. Each device will then have a different number in the last portion of the IP address, for example, 193.14.150.3, 193.14.150.4, or 193.14.150.5.

On the second network, each device will have a number that starts with 193.14.151 because that is the network number. The last number in the IP address changes for each network device, for example, 193.14.151.3, 193.14.151.4, 193.14.151.5, and so forth. No device can have a host number of 0 because that number represents the network and no device can have a host number of 255 because that represents something called the broadcast address. A **broadcast address** is the IP address used to communicate with all devices on a particular network. So, in the example given, no network device can be assigned the IP addresses 193.14.150.0 or 193.14.151.0 because these numbers represent the two networks.

Furthermore, no network device can be assigned the IP addresses 193.14.150.255 or 193.14.151.255 because these numbers represent the broadcast address used with each network. An example of a Class B broadcast is 150.10.255.255. An example of a Class A broadcast is 11.255.255.255. Figure 9.15 shows this configuration.

Figure 9.15: IP Addressing with Two Networks

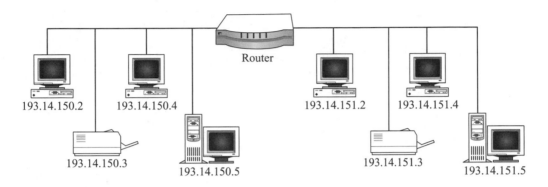

In addition to assigning a computer an IP address, you must also assign a subnet mask. The **subnet mask** is a number that the computer uses to determine which part of the IP address represents the network and which portion represents the host. The subnet mask for a Class A IP address is 255.0.0.0; the subnet mask for a Class B IP address is 255.255.0.0; the subnet mask for a Class C IP address is 255.255.255.0.0. Table 9.4 recaps this important information.

Table 9.4: IP Address Information

Class	First Number	Network/ Host Number	Standard Subnet Mask
A	0-127	N.H.H.H *	255.0.0.0
B	128-191	N.N.H.H *	255.255.0.0
C	192-222	N.N.N.H *	255.255.255.0

Note: * N= Network number; H=Host number

9.8 Configuring Networking

When you install a NIC card in a computer, there are four things that must be configured before connecting to the network:

1. An appropriate driver for the NIC must be installed. The type of driver needed depends on which operating system is being used.

2. You must give the computer a unique name and either a workgroup name (the same name must be used, and this is implemented on a peer-to-peer network), or a domain name (the same name must be used, and this is implemented on a server-based network).

3. You must select the appropriate protocol being used (TCP/IP, IPX/SPX, or Net-BEUI). Contact the network administrator for this information. The majority of businesses and homes use TCP/IP.

4. A network client must be installed. The most common client used in industry is Microsoft's client for Microsoft networks.

There are always other things that could be required depending on the network environment. For example, if the system is a peer-to-peer network, then file sharing (and possibly print sharing) must be enabled. If TCP/IP is configured, some other configuration parameters may be necessary. Exercises at the end of this chapter demonstrate these concepts.

Name a computer using the Network control panel. Each device on the same network must be a unique name. When you double-click on the Network Neighborhood desktop icon, you can view the network device names. It can also be viewed by typing **nbtstat –n** from a command prompt. The command prompt can also be used to access network shares by using the **UNC (Universal Naming Convention)**. For example, a computer called CSchmidt has a network share called TESTS. By typing **\\CSchmidt\TESTS** at the Run prompt, you can access the network share.

To share a folder, use My Computer or Explorer. Locate the folder to be shared and right-click on it. Click on the **Sharing** option. Click on the **Sharing** tab and click in the **Shared As** radio button to enable sharing. In the Share Name text box, type a name for the network share. This name appears in other computers' Network Neighborhood or My Network Places when accessed across the network. In the Access Type section of the window, click on the **appropriate radio button** for the type of access remote users have to the folder. If a password is to be assigned, type it in the text box. Click on the **OK** button and test from a remote computer.

In a network, it is common to map a drive letter to a frequently used network share. To map a drive letter to a network share, right-click on the **Network Neighborhood** or **My Network Places** (Windows 2000) desktop icon. Select the **Map Network Drive** option. Select a drive letter in the **Drive** box by clicking on the down arrow. In the **Folder** or **Path** box (depending on the operating system), type the **UNC** for the network share or use the **Browse** button or **Shared Directories** window (depending on the operating system) to select the network share. The Reconnect at Logon check box allows you to connect to the mapped drive every time you log on.

When configuring TCP/IP, an IP address and subnet mask must be assigned to the network device. The IP address is what makes the network device unique and what allows it to be reached by other network devices. There are two ways to get an IP address:

• Statically define the IP address and mask.
• Use Dynamic Host Configuration Protocol (DHCP).

When an IP address is **statically defined**, that means that someone manually enters an IP address into the computer. This is done through the Network control panel. The appropriate mask must also be entered. The correct IP address and mask can be obtained from the company's network administrator. Entering an IP address that is a duplicate of another network device renders the new network device inoperable on the network. Most support people do not statically define IP addresses unless the device is an important network device such as a web server, database server, network server, router, or switch. Instead, you would use DHCP.

DHCP (Dynamic Host Configuration Protocol) is a method of automatically assigning IP addresses to network devices. A DHCP server (software configured on a network server or router) contains a pool of IP addresses. When a network device has been configured for DHCP and it boots, the device sends out a request for an IP address. A DHCP server responds to this request and issues an IP address to the network device. DHCP makes IP addressing easier and keeps network devices from being assigned duplicate IP addresses.

Another important concept that relates to IP addressing is a default gateway (or gateway of last resort). A **default gateway** is an IP address assigned to a network device that tells the device where to send a packet that is destined for a remote network. The default gateway address is the IP address of the router that is directly connected to that immediate network. A router's job is to find the best path to another network. A router has various network numbers stored in memory. Consider Figure 9.16.

Figure 9.16: A Network with a Default Gateway (or Router)

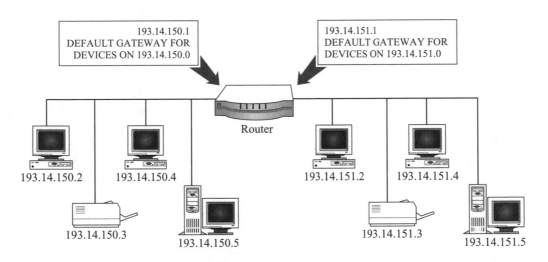

Network devices on the 193.14.150.0 network use the router IP address of 193.14.150.1 as a default gateway address. When a network device on the 193.14.150.0

network wants to send a packet to the 193.14.151.0 network, it sends the packet to the router's IP address that is on the same network (the gateway address). The router, in turn, looks up the destination address (193.14.151.x) in its routing table and sends it out the other interface (193.14.151.1) to the remote network device on the 193.14.151.0 network.

The default gateway address for all network devices on the 193.14.151.0 network is 193.14.151.1, the router's IP address on the same network. Any network device on 193.14.151.0 sending information to the 193.14.150.0 sends it to the default gateway address. For network devices on the 193.14.151.0 network, the gateway address is 193.14.151.1.

Network devices can receive their default gateway address from the DHCP server just like they can an IP address. The DHCP server must be configured for the appropriate default gateway address to give to network devices. An important note is that a DHCP server can give out IP addresses to network devices on remote networks as well as the network to which the DHCP server is directly connected. Default gateway addresses are important for network devices that need to communicate with network devices on other networks. The default gateway address is configured using the Network control panel under the TCP/IP section.

Other elements of TCP/IP information that may need to be configured are one or more DNS server IP addresses and one or more WINS server IP addresses. A **DNS (Domain Name System) server** is an application that runs on a network server that provides translation of Internet names into IP addresses. DNS is used on the Internet, so you do not have to remember the IP address of each site to which you connect. For example, DNS would be used to connect to Scott/Jones Publishing by translating the **URL (Universal Resource Locator)** www.scottjonespub.com into the IP address 167.160.239.173. A computer can receive the DNS server's IP address from DHCP if the DHCP server has been configured for this. You can also manually configure the system for one or more DNS server IP addresses through the Network control panel.

If a DNS server does not know a domain name (it does not have the name in its database), the DNS server can contact another DNS server to get the translation information. Common three-letter codes used with DNS (three letters used at the end of a domain name) are com (commercial sites), edu (educational sites), gov (government sites), net (network-related sites), and org (miscellaneous sites).

A **WINS (Windows Internet Naming Service) server** keeps track of IP addresses assigned to a specific computer name. When connecting to another computer, a user types in a computer's name and not the computer's IP address. The WINS server translates the name to an IP address. The WINS server's IP address can be configured under the Network control panel. WINS is very important especially on computers that receive their IP addresses from DHCP. The IP address can change each time the computer boots because with DHCP, you can configure the DHCP server to issue an IP address for a specific amount of time. In addition, the DHCP server can send the WINS server's IP address to a network device just like the server sends the default gateway address and the DNS address. Another important fact about WINS is that newer DNS servers can now provide the computer name as well as the domain name to IP address translation.

Think of DHCP as a "distributor" of IP addresses and other information. Think of DNS as the phone book. In a phone book, you look up a person's name to get his or her phone number. With DNS, a computer's name, such as **www.somewhere.com**, is looked up and DNS returns its IP address (phone number).

9.9 Troubleshooting Networks

One way to troubleshoot a network is to determine how many devices are affected. For example, if only one computer cannot communicate across a network, it will be handled differently than if several (or all) computers on a network cannot communicate. The easiest way to determine how many devices are having trouble is by using a simple test. Since most computers use TCP/IP, one tool that can be used for testing is the ping command. **Ping** sends a packet to an IP destination (that you determine) and a reply is sent back from the destination device (when everything is working fine). The ping command can be used to determine if the network path is available, if there are delays along the path, and whether the remote network device is reachable.

The ping utility can be used to test the NIC as well as the TCP/IP protocol running on the NIC with the command **ping 127.0.0.1**. The 127.0.0.1 IP address is what is known as a private IP address, which means it cannot be used by the outside world. The 127.0.0.1 is also known as a loopback address. A **loopback address** is not used to check connections to another computer, but is used to test a NIC card's own basic network setup.

If the ping is successful (a message that a reply was received from 127.0.0.1), then the TCP/IP protocol stack is working correctly on the NIC. If the ping responds with a no answer or 100% packet loss error, TCP/IP is not properly installed or functioning correctly on that one workstation.

The ping command can be used to check connectivity all around the network. Figure 9.17 shows a sample network that is used to explain how ping is used to check various network points.

Submarines use a sonar ping to determine if another submarine is in the vicinity. The network **PING** stands for **P**acket **IN**ternet **G**roper.

Figure 9.17: A Sample Network Configuration

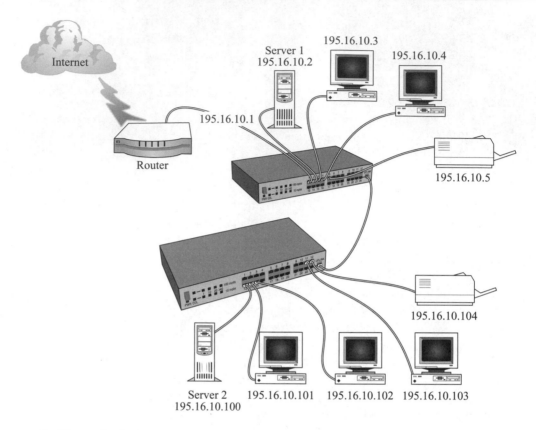

In Figure 9.17, the network consists of various network devices including two servers and two laser printers. The devices connect to one of two switches that are connected together using the uplink port. This port allows two similar devices to be connected together with a standard Ethernet cable or fiber cable. A router connects to the top switch and the router connects to the Internet.

The 195.16.10.3 workstation cannot access a file on Server2 (195.16.10.100). The first in troubleshooting is to ping Server2. If this is successful (the destination reachable), the problem is in Server2 or the file located on the server. If the ping is unsuccessful, there is a problem elsewhere. Right now, the ping is unsuccessful, so ping another device that connects to the same switch. From workstation 195.16.10.3, ping Server1 (195.16.10.2), which connects to the same switch. This ping is successful and tells you the connection between the 195.16.10.3 workstation and the switch is good, the switch is working, the cable connecting to Server1 is fine, and Server1 is functioning. If the ping is unsuccessful, one of these things is faulty.

Now ping workstation 195.16.10.101 (a device other than the server on the remote switch), If the ping is successful, (1) the uplink cable is operational, (2) the second switch is operational, (3) the cable that connects workstation 195.16.10.101 to the switch is good, and (4) the 195.16.10.101 workstation has been successfully configured for TCP/IP. If the ping is unsuccessful, one of these four items is faulty. If the ping is successful, the problems could be (1) Server2's cable, (2) the switch port to which the server connects, (3) server NIC, (4) server configuration, or (5) the file on Server2.

To see the current IP configuration, use the WINIPCFG or IPCONFIG command from a DOS prompt. The **WINIPCFG** command is used with Windows 95 and Windows 98. The **IPCONFIG** command is used with Windows 98, NT Workstation, NT Server, 2000 Professional, and 2000 Server. To access the DOS prompt on Windows 9x, click on the **Start** button, point to the **Programs** option, and click on the **MS-DOS Prompt** option. In NT Workstation or Server, click on the **Start** button, point to the **Programs** option, and click on the **Command Prompt** option. When using Windows 2000 Professional or Server, click on the **Start** button, point to **Programs**, point to the **Accessories** option, and click on the **Command Prompt** option. Figures 9.18 and 9.19 show the switches and output of each command.

Figure 9.18: The WINIPCFG Command

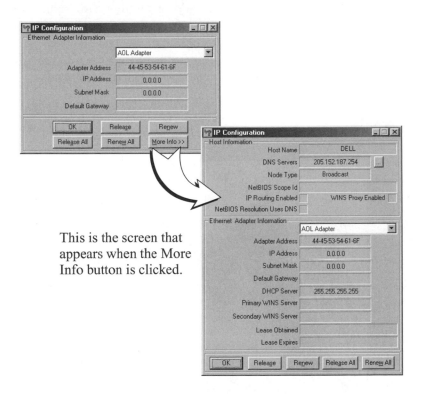

This is the screen that appears when the More Info button is clicked.

Make sure when using WINIPCFG, that you click on the down arrow to select the appropriate NIC.

Use the ping command followed by the name of the device being tested, for example, ping **www.scottjonespub.com**. A DNS server translates the name to an IP address. If the site can be reached by pinging the IP address, but not the name, there is a problem with the DNS server.

A program that helps with DNS server troubleshooting is a tool called **NSLOOKUP**. NSLOOKUP is available on NT Server and 2000 Server. NSLOOKUP allows you to see domain names and their associated IP addresses. When an Internet site (server) cannot be contacted by its name, but can be contacted using its IP address, there is a DNS problem. NSLOOKUP can make troubleshooting these types of problems easier.

The **tracert** command is also a commonly used tool. The tracert command is used to display the path a packet takes through the network. The benefit of using the tracert command is that you can see where a fault is occurring in a larger network.

The **NET command** is also useful in network troubleshooting and configuration. NET DIAG can be used in Windows 98 and 2000 to run a hardware diagnostic program between two computers. Windows 98 and 2000 also have a utility called NET LOGOFF, which breaks the connection between the computer and its connected network resources. The NET USE command can be used to connect or disconnect the computer from a network resource and can be used to display information about network connections. For example, to view all the network connections currently in use, type **NET USE** and press **Enter.** In Windows 98 and 2000, the NET VER command displays the type and version of the network redirector. The NET VIEW command displays a list of computers in a workgroup or a specific computer's shared network resources. A good web site for the NET command is **http://www.computerhope.com/nethlp.htm**.

Figure 9.19: The IPCONFIG Command

```
C:\WINDOWS> ipconfig

Windows 98 IP Configuation

0 Ethernet adapter :

        IP Address . . . . . . . . . : 0.0.0.0
        Subnet Mask. . . . . . . . . : 0.0.0.0
        Default Gateway. . . . . . . :

1 Ethernet adapter :

        IP Address . . . . . . . . . : 0.0.0.0
        Subnet Mask. . . . . . . . . : 0.0.0.0
        Default Gateway. . . . . . . :

2 Ethernet adapter :

        IP Address . . . . . . . . . : 192.168.10.10
        Subnet Mask. . . . . . . . . : 255.255.255.0
        Default Gateway. . . . . . . :

C:\WINDOWS>ipconfig /?
Command line options:
 /All - Display detailed information.
 /Batch [file] - Write to file or ./WINIPCFG.OUT
 /renew_all    - Renew   all adapters.
 /release_all  - Release all adapters.
 /renew   N    - Release adapter N.
 /release N    - Release adapter N.
```

9.10 Configuring a Networked Printer

There are three ways to network a printer:

- Connect a printer to a port on a computer that is connected to the network and share the printer.
- Set up a computer that is designated as a print server. Connect the print server to the network.
- Connect a printer with a network connector installed directly to the network.

Printers can also be password protected on the network. A networked printer is very common in today's home and business computing environments. Networking expensive printers such as laser printers and color printers is cost-effective.

A printer that is connected to a workstation can be shared across the network by enabling File and Print Sharing. An exercise at the end of the chapter explains how to do this. Once File and Print Sharing is enabled, a printer is shared simply by clicking on the **Start** button, pointing to the **Settings** option, clicking on the **Printer** option, right-clicking on the printer to be shared, selecting **Properties**, and clicking on the **Sharing** option.

With Microsoft operating systems, networked printers are much easier to configure than they used to be. To connect and use a networked printer, use the Add Printer wizard. A prompt is available that asks whether the printer is local or networked. A local printer is one that is directly attached to the computer and a networked printer is one attached to another workstation, a print server, or directly connected to the network.

Even though print drivers normally automatically download, sometimes they cause printing problems. The best way to tackle this situation is to manually load the print driver for the networked printer.

9.11 Understanding Dial-Up Networking (DUN)

DUN (Dial-up Networking) is a remote computer that dials into the Internet or a corporation using a modem. Another technology using dial-up networking is virtual private networking. **VPN (Virtual Private Networking)** is a remote computer connecting to a remote network by "tunneling" over an intermediate network such as the Internet or a LAN. Once connected, the remote user can make use of network devices as if they were directly connected to the network. Figure 9.20 illustrates these concepts.

Figure 9.20: DUN and VPNs

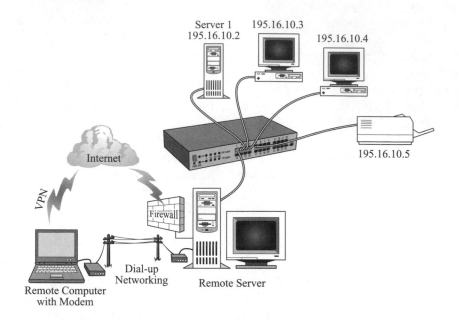

The type of connection, protocol, and settings that you configure on the remote computer depends on the company to which you are connecting. The most commonly used protocol is TCP/IP, but Microsoft operating systems do support IPX/SPX and NetBEUI. A connection protocol used with dial-up networking is PPP. **PPP (Point-to- Point Protocol)** is a connection-oriented, layer 2 protocol that encapsulates data for transmission over phone lines. An older protocol that was used with dial-up networking and was the predecessor to PPP is SLIP (Serial Line Internet Protocol).

In Windows 98, to make a dial-up networking connection, make sure a modem is properly installed. Then access the dial-up networking wizard by double-clicking on the **My Computer** desktop icon, double-clicking on the **Dial-up Networking** folder, and then double-clicking on the **Make New Connection** icon. If the Dial-up Networking folder is not there, you can install the required components using the Add/Remove Programs control panel.

In Windows 2000, click on the **Start** button, access the **Settings** option, and click on the **Network and Dial-up Connections** folder. The Make New Connection wizard is used to setup dial-up networking or configure a VPN connection.

In Windows XP, click on the **Start** button, select **Control Panel**, point to **Network and Internet Connections**, and select **Network Connections**. Under Network tasks, select **Create a new connection**. An area code may have to be entered. Click on the **Next** button.

Select the appropriate type of network connection and click on the **Next** button. Type in a name for the connection and select **Next**. Enter the remote modem's phone number and click on the **Next** button. Ensure the Add a shortcut to this connection to my desktop check box is enabled and click on the **Finish** button.

Before creating a remote connection, you should always determine what parameters are to be entered *before* starting the configuration. Contact the network administrator for exact details on how to configure the remote connection. If the connection is to the Internet via an ISP, detailed instructions are available on the ISP's web site and/or with the materials that come with the Internet package from the ISP.

There are many types of network connections. Dial-up networking normally uses POTS (Plain Old Telephone Service) or ISDN. Businesses use various types of network connections leased from the local phone company or a provider. Table 9.5 shows the types of network connections and speeds.

Table 9.5: Network Connections

Connection Type	Speed
POTS (Plain Old Telephone Service)	2400 bps to 115 Kbps analog phone line
ISDN (Integrated Services Digital Network)	64 Kbps to 1.544 Mbps digital line
Frame Relay	56 K to 1.544 Mbps
56 K point to point	56 K guaranteed bandwidth between two points
T1	1.544 Mbps guaranteed bandwidth between two points
T3	44 Mbps guaranteed bandwidth between two points
DSL (Digital Subscriber Line)	256 Kbps and higher; shares data line with voice line
ATM (Asynchronous Transfer Mode)	Up to 2 Gbps

9.12 Learning about Internet Software

Once a dial-up networking configuration or the LAN configuration tasks have been completed, you can connect to the Internet. Most people use a web browser when connecting to the Internet. A **browser** allows you to view web pages across the Internet. The two most common Internet browsers are Internet Explorer (also known as IE) and Netscape Navigator. Other web browsers include Opera Software's Opera and NeoPlanet, Inc.'s NeoPlanet. Internet Explorer comes with Microsoft operating systems. Netscape Navigator is available from Netscape Communications Corporation (http:// home.netscape.com/)

or free from some ISPs when you enroll with their service. If Internet Explorer is not loaded on the computer, add it using the Add/Remove Programs control panel or go to Microsoft's web site at **www.microsoft.com** to download the latest version.

Keeping the web browser current is important. Internet hackers frequently target Internet browsers and constant updates are provided that help with these attacks. Before upgrading, you should determine the web browser's current version. With any software application, the version is determined by starting the application, clicking on the **Help** menu item, and clicking on the **About** *x*, where *x* is the name of the application. With Internet Explorer, the first two numbers listed are the software version numbers. There is another value called Cipher Strength that is a bit value for encryption. Encryption is the process of changing your transmitted files into data so it cannot be recognized. In the United States, 128-bit encryption is the best.

Internet browsers frequently need plug-ins. A **plug-in** is an application designed to work with the browser. Common plug-ins include Macromedia Flash, Macromedia Shockwave, RealNetwork's RealPlayer, Apple QuickTime, Adobe Acrobat Reader, and WinZip. Macromedia Flash allows web animations to be played. Macromedia Shockwave is for interactive multimedia graphics and audio applications. RealPlayer is for playing streaming audio and video, QuickTime is used for playing video clips. Acrobat Reader is for displaying PDF documents. WinZip is used for compressing and expanding ZIP files.

Another common tool for a web browser is an accelerator. An **accelerator** speeds up downloads and Internet browsing (surfing). Some accelerators are plug-ins for the web browser software and others are standalone applications. Various download and browsing accelerators are available on the Internet. One example is SpeedBit's Download Accelerator Plus; it's available at www.speedbit.com. Two other popular ones are Go!Zilla available from www.gozilla.com and NetSonic available from www.netsonic.com.

Another common Internet software application is an e-mail package. This software allows you to send messages across the Internet. Microsoft operating systems come with Windows Messaging (Inbox). Another popular freeware e-mail software program is Eudora Light. Many Internet providers also have their own e-mail package.

The e-mail service has to be configured. Many settings are configured through the Mail control panel. Two common settings are POP and SMTP server addresses. POP stands for Post Office Protocol and a POP3 server is a server used for retrieving e-mail. **SMTP (Simple Mail Transport Protocol)** is used for sending e-mail. These settings for the e-mail service are available from the network administrator or the ISP in their instructions for configuring dial-up networking.

You must be familiar with troubleshooting browser and e-mail applications. A good place to start is with the userid and password, POP3, and SMTP settings. In Internet Explorer, you need to be familiar with the settings that can be configured under the Internet Options section of the Tools menu item. The Connections tab is a great place to start.

9.13 Understanding Wireless Networks

Wireless networks are networks that transmit data over air using either infrared or radio frequencies. Wireless networks operate at layers 1 and 2 of the OSI model. Most wireless networks in home and businesses use radio frequencies. Wireless networks are very popular both in the home and business computer environments and are great in places that are not conducive to running cable such as an outdoor center, convention center, bookstore, coffee shop, hotel, between buildings, and in between non-wired rooms in homes. You must be familiar with this technology for installation, configuration, and troubleshooting.

The standard for wireless is IEEE 802.11.

There are two main types of wireless networks: ad hoc and infrastructure. An **ad hoc mode** wireless network is also known as a peer-to-peer or IBSS (Independent Basic Service Set) mode or simply IBSS (Independent Basic Service Set). An ad hoc wireless network is when at least two devices such as two computers have wireless NICs (Network Interface Cards) installed. The two devices transmit and receive data.

There are three major types of wireless NICs: PC Card, USB, and PCI (Peripheral Component Interconnect). Note PCI is a type of interface for computer cards, such as video, network and modem, to connect to your computer's motherboard. Figure 9.21 shows a D-link Systems, Inc.'s wireless NIC that could be installed in a laptop computer.

Figure 9.21: D-Link's Wireless PC Card NIC

PCI wireless NICs allow desktop or tower computers to access a wireless network. Figure 9.22 shows D-Link Systems Inc.'s PCI wireless NIC.

Figure 9.22: PCI Wireless NIC PCI Wireless NIC

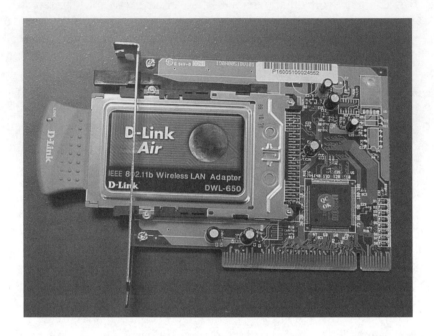

The third most popular type of wireless NIC attaches to the USB port and is quite popular in home networks. Figure 9.23 shows a USB wireless NIC.

Figure 9.23: Linksys USB Wireless NIC

An ad hoc mode wireless network is used when two people want to play a network-based game, two or more computers need to transfer data, or one computer connects to the Internet and the other computer(s) are not wired into the same network. Figure 9.24 shows an ad hoc mode wireless network that consists of two laptops communicating over airwaves.

Figure 9.24: Ad Hoc Wireless Network

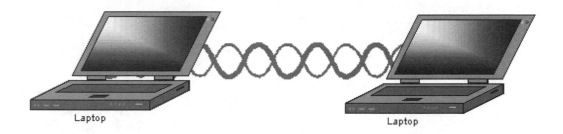

The **infrastructure mode** wireless network connects multiple wireless network devices through an access point. An **access point** is a device that receives and transmits data from multiple computers that have wireless NICs. The easiest way to describe an access point is to think of it as a network hub—it connects the wireless network. Network Figure 9.25 shows an infrastructure mode wireless network with an access point and multiple wireless devices.

Figure 9.25: Infrastructure Mode Wireless Network

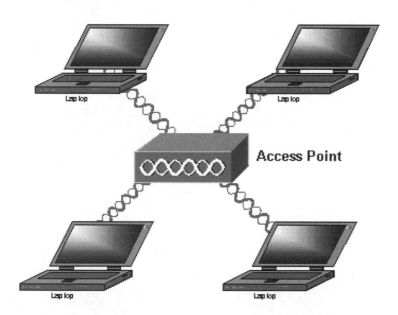

The access point can also be wired to another wireless network or a wired network. The access point can then relay the transmission from a wireless device to another network or to the Internet through the wired network. When multiple devices connect to an access point (whether that access point is wired to a LAN or not), the configuration is known as a **BSS** (**Basic Service Set**). Figure 9.26 shows an infrastructure mode and an access point. The access point connects to a wired network and gives the wireless devices access to the Internet.

Figure 9.26: Wireless Network Connected to Wired Network

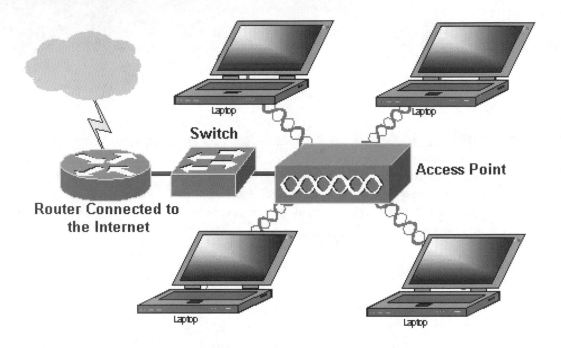

When multiple access points connect to the same main network (known to some as the distribution system), the network design is known as an **ESS** (**Extended Service Set**). Figure 9.27 shows an ESS wireless network.

Figure 9.27: Extended Service Set Wireless Network

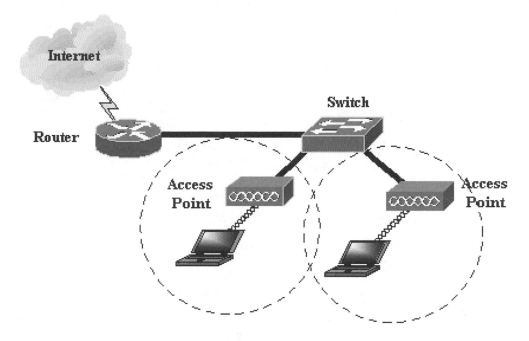

Each access point can handle 60 to 200 network devices depending on vendor, wireless network environment, amount of usage, and the type of data being sent. Each access point is assigned an **SSID** (**Service Set Identifier**). An SSID is a set of 32 alphanumeric characters used to differentiate between different wireless networks. Wireless NICs can automatically detect an access point or be manually configured with the access point's SSID. Some manufacturers refer to the SSID in infrastructure mode as the **ESSID** (**Extended Service Set Identifier**), but this is the same concept as SSID. In situations such as a wireless café, bookstore, or convention center, no SSID may be required on the wireless NIC. However, in a home or business environment, an SSID may be required.

If two access points are used and they connect two different wireless networks, two different SSIDs would be used. Figure 9.28 shows this concept.

Figure 9.28: Two Separate Wireless Networks with Two SSIDs

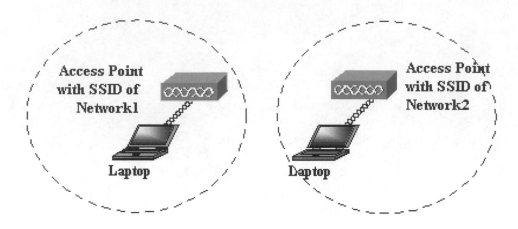

If two access points connect to the same wireless network, the same SSID is used. Figure 9.29 shows this concept.

Figure 9.29: One Extended Wireless Network with the Same SSID on Both Access Points

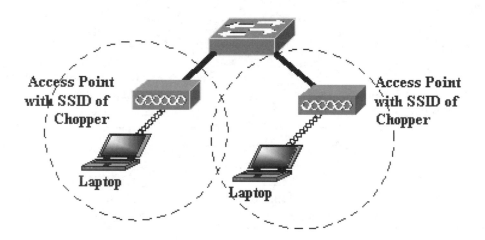

9.14 Installing and Configuring a Wireless Network Card

Before any wireless adapters are installed, the basic configuration parameters should be determined. The following list helps with these decisions.

• Will the wireless adapter be used in an ad hoc environment or infrastructure mode?
• What is the SSID?
• Is WEP enabled? **WEP (Wireless Equivalent Privacy)** encrypts data being transmitted.
• If WEP is enabled what is the key length?
• Is authentication open or shared key?

• If shared key authentication is being used, what is the shared key?
• What is the most current driver for the operating system being used?

Wireless network adapters can be USB, PCI, ISA, or a PC Card. Each of these adapters install like any other adapter of the same type. Not all computers in the wireless network have to have the same type of wireless NIC. For example, a desktop computer could have a PCI wireless NIC installed, a laptop computer in a cubicle office would have an integrated wireless NIC, and another laptop in another cubicle has a PC Card wireless NIC. All three can access the same wireless network and access point.

With most wireless NICs, the manufacturer's software is normally installed before the NIC is installed or attached to the computer. With all wireless NICs, the latest driver for the particular version of Windows should be downloaded from the manufacturer's web site before the card is installed. Once the adapter is inserted or attached and the computer powered on, Windows recognizes that a new adapter has been installed and prompts for the driver, browse to the location of the new downloaded driver. Another method that can be used is to install the driver that comes with the adapter and then upgrade it once installed.

Once the wireless adapter is installed, the options such as SSID and security options can be installed. These parameters are normally configured through a utility provided by the wireless NIC manufacturer or through Windows network control if Windows XP is installed. Figure 9.30 shows the wireless NIC properties screen that is accessible through the Windows XP Network Connections control panel.

Figure 9.30: General Tab of the Wireless NIC Properties Window

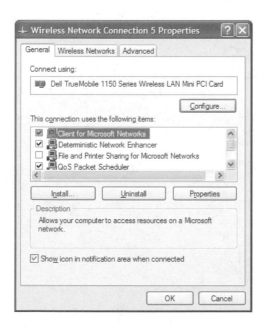

To access the configuration for the wireless network, click on the **Wireless Networks** tab. Figure 9.31 shows this window.

Figure 9.31: Wireless Networks Tab of the Wireless NIC Properties Window

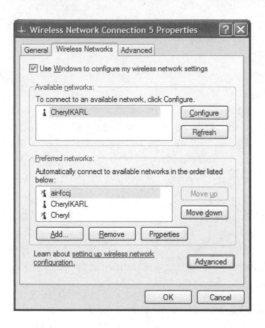

To configure the wireless network adapter for ad hoc mode or infrastructure mode, click on the **Advanced** button at the bottom of the window. Figure 9.32 shows the screen that appears where the selection can be made.

Figure 9.32: Window to Select Ad Hoc Mode or Infrastructure Mode

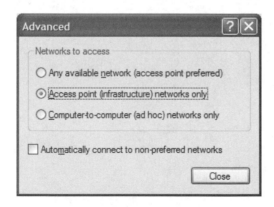

On the Wireless Networks tab, select the **Add** button to configure the wireless NIC for a wireless network. On this screen, the SSID can be input, WEP enabled, and the shared key input. Figure 9.33 shows this window.

Figure 9.33: Windows XP's Wireless NIC Configuration Screen

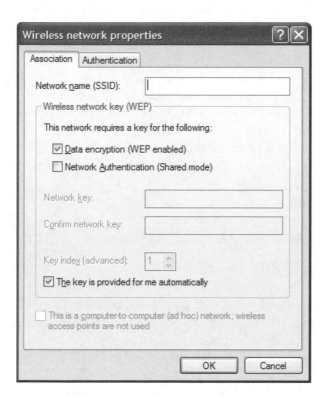

Wireless NICs are very easy to install. The utilities that are provided with the NICs are quite sophisticated and easy to use. Always follow the manufacturer's instructions. All of the screens and configuration utilities have the same type of information. Understanding what the configuration parameters means is important. The hardest part about configuring wireless NICs is obtaining the correct parameters *before* installation begins. Incorrectly inputting any one of the parameters will cause the wireless NIC to not associate with the access point or remote wireless device and not transmit. Planning is critical for these types of cards.

9.15 Understanding Wireless Access Points

Many of the parameters needed for wireless NIC configuration are also needed for access point installation. However, an access point is more involved because it is the central device of the wireless network. The following list of questions help with access point installation. These questions should be answered *before* the access point is installed.

- What is the SSID to be used?
- Is WEP enabled?
- If WEP is enabled what is the key length?
- Is authentication open or shared key?
- If shared key authentication is being used, what is the shared key?
- Is there power available for the access point? Note that some access points can receive power through an in-line switch.
- How will the access point be mounted? Is mounting hardware provided with the access point or does extra equipment have to be purchased?
- Where should the access point be mounted for best coverage of the wireless network area? Perform a site survey to see best performance. Temporarily mount the access point. With a laptop that has a wireless NIC and site survey software, walk around the wireless network area to see the coverage range. The site survey can also be conducted by double-clicking on the network icon on the taskbar; the signal strength is shown in the window that appears. Move the access point as necessary to avoid attenuation and obtain the largest area coverage.
- What Channel ID will be used?
- Will the access point connect to the wired network and, if so, is there connectivity available where the access point will be mounted?

9.16 Troubleshooting Wireless Networks

Troubleshooting wireless networks is sometimes easier than a wired network because of the mobility factor. A laptop with a wireless NIC installed can be used to troubleshoot connectivity, configuration, security, etc. Most wireless network problems stem from inconsistent configuration. The standards deployed must be for the lowest common denominator. For example, if a wireless NIC only supports 64-bit WEP encryption, then that must be what is used even if 128-bit WEP encryption is available on some of the cards. The list that follows are some general wireless networking tips designed to get you started. Most of these tips have been discussed in their previous sections, but it is nice to have a troubleshooting list in one spot.

- Is the SSID correct?
- Is the type of wireless network (ad hoc or infrastructure) correctly configured?
- Is the wireless NIC seen by the operating system? Use Device Manager to check.
- Is WEP enabled? If so, is the WEP key correctly configured? Is the WEP key length correct?
- Is open or shared key authentication being used? Check configuration.
- Can any devices attach to the access point? If not, check the access point.
- Is anything causing interference or attenuation? Check antenna placement.
- Is there a Channel ID overlap problem?
- If a manufacturer's utility is being used and Windows XP is installed, does the Network Properties window have the **Use Windows to configure my wireless network settings**

check box unchecked? If not, uncheck this check box to allow the utility to configure the wireless NIC.

Wireless networking is an emerging technology and will continue to grow in size, technology, and support issues. Today you must be familiar with this technology as corporations and home users install these types of products. Because the technology is reasonably priced, you may want to install your own wireless network for the experience. Enjoy this technology because more wireless technologies are evolving.

Chapter Summary

- A network is comprised of two or more devices that can communicate with one another and share resources.
- The three major types of network topologies are star, ring, and bus.
- Networks require some type of medium to transmit data. This medium is normally wire or wireless. The most common types of cable are twisted-pair and fiber-optic, although some very old networks have coax cable.
- Before a computer can communicate on a network, it must adhere to a set of communication rules to which all computers on the network comply. This set of communication rules is known as a common access method. Ethernet uses a common access method known as CSMA/CD whereas fiber networks and Token Ring use token passing as the common access method.
- The IEEE committee created network standards called the 802 standards. Each standard is given an 802.x number and represents an area of networking.
- ISO has developed the OSI model. The OSI model is a standard for information transfer across the network.
- A network protocol is a data communication language. There are three primary network protocols used: TCP/IP, NetBEUI, and IPX/SPX. TCP/IP is the most common network protocol and is used when accessing the Internet. The most commonly used protocol is TCP/IP.
- When you install a NIC card in a computer, you need to supply an appropriate driver, a unique computer name, a workgroup or domain name, the appropriate protocol, and network client software.
- You can troubleshoot a network by using the ping command. It is the most commonly used networking tool to determine the status of other devices on your network. You can connect a network printer in one of several ways. You can connect a printer to a port on a computer that is connected to the network. You can set up a computer that is designated as a print server. You can then connect the print server to the network. You can connect a printer with a network connector installed directly to the network.
- If you want to access a computer remotely, you can use DUN to dial the Internet or an organization that has a computer connected to a mode. You can also create a VPN to secure your data by implementing a secure "tunnel" through the Internet or an intranet.
- A browser allows you to view web pages across the Internet. Browsers frequently need plug-ins. A plug-in is an application designed to work with the browser. Common plug-

ins include Macromedia Flash, Macromedia Shockwave, RealNetwork's RealPlayer, Apple QuickTime, Adobe Acrobat Reader, and WinZip.

- Another common Internet related software tool is e-mail. It uses the SMTP protocol for sending and receiving e-mail messages.
- Wireless networks are networks that transmit data over air using either infrared or radio frequencies. Wireless networks operate at layers 1 and 2 of the OSI model. Most wireless networks in home and businesses use radio frequencies.
- WEP is used to encrypt data transmitted over wireless networks.

Review Questions

1. A _____ is a group of devices that can share resources in a single area such as a room or a building.
   a) WAN
   b) LAN
   c) IEEE 802.11 standard
   d) OSI

2. A _____ is communication between Local Area Networks.
   a) WAN
   b) LAN
   c) IEEE 802.11 standard
   d) OSI

3. The standard for wireless networks is _____.
   a) WAN
   b) LAN
   c) IEEE 802.11 standard
   d) OSI

4. The IEEE standard for Token Ring is _____.
   a) 802.2
   b) 802.3
   c) 802.4
   d) 802.5

5. The IEEE standard for CSMA/CD is _____.
   a) 802.2
   b) 802.3
   c) 802.4
   d) 802.5

6. A _____ defines how network devices connect.
   a) Protocol
   b) OSI Model
   c) Topology
   d) IEEE 802 standards

7. The most common network topology used today is the _____ .
   a) Ring
   b) Bus
   c) Star
   d) OSI

8. A _____ topology passes a special network packet from one device to another.
   a) Ring
   b) Bus
   c) Star
   d) OSI

9. _____ transmission is data transmitted in both directions but not at the same time.
   a) Full duplex
   b) Duplex
   c) Half duplex
   d) Simplex

10. _____ transmission is data transmitted in both directions but at the same time.
    a) Full duplex
    b) Duplex
    c) Half duplex
    d) Simplex

11. In a _____ topology all devices connect to one cable.
    a) Star
    b) Bus
    c) Ring
    d) Half duplex

12. A _____ server is used for retrieving e-mail.
    a) SMTP
    b) POP
    c) HTTP
    d) TCP/IP

13. A _____ server is used for sending e-mail.
   a) SMTP
   b) POP
   c) HTTP
   d) TCP/IP

14. A _____allows you to view web pages across the Internet.
   a) Web site
   b) Browser
   c) IPX/SPX protocol drive
   d) POP3 Server

15. In wireless communication, when multiple access points connect to the same main network, the network design is known as _____.
   a) BSS
   b) Infrastructure mode
   c) WEP
   d) ESS

16. The _____ is a standard for information transfer across the network.
   a) Half duplex transmission
   b) Topology
   c) BSS
   d) OSI Model

17. A _____, a network device that determines the best path to send a packet, works at layer 3, the network layer.
   a) Hub
   b) Switch
   c) Router
   d) Ethernet

18. A _____ of data is transmitted at the network layer of the OSI Model.
   a) Frame
   b) Bit
   c) Packet
   d) Byte

19. The _____ defines how the bits are transferred and retrieved across the network media. This layer is where connectors, cable, and voltage levels are defined.
   a) Presentation
   b) Session
   c) Network
   d) Physical

20. A _____ is guaranteed to be 1.544 Mbps bandwidth between two points.
   a) T1
   b) T2
   c) T3
   d) DSL

Lab Projects

Lab Project 1

The goal of this project is for you to be able to install and configure a NIC in a Windows 9x computer. You will need a computer with Windows 9x installed, and a NIC card with driver. The method used to install a NIC in Windows 9x depends on whether the NIC is a plug and play device or a non-plug and play or legacy device.

Installing a Plug and Play NIC

1. With the computer turned off, remove the **computer cover**.

2. Using proper ESD precautions, insert the **NIC** in a compatible bus slot and secure with a screw.

3. Turn the computer on and verify that Windows 9x loads.

4. Log on to Windows 9x using the userid and password provided by the instructor or lab assistant.

5. Windows 9x automatically detects and installs the NIC. If Windows 9x does not detect a driver for the NIC, you will be prompted for a driver location. If this is the case, insert the driver disk and enter the path to the driver. Proceed to the **Checking the Installation** section.

Installing a Non-Plug and Play or Legacy NIC

6. Using jumpers or a software configuration utility, configure the NIC so it will use system resources that do not conflict with any other device.

7. With the computer turned off, remove the **computer cover**.

8. Using proper ESD precautions, insert the **NIC** in a compatible bus slot and secure with a screw.

9. Turn the computer on and verify that Windows 9x loads.

10. Log on to Windows 9x using the userid and password provided by the instructor or lab assistant.

11. Click on the **Start** button, point to the **Settings** option, and double-click on the **Control Panel** option. The Control Panel window opens.

12. Double-click on the **Add/Remove Hardware** icon. The Add/Remove Hardware wizard opens. Select the **Next** button twice.

13. Windows searches for new plug and play devices. When Windows does not find any, you are given the option to allow Windows to search for non-plug and play devices or you can select the hardware from a list. Choose **No, I want to select the hardware from a list** and click on the **Next** button.

14. Scroll down and select **Network Adapters** and then click on the **Next** button.

15. Select the **NIC Manufacturer and Model** from the list. If the NIC is not listed, select **Have Disk**, enter a path to the driver files, and click on **OK**.

16. After selecting the proper NIC, click on the **Next** button.

17. Select **Finish** to continue the installation. If prompted, enter the proper configuration information for the NIC and click on **OK**.

18. Restart the computer.

Checking the Installation

19. From the **Start** menu, point to **Settings,** and then click on the **Control Panel** option.

20. From the Control Panel window, double-click on the **System** icon, and then select **Device Manager**.

21. Expand **Network Adapters**, select the **network adapter** installed in the computer, and then click on **Properties**.

22. Click on the **General** tab.

    What is the device status of your NIC?

    _____

23. Select the **Driver** tab.

    What is the driver version number of the NIC?

    _____

24. Select the **Resources** tab.

    What resources are being used by the NIC?

    _____

    Are any devices conflicting with your NIC? If so, list them below.

    _____

25. Click on **OK** to close the Network Adapters Properties window.

Lab Project 2

The goal of this project is for you to be able to properly install and configure a NIC using Windows NT Workstation. You will need a computer with Windows NT Workstation loaded, a NIC card, and a NIC driver disk. Installing a NIC using NT Workstation is different from using Windows 95, 98, or 2000 Professional because NT Workstation is not a plug and play operating system.

1. With the computer turned off, remove its cover.

2. Using proper ESD precautions, insert the **NIC** in a compatible bus slot and secure with a screw.

3. Turn the computer on and verify that Windows NT Workstation loads.

4. Log on to NT Workstation using the userid and password provided by the instructor or lab assistant.

   What rights are required to be able to install a NIC in Windows NT Workstation?

   _____

5. Right-click on the **Network Neighborhood** desktop icon and select **Properties**. The Network window opens.

   What alternate method can be used to open the Network window?

   _____

6. Select the **Adapters** tab, the Adapters Installation and Configuration window opens.

7. Click on the **Add** button and the Select Network Adapter window opens.

8. If the NIC that is installed in the computer is listed, click on it from the list. If the proper NIC is not listed, insert the NIC driver disk, click on the **Have Disk** button, and enter the path to the driver.

9. If prompted, insert the Windows NT Workstation CD, or enter the path to the installation files, and click on the **Continue** button.

10. If prompted, enter configuration information such as Ethernet ID, bus type, and slot number and click on the **OK** button.

11. Windows NT Workstation copies and installs the NIC driver files.

12. Click on the **Close** button to exit the Adapter Installation and Configuration window.

Lab Project 3

The goal of this project is for you to be able to properly install and configure a NIC using Windows 2000 Professional. You will need a computer with Windows 2000 Professional installed, a NIC card, and a NIC driver. The method used to install a NIC in Windows 2000 Professional depends on whether the NIC is a plug and play device or a non-plug and play device (also known as a legacy device).

What type of NIC is to be installed into the computer: plug and play or non plug and play?

_____

Contact your instructor or lab assistant if unsure. Once the type of NIC is determined, follow the directions appropriate for the type of NIC: Installing a Plug and Play NIC or Installing a Non-Plug and Play or Legacy NIC.

Installing a Plug and Play NIC

1. With the computer turned off, remove the **computer cover**.

2. Using proper ESD precautions, insert the **NIC** in a compatible bus slot and secure with a screw.

3. Turn on the computer and verify that Windows 2000 Professional loads.

4. Log on to Windows 2000 Professional using the userid and password provided by the instructor or lab assistant.

5. Windows 2000 Professional automatically loads the drivers and configures the NIC. If Windows 2000 Professional does not have a driver for the NIC, you will be prompted for a driver location. If this is the case, insert the driver disk into the floppy drive and enter the path to the driver (A:).

6. Go to the section labeled **Checking the Installation**.

Installing a Non-Plug and Play or Legacy NIC

7. With the computer turned off, remove the **computer cover**.

8. Using proper ESD precautions, insert the **NIC** in a compatible bus slot and secure with a screw.

9. Turn on the computer and verify that Windows 2000 Professional loads.

10. Log on to Windows 2000 Professional using the userid and password provided by the instructor or lab assistant.

11. Click on the **Start** button.

12. Point to the **Settings** option.

13. Click on the **Control Panel** option. The Control Panel window opens.

14. Double-click on the **Add/Remove Hardware** icon. The Add/Remove Hardware wizard opens.

15. Click on the **Next** button.

16. Choose the **Add/Troubleshoot a device** option and select **Next**.

17. Windows searches for plug and play devices. When the search is over, select the **Add a new device** option from the **Choose a Hardware Device** window, and then click on the **Next** button.

18. Choose the **Yes, search for new hardware** option and click on the **Next** button.

19. Windows searches for non-plug and play hardware and displays devices found. Choose your NIC from the list and select **Next**.

20. In the **Found New Hardware** wizard window select the **Resources** button.

21. Ensure the resources assigned to the NIC are correct. Make any necessary changes and click on the **OK** button to return to the Found New Hardware Wizard window.

22. Select the **Finish** button twice to complete the installation.

23. Continue to the **Checking the Installation** section.

Checking the Installation

24. Click on the **Start** button.

25. Point to the **Programs** option.

26. Point to the **Administrative Tools** option.

27. Click on the **Computer Management** option. The Computer Management window opens.

28. In the left window, select **Device Manager**.

29. In the right window, select the **+ (plus sign)** next to **Network adapters**.

30. Right-click on the NIC you just installed and select the **Properties** option from the menu. The Properties page opens.

31. Click on the **General** tab.

    What is the device status of the NIC selected?

    _____

32. Select the **advanced** tab (if available).

    Are any properties listed? If so, list one property and its value.

    _____

33. Select the **Driver** tab.

    What is the driver version number of your NIC's driver?

    _____

34. Select the **Resources** tab.

    What resources are being used by your NIC?

    _____

    Are any devices conflicting with your NIC? If so, list them below.

    _____

35. Click on **OK** to close the Properties page.

Lab Project 4

The goal of this project is to install the Microsoft Client on a Windows 9x computer. You will need a Windows 9x computer with NIC installed and configured. The Microsoft Client for Windows 9x enables a client computer to take advantage of the built-in Microsoft networking services in a Microsoft peer-to-peer network. It also allows a Windows 9x computer to access a Windows domain.

1. Turn on the computer and verify that Windows 9x loads.

2. If necessary, log on to Windows 9x using the userid and password provided by the instructor or lab assistant.

3. From the Start menu, point to the **Settings** option, click on the **Control Panel** option, and then double-click on the **Network** icon. The Network Properties window opens.

4. Click on the **Add** button. The Select Network Component Type window opens.

5. From the Select Network Component Type window, choose **Client** and then select **Add**. The Select Network Client window opens.

6. Highlight the **Microsoft** option, select **Client for Microsoft Networks**, and then click on the **OK** button.

   Which Microsoft network clients appear as available in the Select Network Client window?

   _____

7. From the Network Properties window, verify that **Client for Microsoft Networks** appears in the Installed Components window, and then click on the **OK** button.

8. If prompted, insert the Windows 9x CD-ROM in the drive or enter a path to the installation files.

9. When the Client for Microsoft Networks installation finishes, reboot the computer for the new client to take effect.

Lab Project 5

The goal of this project is to install networking on a computer that uses NT Workstation. You will need an NT Workstation computer with NIC installed and configured. The instructor or lab assistant must be prepared to answer questions such as these:

• Should the student select Workgroup or Domain network model?

• Is DHCP being used in the lab?

You must configure NT networking a bit differently than the other operating systems. The Network Setup wizard takes you through the installation process.

1. Turn on the computer and verify that NT Workstation loads.

2. Log on to Windows using the userid and password provided by the instructor or lab assistant.

3. From the **Start** menu, point to the **Settings** option, click on the **Control Panel** option, and then double-click on the **Network** icon. A dialog box appears stating that networking is not installed and asks if you want to install it. Click on the **Yes** button.

4. The next prompt asks if you are wired to the network or if you are going to use a modem to connect. In a lab environment, you are probably wired to the network. Look at the back of the computer and see if a NIC is installed and a network cable connects to the NIC. If so, click on the **Wired to the network** check box and click on the **Next** button.

5. On the next screen, click on the **Start Search** button so the operating system looks for the installed NIC. This exercise assumes the NIC is installed, but if it hasn't been, you can click on the **Select from list** button and install the NIC drivers and then continue. The NIC appears in the window. Click on the **Next** button.

6. A list of protocols appears. Ensure the **TCP/IP Protocol** is selected and click on the **Next** button.

7. The Network Services screen appears. These default services are what allow your computer to participate in a peer-to-peer network or in a server-based network. Ensure that **RPC Configuration**, **NetBIOS Interface**, **Workstation**, and **Server** are all checked and click on the **Next** button.

8. A message appears that NT is going to install the components. Click on the **Next** button. Another screen appears that allows you to change your binding order. Simply click on the **Next** button. You may be asked to insert the NT installation CD or be prepared to type in the path to where the programs are stored. Contact your instructor or lab assistant if you are unsure what to do.

9. Since TCP/IP was selected, you will be asked if there is a DHCP server connected to the network. This is lab-dependent. Most schools have a DHCP server, but contact the instructor or student assistant if you are unsure. If you select **No**, you must enter the IP address, mask, and default gateway information. If you select **Yes**, the computer will be assigned this information by the DHCP server.

10. Click on the **Next** button to start the NT networking services.

11. You are asked to give the computer a name and determine if the computer participates in a peer-to-peer network (Workgroup option) or a server-based network (Domain option); either way, you will have to enter either a workgroup name or a domain name. Contact the instructor or lab assistant for the correct names if you are unsure. Click on the **Next** button after all information has been entered.

    What is the network name of your computer?

    _____

    Is the computer participating in a peer-to-peer network or a server-based network?

    _____

    Is DHCP being used?

    _____

12. Click on the **Finish** button to complete the installation.

13. The computer must reboot in order for the setting to take effect. Click on the **Yes** button to restart the computer.

14. After restarting and logging in, double-click on the **Network Neighborhood** desktop icon.

    How many other computers do you see on the network?

    _____

Lab Project 6

The goal of this project is to correctly install Microsoft Client on a Windows 2000 Professional computer. You will need a computer with Windows 2000 Professional Workstation with a NIC installed and configured (Client for Microsoft Networks is not installed.)

Microsoft Client enables a computer to take advantage of the built-in Microsoft networking services in a Microsoft peer-to-peer network. It also allows a computer to access a Windows NT domain.

1. Turn on the computer and verify that Windows 2000 Professional loads.

2. Log on to Windows 2000 Professional using the userid and password provided by the instructor or lab assistant.

3. Right-click on the **My Network Places** desktop icon, and select the **Properties** option. The Network and Dial-up Connections window opens.

4. Right-click on the **Local Area Connection** icon and select the **Properties** option. The Local Area Connections window opens.

   Which installed network components are being used by this connection?

   _____

5. Select the **Install** button. The Select Network Component Type window opens.

6. Choose the **Client** option, and then click on the **Add** button. The Select Network Client window opens.

   What network clients are listed as available?

   _____

7. Select the **Client for Microsoft Networks** option, and click on the **OK** button.

8. If prompted, insert the Windows 2000 Professional installation CD-ROM or enter a path to the installation files.

9. When prompted, reboot the workstation for the new network settings to take effect.

Lab Project 7

The goal of this project is for you to be able to install and configure the TCP/IP protocol on a Windows 9x computer. You will need a Windows 9x computer with a NIC installed and configured. The TCP/IP protocol is a routable protocol. It is the protocol that powers the Internet, so it is important that you understand how it is installed and configured.

1. Turn on the computer and verify that Windows 9x loads.

2. If necessary, log on to Windows 9x using the userid and password provided by the instructor or lab assistant.

3. Right-click on the **My Network Places** desktop icon, and then select **Properties**. The Network Properties window opens.

   What other method can be used to access Network Properties?

   _____

4. Click on the **Add** button, and the Select Network Component Type window opens.

   What network component types are listed as available?

   _____

5. Click on the **Protocol** item, and then click on the **Add** button. The Select Network Protocol window opens.

6. Click on the **Microsoft** option in the left window. In the right window, click on the **TCP/IP** option and then click on the **OK** button.

7. From Network Properties, scroll down and choose **TCP/IP** and then select **Properties**.

8. If you are using DHCP on your network, choose **Obtain an IP address automatically**. If you are not using DHCP on your network, choose **Specify an IP address**, and enter an **IP address** and **subnet mask**. Contact the instructor or a lab assistant if you are unsure which option to use.

9. If needed, select **DNS configuration** and enter DNS information, select **Gateway** and enter gateway information, and select **WINS configuration** and enter WINS information. Again, contact the instructor or lab assistant if you are unsure which option to use.

   Which of the following is responsible for host name to IP address resolution: DNS, gateway, or WINS?

   _____

   Which of the following is responsible for NetBIOS name to IP address resolution: DNS, gateway, or WINS?

   _____

10. Click on the **OK** button and if prompted, insert the Windows 9x CD-ROM or enter the path to the installation files.

11. Reboot the computer for the new settings to take effect.

Lab Project 8

The goal of this project is for you to be able to install and configure the TCP/IP protocol on a Windows NT Workstation computer. You will need a Windows NT Workstation computer with a NIC installed and configured. The TCP/IP Protocol is a routable protocol. It is the protocol that powers the Internet, so it is important that you understand how it is installed and configured.

1. Turn on the computer and verify that NT Workstation loads.

2. Log on to NT Workstation using the userid and password provided by the instructor or lab assistant.

3. From the Start menu, point to the **Settings** option, click on the **Control Panel** option, and then double-click on the **Network** icon. The Network Installation and Configuration window opens.

   What alternate method can be used to access the Network Installation and Configuration window?

   _____

4. Click on the **Protocols** tab. The Protocols window opens.

   Which protocols are already installed on your computer?

   _____

5. Click on the **Add** button. The Select Network Protocol window opens.

   List the protocols that are available for installation.

   _____

6. Select the **TCP/IP Protocol** option and click on the **OK** button.

7. If DHCP is used on the network, select **Yes** to use DHCP, otherwise select **No**. Contact the instructor or lab assistant if you are unsure which option to select.

8. If prompted, insert the Windows NT Workstation CD-ROM or enter the path to the installation files and click on the **Continue** button. When TCP/IP finishes installing, select **Close**. The TCP/IP Properties page opens.

9. If DHCP is used on the network, select **Obtain an IP address from a DHCP server**. If DHCP is not used, select **Specify an IP address** and enter the **IP address, subnet mask**, and **default gateway** information provided by the instructor or lab assistant.

   When DHCP is not used, which one of the following is optional: IP address, subnet mask, or default gateway?

   _____

10. If directed by the instructor, click on the **DNS** tab and enter the provided DNS information. Click on the **WINS** tab and enter the provided WINS information.

11. When all TCP/IP configuration information has been entered, click on the **OK** button. NT Workstation goes through a bindings process, and you will be prompted to restart the computer. Restart the computer for the new settings to take effect.

Lab Project 9

The goal of this project is for you to be able to install and configure the TCP/IP protocol on a Windows 2000 Computer. You will need a Windows 2000 Professional computer with a NIC installed and configured. The TCP/IP protocol is a routable protocol. It is the protocol that powers the Internet, so it is important that you understand how it is installed and configured.

1. Turn on the computer and verify that Windows 2000 Professional loads.

2. Log on to Windows 2000 Professional using the userid and password provided by the instructor or lab assistant.

3. Right-click on the **My Network Places** desktop icon, and then select **Properties**. The Network and Dial-up Connections window opens.

4. Right-click on the **Local Area Connection** icon, and then select **Properties**. The Local Area Connections page opens.

5. Choose **Install.** The Select Network Component Type window opens.

   What types of network components are available?

   _____

6. Choose **Protocol** and then select **Add**. The Select Network Protocol window opens.

   Which network protocols are available for installation?

   _____

7. Choose the **TCP/IP** protocol and then select **OK**.

8. If prompted, insert the Windows 2000 Professional CD-ROM into the drive, or enter the path to the installation files.

9. From the Local Area Connection window, highlight **TCP/IP**, and then select **Properties**. The TCP/IP Properties window opens.

10. If you are using DHCP, select the **Obtain an IP Address Automatically** option. If you are not using **DHCP**, select the **Use the Following IP Address** option and enter an **IP address**, a **subnet mask**, and the **default gateway** information provided by the instructor or lab assistant.

    Which of the following is optional: IP address, subnet mask, or default gateway?

    _____

11. When you are finished entering TCP/IP configuration information, click on the **OK** button and close the Local Area Connections Properties window.

Lab Project 10

The goal of this project is to create a functional CAT 5 UTP network cable. You will need the following parts:

- Category 5 UTP cable
- RJ-45 connectors
- CAT 5 stripper/crimper tool
- UTP cable tester

Note: Standard Ethernet networks are cabled with either CAT 5 UTP cable or RG-58 coaxial cable. In this exercise, you create a standard CAT 5 cable for use with either 10BaseT or 100BaseT networks connected through a central hub or switch.

1. Category 5 UTP cable consists of four twisted pairs of wires, color-coded for easy identification. The color-coded wires are as follows:

   Pair 1: White/Orange and Orange

   Pair 2: White/Blue and Blue

   Pair 3: White/Green and Green

   Pair 4: White/Brown and Brown

2. Using the **stripper/crimper tool**, strip approximately **1/2 inch** of the protective outer sheath to expose the four twisted pairs of wires. Most strippers have a strip gauge to ensure stripping the proper length. (See Figure 9.34.)

   Note: In order to make it easier to sort the wire pairs, the sheathing can be stripped further than 1/2 inch, then the wires can be sorted properly and trimmed to the proper length.

Figure 9.34: Using a Stripper/Crimper Tool

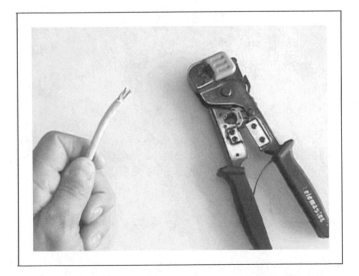

3. Untwist the exposed wire pairs. Be careful that you do not remove more twist than necessary. Sort the wires according to the following:

Wire 1: White/Orange Wire 5: Blue

Wire 2: Orange Wire 6: Green

Wire 3: White/Green Wire 7: White/Brown

Wire 4: White/Blue Wire 8: Brown

Ethernet cabling utilizes wires 1, 2, 3, and 6. Using the above wiring scheme means that the cable will use the White/Orange-Orange and White/Green-Green wire pairs. (See Figure 9.35.)

Will both ends of the cable need to follow the same wiring schematic?

Figure 9.35: Creating the Correct Cable Scheme

4. Insert the sorted and trimmed **cable** into an **RJ-45 connector**. The RJ-45 connector's key should face downward. Verify that all eight wires fully insert into the RJ-45 connector and that they are inserted in the proper order. (See Figure 9.36.)

Figure 9.36: The Inserted Cable Ends

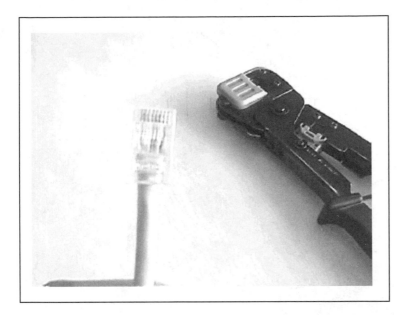

5. Insert the **cable-connector assembly** into the **stripper/crimper tool** and crimp the connector firmly. (See Figure 9.37.)

Figure 9.37: Crimping the Cable

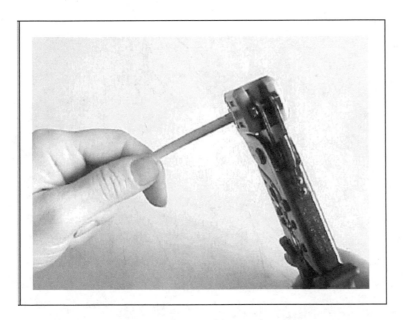

6. Remove the **cable/connector assembly** from the **stripper/crimper tool** and verify that the wires fully insert into the connector and that they are in the proper order. (See Figure 9.38.)

Figure 9.38: Removing the Cable

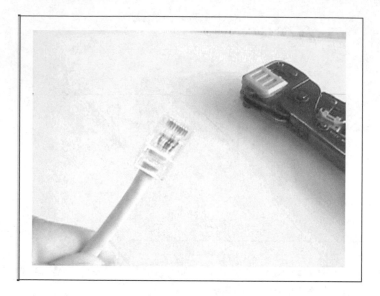

7. Repeat steps 2 through 6 for the other end of the CAT 5 UTP cable. (See Figure 9.39.)

Figure 9.39: The Newly Created Cable

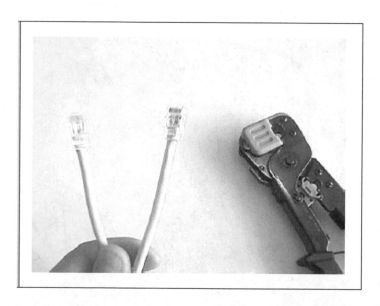

Can the cable be used at this point?

8. Before using the cable, it should be tested with a cable tester. This verifies that you have end-to-end continuity on individual wires and proper continuity between wire pairs. Insert the **RJ-45 connector** into the proper **cable tester receptacle** and verify that the cable is functional. (See Figure 9.40.)

Figure 9.40: Testing the Cable

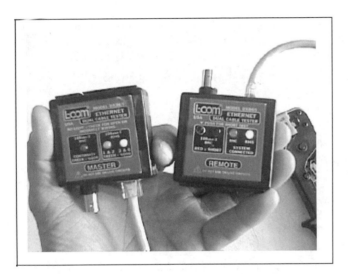

Lab Project 11

The goal of this project is to create a functional CAT 5 UTP crossover cable. You will need the following parts:

• Category 5 UTP cable

• RJ-45 connectors

• Stripper/crimper tool

• UTP cable tester

Note: In normal situations, straight-through CAT 5 UTP cabling is used to connect to a central hub or switch. In this exercise, you create a crossover CAT 5 cable for use when connecting two network devices—computers without using a central hub or switch.

1. Category 5 UTP cable consists of four twisted pairs of wires that are color-coded for easy identification. The color-coded wires are as follows:

   Pair 1: White/Orange and Orange

   Pair 2: White/Blue and Blue

   Pair 3: White/Green and Green

   Pair 4: White/Brown and Brown

2. Using the **CAT stripper/crimper tool**, strip approximately **1/2 inch** of the protective outer sheath to expose the four twisted pairs of wires. Most tools have a strip gauge to ensure stripping the proper length.

   Note: In order to make it easier to sort the wire pairs, the sheathing can be stripped further than 1/2 inch. The wires can then be sorted properly and trimmed to the proper length.

3. Untwist the exposed wire pairs. Be careful that you do not remove more twist than necessary. Sort the wires as follows:

   Wire 1: White/Orange Wire 5: Blue

   Wire 2: Orange Wire 6: Green

   Wire 3: White/Green Wire 7: White/Brown

   Wire 4: White/Blue Wire 8: Brown

   Ethernet networks utilize wires 1, 2, 3, and 6. Using the above wiring scheme means the cable will use the White/Orange-Orange and White/Green-Green wire pairs.

   When making a crossover cable, will both ends of the cable need to follow the same wiring schematic?

   _____

4. Insert the sorted and trimmed **cable** into a **RJ-45 connector**. The RJ-45 connector's key should face downward. Verify that all eight wires fully insert into the RJ-45 connector, and that they are inserted in the proper order.

5. Insert the **cable-connector assembly** into the **CAT 5 stripper/crimper tool** and crimp the connector firmly.

6. Remove the **cable/connector assembly** from the **CAT 5 stripper/crimper** tool and verify that the wires are fully inserted into the connector and that they are in the proper order.

7. To create the crossover cable, the wire pairs must be put in a different order. To accomplish this, repeat steps 2 through 6 on the opposite end of the cable, but when sorting the wire pairs, use the following color codes:

   Wire 1: White/Green Wire 5: Blue

   Wire 2: Green Wire 6: Orange

   Wire 3: White/Orange Wire 7: White/Brown

   Wire 4: White/Blue Wire 8: Brown

   Can the crossover cable be used at this point?

   _____

8. Before using the crossover cable, it should be tested with a cable tester. This verifies that you have end-to-end continuity on individual wires and proper continuity between wire pairs. Insert the **RJ-45 connector** into the proper **cable tester receptacle** and verify that the cable is functional.

   Note: Your cable tester must have the capability to test crossover cables.

Lab Project 12

The goal of this project is for you to be able to share a local printer on a Windows 9x computer so it will be available to other workstations. You will need a Windows 95 or 98 computer with a NIC installed and configured, and a printer physically attached and configured.

Note: A printer that is physically attached (local) to a networked workstation can accept and process print jobs from other workstations on the network. Before this can happen, the local printer must be shared on the network. Before a printer can be shared in Windows 9x, Printer Sharing must be installed.

Installing Printer Sharing

1. Turn on the computer and verify that Windows 9x loads.

2. If necessary, log on to Windows 9x using the userid and password provided by the instructor or lab assistant.

3. Right-click on the **Network Neighborhood** desktop icon and select the **Properties** option. The Network Properties window opens.

4. Choose **File and Print Sharing**, select **I want to be able to allow others to print to my printers**, and then click on the **OK** button.

   What will this setting allow you to share?

   _____

5. From the Network Properties window, click on the **OK** button.

6. If prompted, insert the Windows 9x installation CD-ROM or enter a path to the installation files.

7. Reboot the computer when prompted.

Sharing a Windows 9x Printer

8. From the **Start** menu, point to **Settings**, and then click on the **Printers** option. The Printers folder opens.

9. Right-click on a specific printer that is attached to the computer, and then select the **Sharing** option. The Printer Sharing window opens.

10. Select the **Shared As** radio button, and enter a share name of **TESTPRINT** in the Share Name field.

11. In the **Comment** field, enter a user-friendly description of this printer.

12. In the Password field, type the word **password**.

    What effect will setting a password have?

    _____

13. Click on the **Apply** button to save your sharing settings, re-enter the password **(password)** when prompted, and then click on the **OK** button.

14. Click on the **OK** button to exit the Printer Sharing window. Your printer is now shared and available on the network.

Lab Project 13

The goal of this project is for you to be able to share a local printer on a Windows NT Workstation so it will be available to other workstations on the network. You will need a Windows NT Workstation with a NIC installed and configured, and a printer physically attached and configured.

Note: A printer that is physically attached (local) to a networked workstation can accept and process print jobs from other workstations on the network. Before this can happen, the local printer must be shared on the network.

1. Turn on the computer and verify that NT Workstation loads.

2. Log on to NT Workstation using the userid and password provided by the instructor or lab assistant.

3. Click on the **Start** button, point to **Settings**, and then click on the **Printers** option. The Printers folder opens.

   What other method can be used to access the Printers folder?

   _____

4. Right-click on the name of the printer that is attached to the workstation. Select the **Sharing** option. The Printer Sharing window opens.

   What other method can be used to access the Printer Sharing window?

   _____

5. Click on the **Shared** radio button and enter a name in the Share Name field.

   What name did you assign to the printer?

   _____

6. Click on the **OK** button to return to the Printers folder. The printer is now shared.

   How can you verify that the printer has been shared?

   _____

Lab Project 14

The goal of this project is for you to be able to share a local printer on a Windows 2000 Professional Workstation so it will be available to other workstations. You will need a Windows 2000 Professional Workstation with a NIC installed and configured, and a printer physically attached and configured.

Note: A printer that is physically attached (local) to a networked workstation can accept and process print jobs from other workstations on the network. Before this can happen, the local printer must be shared on the network.

1. Turn on the computer and verify that Windows 2000 Professional loads.

2. Log on to Windows 2000 Professional using the userid and password provided by the instructor or lab assistant.

3. Click on the **Start** button, point to the **Settings** option, and then click on the **Printers** option. The Printers folder opens.

4. Right-click on the local printer attached to the workstation and select the **Properties** option. The printer's Properties window opens.

5. Click on the **Sharing** tab. From the Sharing window, you can share the printer, give it a share name, and install additional drivers for each type of Windows operating system connected to the network that will use the printer. You can also publish the printer in Active Directory if the workstation is part of a Windows 2000 domain.

6. Choose the **Shared as** option and type **TestShare** in the Share Name field.

7. Choose **Apply** and then click on the **OK** button. The printer's Properties window closes and then returns to the Printers folder. The local printer is now shared and is available to other workstations on the network.

   How can you tell the printer has been shared?

Lab Project 15

The goal of this project is for you to understand how to connect to and use a networked printer on a Windows 9x Computer. You will need a Windows 9x computer with a NIC installed and configured, and local printer installed and shared on the network.

Note: A printer that is physically attached (local) to a networked computer and shared on the network can accept and process print jobs from remote computers on the network. Before this can happen, the remote computers must connect to the shared printer and install the proper printer driver.

1. Turn the computer on and verify that Windows 9x loads.

2. Log on to Windows 9x using the userid and password provided by the instructor or lab assistant.

3. Click on the **Start** button, point to the **Settings** option, and then click on the **Printers** option. The Printers folder opens.

4. Double-click on the **Add Printer** icon and the Add Printer wizard runs.

5. Click on the **Next** button, select the **Network printer** option, and then click on the **Next** button.

6. Locate the shared printer, highlight the printer, and then click on the **OK** button. Contact the instructor or lab assistant if you cannot locate the shared printer. The printer's UNC name appears in the Network path or Queue Name field.

   What does UNC stand for?

   _____

7. Choose **No** to the Do you print from MS-DOS based programs prompt, and click on the **Next** button.

8. Enter the name **LABTEST** for this printer in the Printer Name field.

   Where does this printer name appear?

   _____

9. Select the **Yes** option in order to have Windows use this printer as the default printer and then click on the **Next** button.

10. Choose **Yes** to print a test page and then select **Finish**.

11. The printer driver downloads and installs on your local computer.

12. To complete the connection, type **password** for the printer share password and then click on the **OK** button.

    Can the printer be used across the network without a network user supplying the password?

    _____

13. If the printer connection and driver installation were successful, a printer test page prints.

Lab Project 16

The goal of this project is for you to be able to connect to and use a networked printer in Windows NT Workstation. You will need a Windows NT Workstation with a NIC installed and configured, and a local printer installed and shared on the network.

Note: A printer that is physically attached (local) to a networked computer and shared on the network can accept and process print jobs from remote computers on the network. Before this can happen, the remote computers must connect to the shared printer and install the proper printer driver.

1. Turn the computer on and verify that Windows NT Workstation loads.

2. Log on to Windows NT Workstation using the userid and password provided by the instructor or lab assistant.

3. Click on the **Start** button, point to the **Settings** option, and then click on the **Printers** option. The Printers folder opens.

4. Double-click on the **Add Printer** icon and the Add Printer wizard runs.

5. Choose **Network Printer Server** and then click on the **Next** button. The Connect to Printer window opens.

6. From the Connect to Printer window, browse through the available computers and shared printers until you locate the appropriate shared printer. After several minutes of browsing, contact the instructor or lab assistant if you cannot locate the shared printer.

7. Click on the **appropriate shared printer**. The printer's UNC name appears in the Printer field. Click on the **Next** button.

   The UNC name is made up of two parts. What do these two parts represent?

   _____

8. Choose **Yes** for Windows applications to use this printer as the default printer, and then click on the **OK** button.

9. The printer driver downloads and installs. Click on the **Finish** button to exit the Add Printer wizard. You have now connected to and installed the driver for a networked printer.

   How can you tell the printer has been connected to a shared printer?

   _____

Lab Project 17

The goal of this project is to connect to and use a networked printer in Windows 2000 Professional. You will need a Windows 2000 Professional computer with a NIC installed and configured, and a local printer installed and shared on the network.

Note: A printer that is physically attached (local) to a networked workstation and shared on the network can accept and process print jobs from remote workstations on the network. Before this can happen, a remote workstation must connect to the shared printer and install the proper printer driver.

1. Turn the computer on and verify that Windows 2000 Professional loads.

2. Log on to Windows 2000 Professional using the userid and password provided by the instructor or lab assistant.

3. From the **Start** menu, point to **Settings**, and then click on the **Printers** option. The Printers folder opens.

4. Double-click on the **Add Printer** icon and the Add Printer wizard opens.

5. Click on the **Next** option, and the Local or Network Printer window opens.

6. Choose **Network Printer** and then select **Next**. The Locate Your Printer window opens.

7. If the workstation is part of an Active Directory domain, you could choose the **Find a printer in the directory** option or select the **Connect to a printer on the Internet or on your intranet** option and enter the URL for the printer. Contact the instructor or lab assistant if you are unsure about which option to choose.

   What does the acronym URL stand for?

   _____

8. Choose the **Type the printer name or click on Next to browse for a printer** option and then click on the **Next** button. The Browse for Printer window opens.

9. From the Browse for Printer window, browse through the available computers and shared printers until you find the appropriate shared printer. Contact the instructor or lab assistant if you are unsure about which printer to choose.

10. Highlight the **shared printer**. The printer's UNC name displays in the Printer Name field. Click on the **Next** button.

    The UNC name is made up of two parts. What do these two parts represent?

    _____

11. From the Default Printer window, choose **Yes** for Windows to use this printer as your default printer, and then select **Next**.

12. From the Completing the Add Printer Wizard window, review the settings and click on **Finish**.

13. The printer driver automatically downloads from the host workstation and you return to the Printers folder after the driver downloads. You have now connected to and installed the driver for a networked printer.

How can you tell the printer has been connected to a shared printer?

Lab Project 18

The goal of this project is for you to understand how to create a dial-up connection using Windows 98. You will need a Windows 98 computer with a modem and Dial-up Networking installed and configured. You will also need a phone number of a dial-up server.

Note: The Windows Dial-up Networking (DUN) utility allows you to create and configure dial-up connections to dial-up access servers. In this exercise, you create a dial-up connection using Windows 98.

1. Turn the computer on and verify that Windows 98 loads.

2. Log on to Windows 98 using the userid and password provided by the instructor or lab assistant.

3. Double-click on the **My Computer** desktop icon and then double-click on the **Dial-up Networking** folder. The Dial-up Networking folder opens.

   Can you create a new connection if a modem has not been installed?

   _____

4. Double-click on the **Make New Connection** icon. The **Make New Connection** window opens.

5. Type **Test** in the Connection Name field and from the **Select a device** drop-down menu select the modem to use for this connection. Click on the **Next** button.

6. Enter the **area code** and **phone number** of the remote dial-up server to be dialed, select the **Country or Region code** from the drop-down menu, and click on the **Next** button. Contact the instructor or lab assistant for this number.

7. Click on the **Finish** button to create the Test connection.

8. Close the **Make New Connection** window. The Test Connection icon appears in the Dial-up Networking folder.

   Can you modify the dialing properties of the Test connection after it has been created?

   _____

   To use the Test connection, follow these steps:

9. Double-click on the **My Computer** desktop icon and then double-click on the **Dial-up Networking** folder. The Dial-up Networking folder opens.

10. Double-click on the **Test Connection** icon. The Connect to window opens.

11. Enter a **username** and **password** for the connection, verify the proper **phone number** is listed, and click on the **Connect** button. The Dial-up Networking utility will complete the connection to the remote dial-up server.

Lab Project 19

The goal of this project is to use the Dial-up Networking utility to create a dial-up connection in Windows NT Workstation. You will need a computer with Windows NT Workstation and a modem and Dial-up Networking installed.

Note: Windows NT Workstation comes with Dial-up Networking to enable you to create a dial-up connection to a remote dial-up access server.

1. Turn the computer on and verify that Windows NT Workstation loads.

2. Log on to NT Workstation using the userid and password provided by the instructor or lab assistant.

3. Double-click on the **My Computer** desktop icon and then double-click on the **Dial-up Networking** icon. The Dial-up Networking window opens.

4. To create a new dial-up connection, select **New**. The New Phonebook Entry wizard starts.

5. Type **Test** in the **Name the phonebook entry** field and click on the **Next** button.

6. From the Server window, select the type of dial-up connection you are configuring.

   Which type of connection would you choose for browsing the web?

   _____

7. Choose **I am calling the Internet** and click on **Next**.

8. Enter the phone number of the Internet Service Provider (ISP) you are calling and click on **Next**.

9. Click on **Finish** to complete the creation of the Test connection.

   Where will the new connection appear?

   _____

10. To use the Test connection, double-click on the **My Computer** desktop icon and then double-click on the **Dial-up Networking** icon.

11. From the Phonebook Entry drop-down menu, select the **Test** connection.

12. Select **Dial**, enter a **user name**, **password**, and **domain** (if required), and click on **OK**. Dial-up Networking dials the Internet Service Provider and completes the connection.

Lab Project 20

The goal of this project is to use the Dial-up Networking utility to create a dial-up connection in Windows 2000 Professional. You will need a computer with Windows 2000 Professional and a modem installed and a phone number of a dial-up server.

Note: Windows 2000 Professional comes with Dial-up Networking to enable you to create a dial-up connection to a remote dial-up access server.

1. Turn the computer on and verify that Windows 2000 Professional loads.

2. Log on to Windows 2000 Professional using the userid and password provided by the instructor or lab assistant.

3. From the Start menu, choose **Settings**, and then select **Network and Dial-up Connections**.

   What other method can be used to access Network and Dial-up Connections?

   _____

4. Double-click on the **Make New Connection** icon. The Network Connection wizard starts.

5. Click on the **Next** button.

6. From the Network Connection Type window, you can select the **type of connection** you are making.

   Which connection type would you select to allow your computer to act as a Remote Dial-up Access Server?

   _____

7. Select **Dial-up to Private Network** and click on the **Next** button.

8. Enter the **Phone number** of the Remote Dial-up Access Server. Contact the instructor or lab assistant for the number. Click on the **Next** button.

9. Select **Create this connection for all users** and click on the **Next** button.

10. Enter **Test in the Connection Name** field and click on the **Finish** button.

11. A new Dial-up connection appears in the Network and Dial-up Connection window.

12. Double-click on the **Test connection** icon, enter a User name and Password, and click on **Dial**. Dial-up Networking places the call and completes the connection to the remote dial-up access server.

Lab Project 21

The goal of this project is to have you share a folder within Windows XP and view the results of this shared folder.

1. Boot the Windows XP computer.

2. Create a folder on the hard drive or floppy drive named "OS Book."

3. Right-click on the folder, click **Properties,** and then click **Sharing**. The Sharing properties for the folder appear.

4. Click **Share this folder** and then click **OK**. Note that the folder name and the share name are the same in this case but can be different.

5. Refer to Figure 9.41 for a sample screenshot.

Figure 9.41: Sharing a Folder

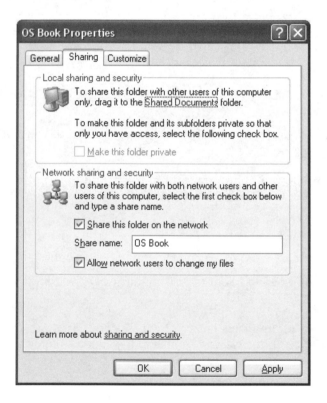

6. To verify the folder is shared, click **Start** and then click **Run**. The Run dialog box appears.

7. In the Run dialog box, type *computer-name**share-name,* where *computer-name* is the name of your computer and *share-name* is the share name. For example, in Figure 9.42, you can see that the computer name is XP2 and the share name is OS Book.

8. Close all windows.

Figure 9.42: Results of Sharing a Folder

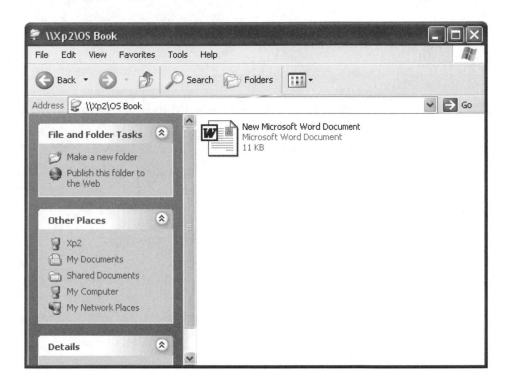

Lab Project 22 Challenge

The goal of this project is to set up a Windows XP computer as a firewall to protect it and other computers from hacker attack. This project assumes at least two computers connected using either a crossover cable or a straight-through cable with a switch/hub. Additionally, the XP computer will be the only one connected to the Internet and to the other computer. The second computer will not be connected to the Internet.

1. Boot the XP computer (the one connected to both the Internet and the second computer).

2. Click **Start**, point to **My Network Places**. The **My Network Places** window appears.

3. Under the Network Tasks section in the left windowpane, click **View Network Connections**. The **Network Connections** window appears.

4. Right-click the Local Area Connection icon that is connected to another PC and then click **Properties**. The **Properties** page appears.

5. Click **Advanced**. A window such as the one in Figure 9.43 appears.

Figure 9.43: The Advanced Page of a Local Area Connection's Properties

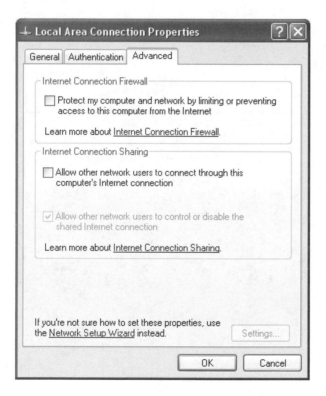

6. Attempt to ping the second computer.

   Did it work? (It should!)

   _____

7. To protect your computer and network by limiting or preventing access, check **Protect my computer and network by limiting or preventing access to this computer from the Internet**.

8. Once you check this box, the Settings button is highlighted.

9. Click the **Settings** button.

10. List the services that can be blocked.

11. Attempt to ping from the second computer.

    Did it work? (It should not due to firewall protection!)

    _____

12. To allow other users to use this computer to connect to the Internet, check **Allow other network users to connect through this computer's Internet connection**.

13. What IP address was selected for this connection?

    _____

14. Close all windows.

Lab Project 23 Challenge

The goal of this Challenge Lab Project is to connect two computers together using a CAT 5 crossover network cable. You will need to have completed Lab 11 prior to beginning this Challenge Lab Project. You will also need two Windows 9x computers with network cards installed and configured.

Note: In normal situations on Ethernet networks, all networked computers connect to a central hub or switch using CAT 5 straight-through cables. In this exercise, you connect two Windows 9x computers using a CAT 5 crossover cable.

1. Plug the **CAT 5 crossover cable** directly into the network cards' RJ-45 ports on the two Windows 9x computers.

   Can you connect more than two computers together using a CAT 5 crossover network cable?

   _____

   Why or why not?

   _____

2. Turn the computers on and verify that Windows 9x loads.

3. Log on to both Windows 9x computers using the userid and password provided by the instructor or lab assistant.

4. Right-click on the **Network Neighborhood** desktop icon and select **Properties**. The Network Properties window opens.

5. From the General tab, select the **File and Print sharing** button. Verify that the **I want to be able to give others access to my files** check box is selected, and click on **OK**.

6. Select the **Identification** tab. Verify that both computers are members of the same **workgroup**.

7. Click on **OK** to close the Network Properties window.

8. If prompted, insert the Windows CD into the CD-ROM and restart the computers.

9. From the My Computer desktop icon, right-click on the **C:** drive and select **Sharing**.

10. From the Sharing window, select **Shared as**, enter a share name, and click on **OK.**

    What is the significance of a share name?

    _____

11. After finishing sharing both C: drives, double-click on the **Network Neighborhood** desktop icon.

12. If the crossover cable connection is working, both computers should appear in the Network Neighborhood browser window. If it is not working, check the NIC configuration settings, the crossover cable, or redo the steps in this exercise. Show the instructor or lab assistant the two computer names in the Network Neighborhood browser window.

Internet Discovery

Internet Discovery Lab 1

The goal of this Internet Discovery Lab is to access the Internet to obtain specific information regarding a computer or its associated parts. You will need access to the Internet.

1. On an HP BRIO computer and after installing a 10/100 BT PCI Ethernet adapter, NT Workstation displays the error message, "At least one service failed to start." What is the problem, solution, and at what Internet address did you find the solution?

2. What does the term "Wake on LAN" mean and at what URL did you find the answer?

3. On a clone computer running Windows 95 and with a 3Com 3C359B adapter installed, the "Divide by zero error R6003" error appears. The computer is upgraded to Windows 98 and the problem does not change. What is the solution? List the URL where the answer was found.

4. How can you tell if an infrared device is within range on a Windows 2000 Professional computer? Write the answer and the URL in the space below.

5. Find an Internet site that explains the differences between CAT 5 and CAT 5E UTP cable. Write one difference and the URL in the space below.

Internet Discovery Lab 2 Challenge

The goal of this Internet Discovery Lab is to access the Internet to obtain specific information regarding networking. You will need access to the Internet.

1. Research the Internet and locate two companies that sell T1 technology. Compare their prices. You may have to call the companies.

2. Research the Internet and locate at least three companies, including their web sites, that produce the following:

   a) routers
   b) switches
   c) hubs
   d) fiber

3. Research the Internet, locate and download pictures of the following in use:

   a) star topology
   b) bus topology
   c) ring topology
   d) mesh topology

Soft Skills: Help Desk Support

1. A customer calls you and says she cannot access the Internet. What types of questions do you ask to help troubleshoot the problem?
2. A user is attempting to download a file and it seems to be taking too long. What types of issues could be causing this apparent slowness?

Critical Thinking

1. Discuss the differences between a crossover cable and a regular straight-through cable. Prove why they cannot be interchanged.
2. Research the Internet or other sources for information on what the uplink toggle switch is used for on a switch/hub. Can you interchange a crossover and straight-through cable by toggling this switch?
3. Discuss the benefits of using a star topology over a mesh topology. Now, discuss the benefits of the mesh topology over a star topology.
4. How long will it take to transfer a 2 GB file over a T1 assuming you have the maximum transfer rate for your file transfer?
5. How long will it take to transfer the same 2 GB file over POTS assuming the low end of the speed range?

Study Skills

Test Preparation

Test preparation occurs over time—not the night before the test.
- Know what you will be tested on.
- Know what time the test begins.
- Bring all necessary test materials such as pencils and paper. If the test is open book and open note, bring those to class.
- Review all necessary material particularly instructor lecture notes, labs, and chapter highlights such as the summary.
- Go over the material at least three times.
- Get a good night's rest before the test.
- Eat something before the test.
- Get to the testing area on time.
- Do not cheat off another's test paper.
- Study daily! Spend at least 30 minutes a day reviewing material.
- When you get the test, do a "brain dump" of key terms so they will be readily accessible when you need them.
- Have confidence—if you have kept up with your studies, you should have no surprises on the test.

Self-Study Question(s)
1. Did you prepare for a test today? You should prepare a little each day.
2. Identify at least two test preparation Study Skills you did this week.

Glossary

% The character used for passing positional parameters to a batch file.

The prompt for the UNIX/Linux root user.

$ The prompt for a UNIX/Linux user.

$HOME A variable that represents a user's home directory.

%systemdrive% A variable within Microsoft Windows operating systems that represents the root directory. For example C:\.

%systemroot% A variable within Microsoft Windows operating systems that represents the boot partition. For example C:\WINNT.

* The wildcard character used to match all characters.

. Represents the current directory.

.. Represents the parent directory.

? The wildcard character used to match specific character positions.

[...] A wildcard technique used in UNIX/Linux where wildcarding occurs on multiple characters. For example [abc] would match on the characters a, b or c.

| The pipe symbol used to filter output from one command as input to another.

< The redirection symbol used to redirect input.

> The redirection symbol used to redirect output and a new file is created.

>> The redirection symbol used to redirect output and append to a file.

16-bit mode Also known as real mode. See real mode.

32-bit mode Also known as protected mode. See protected mode.

8.3 rule The rule used by DOS for creating files. There can be a maximum of 8 characters in the file name followed by 3 characters in the extension.

absolute path Another name for full path.

access point An access point is a device that receives and transmits data from multiple computers that have wireless NICs.

accumulator A register that contains the accumulated value of operations such as addition.

active partition A partition that has been designated as "active" is one that will be booted. A partition is made active by using the FDISK command.

ad hoc mode An ad hoc wireless network is when at least two devices such as two computers have wireless NICs (Network Interface Cards) installed.

Add Printer An icon allowing you to add a printer to your computer.

Add/Remove Hardware A Microsoft Windows control panel icon allowing you to manage hardware.

Add/Remove Programs A Microsoft Windows control panel icon allowing you to manage programs.

Administrative Tools A Microsoft Windows program allowing you access to administrative programs.

algorithm A set of instructions to perform a task.

allocation unit The Microsoft suggested term for a cluster. An allocation unit is equal to a specific number of sectors on disk.

American Standard Code for Information Interchange (ASCII) The character set used on PC based computer systems.

Application Program Interface (API) A standard used by programmers to guarantee standards among applications.

application software Software used to perform office work. Examples include word processing, spreadsheet and database software.

Apply button A Microsoft Windows button that applies changes immediately to an option once it is clicked.

Archive attribute A Microsoft DOS and Windows attribute that indicates a file needs to be backed up the next time a backup is scheduled. All new files and modified files will have this attribute set by the operating system.

Arithmetic Logic Unit (ALU) The ALU is the portion of the CPU where all the arithmetic and logic functions of the computer are done.

assembling The process of converting the assembly language statements and data into machine-readable (known as binary code or executable code) form.

assembly language The computer language that uses mnemonic statements and is native to each processor (CPU).

AT&T System Release V (SR5) A version of UNIX developed by AT&T.

ATTRIB A Microsoft DOS and Windows command that allows you to change the attributes of a file or directory (folder).

auditing A technique allowing you to track changes to files and folders.

Auto Arrange A Microsoft Windows option forcing your icons on your desktop to be automatically arranged towards the left hand side of the screen.

AUTOEXEC.BAT A Microsoft Windows and DOS customizable batch file. The commands in this file will be executed upon system boot.

AUTOEXEC.NT A file on the ERD that emulates AUTOEXEC.BAT.

Automated System Recovery (ASR) This software is used in Windows XP and replaces the ERD used in Windows NT and 2000.

Automatic Skip Driver Agent (ASD) A Microsoft Windows program that helps determine which drivers failed to load during startup.

Autorun A Microsoft Windows feature that allows a software program on a CD to be automatically started when inserted into the CD drive.

awk A UNIX/Linux command that allows you to perform pattern matching on text within a file.

bad-cluster remapping A technique where bad clusters on a disk are marked so they won't be used in the future.

Base memory address range An area of memory allocated to a device for its device drivers.

Baseline A snapshot of your computer's performance during normal operations.

bash shell A UNIX/Linux command that provides the shell prompt.

Basic Input/Output System (BIOS) A set of instructions permanently stored on a chip that handles the basic input and output functions of the computer.

Examples include keyboard and video functionality.

Basic Service Set (BSS) When multiple devices connect to an access point (whether that access point is wired to a LAN or not), the configuration is known as a BSS.

batch file A Microsoft DOS and Windows file that contains other commands.

Berkeley Source Distribution (BSD) A version of UNIX developed by The University of California at Berkeley.

binary The language of the computer consisting of zeroes and ones.

binary program Also known as an executable program, a binary program is a set of instructions in computer form that can be executed, or run, by the computer.

Bit A binary digit, either a zero or a one.

Block A block is also known as a cluster. Linux uses the term block to represent a collection of sectors on disk.

blue screen or Blue Screen of Death The blue colored screen you receive when a Microsoft Windows operating system crashes.

boot partition Under Microsoft Windows NT, 2000 and XP, the boot partition is where the operating system itself is located.

BOOT.INI A file used to determine which operating system is loaded during boot up.

booting The process of starting the computer. This includes loading the operating system into memory.

BOOTSECT.DOS A Microsoft Windows NT file used for dual-booted systems.

broadcast address A special IP address used to communicate to all hosts on a specific network.

bus topology A topology where each devices is connected by a single cable.

byte A series of 8 bits taken collectively.

C:\WINDOWS The default installation folder for Windows 2000 and XP.

C:\WINNT The default installation folder for Windows NT.

cache A form of high speed memory. Programs and data that are frequently used are held in cache memory.

cal A UNIX/Linux command that allows you to display a calendar.

CALL A Microsoft DOS and Windows command that calls one DOS batch program within another.

Cancel button A Microsoft Windows button that returns options to their original state when it is clicked.

Carrier Sense Multiple Access/Collision Avoidance (CSMA/CA) A technique used by wireless networks and Apple networks allowing devices to communicate.

Carrier Sense Multiple Access/Collision Detect (CSMA/CD) A technique used by Ethernet networks allowing devices to communicate.

cat A UNIX/Linux command that allows you to display the contents of a file.

cd .. A UNIX/Linux command that allows you to parent directory.

CD .. A Microsoft DOS and Windows command that allows you to change to your parent directory in DOS.

cd / A UNIX/Linux command that allows you to change to the root directory.

**CD ** A Microsoft DOS and Windows command that allows you to change to the root directory in DOS.

CD A Microsoft DOS and Windows command allowing you to change directory positions within the directory tree structure.

cd A UNIX/Linux command that changes directory locations within the directory structure.

CDFS A file system used for CD-ROMs.

Cells A combination of rows and columns used in a spreadsheet to identify data.

Central Processing Unit (CPU) The part of the computer system used perform arithmetic, logic and control functions of the computer system.

centralized processing This term is typically used when discussing the operations of a mainframe computer. With centralized processing, the processing and storage of a computer system is performed in a central location.

character set The set of characters on the keyboard and their binary equivalent.

chassis Also known as the system unit.

CHDIR The same as the Microsoft DOS and Windows command CD command. See CD.

check box A Microsoft Windows box where you can turn on or off an option.

child directory A directory contained within a parent.

CHKDSK A DOS and Microsoft Windows command that checks the status of a disk.

chmod A UNIX/Linux command that allows you to modify permissions of a file or directory.

CHOICE A Microsoft DOS and Windows command that waits for the user to choose one of a set of choices in a batch file.

CIPHER The Microsoft Windows command allowing you to encrypt and decrypt a file.

clean install Installing an operating system on a computer that does not have an existing operating system.

client In a network environment, a computer that uses resources on a server.

client/server model In networking, the client server model, is where a few server computers serve many client computers.

Close button A Microsoft Windows button, located in the upper right hand corner of a window, allowing you to close a window.

Close button The X located in the upper right hand corner of a window allowing you to close it.

CLS A Microsoft DOS and Windows command used to clear the screen.

cluster A cluster is equal to a specific number of sectors on disk.

cluster address The hexadecimal address of a disk cluster.

cmd The command that allows you to access the DOS command line in Microsoft Windows NT, 2000 and XP.

coaxial cable (or coax) Coax cable has a center copper conductor surrounded by insulation. Outside the insulation is a shield of copper braid, a metallic foil, or both, that protects the center conductor from EMI.

command The command that allows you to access the DOS command line.

command line A text area when commands are entered and processed by a command processor.

Command Line Interface (CLI) A text area when commands are entered and processed by a command processor.

command prompt The prompt, such as C:\ in DOS or $ in UNIX/Linux, that allows you enter commands.

COMMAND.COM The DOS command interpreter. This is what gives you the

A:\> or C:\> prompt and either executes a command or displays an error.

Common User Access (CUA) A standard developed by IBM and Microsoft. The CUA makes sure Microsoft Windows programs are consistent.

COMPACT A Microsoft Windows command allowing you to compress or uncompress files.

compiling The process of converting the programming language statements and data into machine-readable (known as binary code or executable code) form.

Complex Instruction Set Computer (CISC) A computer that contains a large number of instructions in its instruction set. Your PC is a CISC computer.

compression A technique where redundant data is removed from a file thereby making a smaller file. Compression allows a file to occupy less disk space. Compression also reduces the amount of time a file will transfer or download.

Computer Management Console A Microsoft Windows program allowing you to manage shared folders, disk drives, stop and start services, view performance logs and system alerts and troubleshoot hardware.

condition A technique used to test criteria and then perform commands based upon the criteria.

CONFIG.NT A file on the ERD that emulates CONFIG.SYS.

CONFIG.SYS A DOS configuration file that has entries for device drivers such as memory, the mouse, and the CD-ROM.

contiguous The term contiguous means adjacent and is typically used to refer to how a file is stored on disk. If a file is stored in contiguous sectors, retrieval of the file is usually faster than if the file were not stored in contiguous sectors.

Control Panel A Microsoft Windows method of configuring various Microsoft Windows components.

Control Unit (CU) The CU is the part of the CPU, which handles the main activity.

conventional memory Also known as base memory, conventional memory is in the range of 0 to 640 KB. DOS and applications reside here.

CONVERT A Microsoft DOS and Windows command allowing you to convert from FAT16 to FAT32.

cooperative multitasking A form of multitasking where an application takes control of the system resources.

COPY A Microsoft DOS and Windows command allowing you to copy a file.

cp A UNIX/Linux command that allows you to copy files.

crosstalk A problem that occurs when signals on wires interfere with one another.

cut A UNIX/Linux command that allows you remove text from files.

cycle A unit of measure for CPU speed.

data Raw facts.

database A storage area for data.

DATE A Microsoft DOS and Windows command used to display or change the date.

date A UNIX/Linux command that allows you to display the current date.

default gateway A device that forwards a packet to a remote network.

defragmentation The process of placing a file into contiguous areas of a disk for quick retrieval.

DEL A DOS command used to delete a file.

DELTREE A Microsoft DOS and Windows command used to remove a tree, which includes files and directories. This is not available in Microsoft Windows XP and 200. In Microsoft Windows 2000 and XP, you use the **rd /s** command.

desktop The area of a Microsoft Windows screen where users interact with the operating system.

DEVICE The DOS command allowing you to specify a device driver. For example DEVICE=mouse.sys.

device driver Software that allows a device to operate.

Device Manager A Microsoft Windows software program allowing you to manage devices.

df A UNIX/Linux command that allows you display the amount of free disk space on a partition.

dialog box A Microsoft Windows box that allows you to interact with the operating system when configuring software.

Dial-up Networking (DUN) A Microsoft product where a computer dials the Internet using a modem.

diff A UNIX/Linux command that allows you display the difference between two files.

DIR A Microsoft DOS and Windows command that displays both files and directories.

Direct Memory Access (DMA) A technique where devices bypass the CPU and access memory directly.

directory A storage location for other directories and files. A directory is also known as a folder.

Disk Operating System (DOS) A Microsoft operating system that is disk based and command oriented.

disk quotas A technique where users are given a limited amount of disk storage space.

dispatch table A dispatch table is also known as an Interrupt Vector Table. It is used to point to where device drivers are stored in memory.

distributed processing A technique used in a computer environment where process is spread over multiple computers. The advent of the PC facilitated distributed processing.

Domain Name Service (DNS) A technique where TCP/IP names are resolved to IP addresses.

DOSKEY A Microsoft DOS and Windows command that allows you to scroll up and down through the list of commands that you've entered.

Dr. Watson A Microsoft Windows program that assists with debugging applications.

drive specification A DOS term comprised of a drive letter followed by a colon. Examples include A:, C: and D:.

driver roll back A feature in Windows XP where you can reinstall on old driver when a new driver causes problems.

drop-down menu A Microsoft Windows box that contains a down arrow allowing you to select from a list of choices for an option.

DRVSPACE.BIN A Microsoft DOS and Windows command used for disk compression.

DSL A technique used to transfer data over a WAN.

dual-boot A technique where multiple operating systems are installed on a computer and a choice is made as to which one to boot.

DualView A feature where you can display output on multiple monitors.

dynamic disks An NTFS technique allowing you to extend a partition from unallocated free space.

Dynamic Host Configuration Protocol (DHCP) DHCP is a method of automatically assigning IP addresses to network devices.

Dynamic Link Library (DLL) An executable file that contains other programs, called functions, and data.

ECHO A Microsoft DOS and Windows command used to display text on the screen.

echo A UNIX/Linux command that allows you to display text.

EDIT The Microsoft DOS and Windows command editor.

Electronic Magnetic Interference (EMI) Electronic noise generated by electrical devices.

Emergency Repair Disk (ERD) A disk used to repair a Microsoft Windows operating system.

emulated environment When a 32-bit application can run a 16-bit application.

emulation software Software that emulates (or imitates) characteristics of a device.

encryption A technique where data is converted into an unreadable form for security reasons.

Encryption File System (EFS) A Microsoft Windows 2000 and XP feature where you can encrypt files.

End Of File (EOF) A special character indicating to the operating system that a file contains no more data.

Escape key The key you use in the UNIX/Linux vi command that toggles between command and text mode.

Ethernet A network access mechanism used in star and bus topologies.

Event Viewer A Microsoft Windows program allowing you to manage events.

executable program A binary program is a set of instructions in computer form that can be executed, or run, by the computer.

execute permission For a directory, this UNIX/Linux permission allows you to list files and directories in the directory with this permission. For a file, this UNIX/Linux permission allows you to execute (or run) a program or shell script.

Execution time (E-time) The second part of the two-part instruction cycle where an instruction is executed.

Expanded Memory Specification (EMS) EMS uses a technique where 16 KB blocks of data are transferred in and out of a reserved 64 KB section of upper memory.

expansion slot A hardware component where devices connect to the computer.

ext2 A Linux file system.

ext3 A newer version of the ext2 file system that includes an on-disk journal that keeps track of changes.

Extended Memory Specification (XMS) XMS is memory that can exceed the 640 KB memory limit.

extended partition An extended partition is a partition that can be further divided into multiple logical partitions, called logical drives.

Extended Service Set (ESS) When multiple access points connect to the same main network (known to some as the distribution system), the network design is known as an ESS.

Extended Service Set Identifier (ESSID) Some manufacturers refer to the SSID in infrastructure mode as the ESSID, but this is the same concept as SSID.

external commands External commands can be seen when viewing files on a disk or a hard drive. External commands execute slower than internal commands because the external commands must be retrieved and loaded from the disk or hard drive. External commands reside in a system directory such as C:\WINDOWS.

FAT12 A File Allocation Table type used for floppies.

FAT16 A File Allocation Table type used for file storage.

FAT32 A File Allocation Table type used for file storage.

fault tolerance A technique where the failure of one part does not prevent the failure of the whole.

FDISK A Microsoft DOS and Windows command to create primary, extended and logical partitions.

fiber-optic cable A type of cable that uses glass or plastic fiber to carry light pulses.

field A set of related characters. For example, Social Security Number or Employee Name are considered fields.

file A storage area for data.

File Allocation Table (FAT) A technique for storing and retrieving data.

file attributes File settings such as Read, Archive and Hidden.

file extension A three character part of a file name that is used to associate an application to a file.

file level security The capability of setting security permissions, such as Read or Write, upon files.

file name A file name is used to specify a file. In DOS, it can be 8 characters followed by a period and a three character extension.

file system Defines how the files and folders (directories) are stored on disk.

file type association A technique where a file's extension is associated with a program.

FILES A Microsoft DOS and Windows command allowing you to specify the number of concurrent open files.

FIND A Microsoft DOS and Windows command allowing you to search for text within a file.

find A UNIX/Linux command that allows you to search for files or directories within the directory tree structure.

Firmware A hardware component that contains specific software instructions.

FIXBOOT A Microsoft Windows program that will fix the boot sector.

FIXMBR A Microsoft Windows program that will fix the MBR.

folder A storage location for other folders and files.

FOR A Microsoft DOS and Windows command used to perform looping.

format The process of preparing a disk for use by the operating system.

FORMAT A Microsoft DOS and Windows command that allows you to format a drive.

fragmentation A concept where a file is spread over non-contiguous areas of disk.

frame Data at the Data Link layer of the OSI model.

fsck The UNIX/Linux command to verify a file system. This is similar to the DOS CHKDSK or SCANDISK command.

full duplex Full duplex transmission is data transmitted in both directions on a cable simultaneously.

full path A full path always begins from the root directory of a partition.

functions Small programs that perform a specific purpose.

GDL.EXE A Microsoft Windows system file that provides support for the GUI environment.

General Protection Fault (GPF) An error that occurs when a program tries to access the same area of memory as another program.

Gigabyte (GB) 1 billion bytes.

GOTO A Microsoft DOS and Windows command that directs processing to a specific location in a batch program.

Graphical User Interface (GUI) A user-friendly software interface.

grep A UNIX/Linux command that allows you to search for text within files.

half duplex Half duplex transmission is where data transmitted in both directions on a cable, but not at the same time.

Hardware Abstraction Layer (HAL) A layer of a Microsoft Windows operating system between the operating system and the hardware.

hardware interrupt An interrupt originating from hardware.

Help Desk Representative The Help Desk Support Representative, or Help Desk Technician, typically works in a telephone support environment where users of all types call in with problems.

Hertz A measure of processor speed.

High Memory Area (HMA) The first 64 KB of extended memory used by DOS.

high-level programming language A language such as COBOL, C or Visual BASIC that is "English-like" in nature.

HIMEM.SYS A device driver used for extended memory.

Hidden attribute A Microsoft DOS and Windows attribute that indicates a file is hidden for normal view. You can see a hidden file in DOS if you enter the ATTRIB command. You can change the File Options to see a hidden file in Windows Explorer.

hive A Registry term used to represent a set of keys and values.

host In TCP/IP terms, a computer that has an IP address. In mainframe terms, a device that has a keyboard, terminal and a link to a mainframe for processing.

host number The portion of an IP address that represents a specific computer on the network.

hub A device that connects other devices, such as computers or printers, on a network.

icon A graphical representation of items such as programs, files or folders.

IEEE 802 standards Standards developed in February of 1980 by the IEEE.

IF A Microsoft DOS and Windows command allowing you to perform conditional processing. Typically used in a batch file.

IF ERRORLEVEL A Microsoft DOS and Windows command used in a DOS batch file to test if a key was pressed.

information Data that has been processed.

infrastructure mode The infrastructure mode wireless network connects multiple wireless network devices through an access point.

inode A unique number representing a file in the Linux operating system.

input device A device that facilitates input into the computer system.

Input/Output The term used to describe the input and output functions in a computer system.

Input/Output device (I/O) The term used to describe input and output devices such as keyboards, disk drives, and monitors.

Input/Output port A memory address through which data is transferred between a device and the processor.

Institute for Electrical and Electronics Engineers (IEEE) The IEEE committee created network standards called the 802 standards.

Instruction Register A register that contains the current instruction that is being executed by the CPU.

instruction set The set of instruction statements a processor can understand and use.

Instruction time (I-time) The first part of the two-part instruction cycle where an instruction is fetched from memory to the processor's registers.

internal commands Microsoft DOS and Windows commands that exist in the file named COMMAND.COM.

International Standards Organization (ISO) ISO has developed a model for network communications known as the OSI (Open Systems Interconnect) model.

Internet conferencing Software allowing you to communicate with other users on the Internet.

Internet Connection Sharing (ICS) Software allowing multiple computers to access the Internet via one computer.

Internet Explorer (IE) A web browser for connecting to the Internet.

Internet Service Provider (ISP) An organization through which you connect to the Internet.

Internetwork Packet Exchange/ Sequenced Packet Exchange (IPX/ SPX) IPX/SPX is used when connecting to a Novell network, but Novell networks now use TCP/IP as their standard protocol.

interpreting The process of converting programming language statements to executable code and then executing them.

interrupt A signal sent by a device to the processor indicating that the device needs the processor's assistance.

interrupt handler Software that handles processor interrupts.

Interrupt Request (IRQ) A signal sent by a device to the processor indicating that the device needs the processor's assistance.

Interrupt Vector Table A table used to point to where device drivers are stored in memory.

IO.SYS A DOS system hidden file that loads drivers.

IP address A 32 bit software address assigned to a computer running TCP/IP.

IP Security (IP Sec) A method of encrypting IP packets.

IPCONFIG command A Microsoft command that displays IP settings on Windows 98, NT, 2000 and XP.

ISDN A technique used to transfer data over a WAN.

job A running program. Also known as a task or process.

kernel The core of the operating system that remains in memory.

keys Registry folders.

Kilobyte (KB) 1 thousand bytes.

LABEL A Microsoft DOS and Windows command allowing you to change the disk label.

Last Known Good Configuration A setting you can choose during the Startup menu that allows you to access the last known good configuration of the operating system. Commonly used when you load a device driver that does not work.

legacy device A non-Plug and Play device.

linkage editor Software used to create a program. Once a program is assembled, it is then linked.

linker Software used to create a program. Once a program is assembled, it is then linked.

Linux A multi-user, multitasking operating system.

Local Area Network (LAN) A group of devices that can share resources in a single area, such as a room or a building.

log file A file that records message information about events that have occurred on the system.

logical partition A logical partition is where drive specifications, such as A: or C:, refer to a physical partition.

Long File Name (LFN) A Microsoft Windows file name that can contain up to 255 characters.

loop A technique used when program statements are repeated.

loopback address The address, 127.0.0.1, of the local NIC in a computer.

low-level programming language A language such as Assembly that is the language of the CPU.

ls The UNIX/Linux command that displays files and directories.

MAC OS An operating system developed by Apple Computer.

mainframe operating system Operating systems that are designed to handle the Input/Output (I/O), processing and storage requirements for a lot of users.

MAKEDISK.BAT A Microsoft Windows batch file that allows you to create an anti-virus disk.

man A UNIX/Linux command that allows you to display the manual help pages on a command.

Master Boot Record (MBR) The Master Boot Record (MBR) is a record that tells the operating system about the partitions. The MBR looks for the active partition and boots it.

Master File Table (MFT) NTFS uses a Master File Table (MFT), which is analogous to the table used by FAT16 and FAT32. The MFT is the first file on an NTFS volume that contains information about each file and folder on each volume.

Maximize button A Microsoft Windows button, located in the upper right hand corner of a window, allowing you to maximize your window.

MD A Microsoft DOS and Windows command allowing you to create a directory.

Media Access Control (MAC) address The physical hardware address associated with a NIC. A MAC address is 48 bits and is unique.

Megabyte (MB) 1 million bytes.

MEM A Microsoft DOS and Windows command that displays memory statistics.

memory address range A section of memory address.

mesh topology A topology where all devices connect to each other for redundancy.

Microsoft DOS A Microsoft operating system that is disk based and command oriented.

Microsoft System Information (MSINFO) A Microsoft Windows program allowing you to display information about system resources and can be used to detect conflicts between devices.

Microsoft Windows 2000 A Microsoft operating system that is graphically oriented.

Microsoft Windows 3.1 A Microsoft application used to interact with the operating system. Microsoft Windows 3.1 required an actual operating system such as MS-DOS.

Microsoft Windows 98 A Microsoft operating system that is graphically oriented.

Microsoft Windows Explorer A Microsoft application used to create, copy, and move files and folders.

Microsoft Windows NT A Microsoft desktop and server operating system.

Microsoft Windows XP A Microsoft desktop operating system.

midrange operating system Midrange operating systems are handle the Input/Output (I/O), processing and storage requirements and are generally used for medium-sized organizations. They operate in a centralized manner with terminals and PCs using emulation software, accessing applications remotely.

Minimize button A Microsoft Windows button, located in the upper right hand corner of a window, allowing you to minimize your window.

MKDIR The Microsoft DOS and Windows command that is the same as the MD command. See the MD command.

mkdir A UNIX/Linux command that allows you to create a directory.

mnemonic Another name for an assembly language statement is mnemonic.

modem A device that converts (or modulates) digital signals to analog and then demodulates the analog signals back to digital. A modem allows you to connect to a computer with another

modem, say one at your Internet Service Provider. Modem comes from the terms "modulate/demodulate".

MORE A Microsoft DOS and Windows command that allows you to display a screen of text at once.

more A UNIX/Linux command that allows you to display a screen of text.

motherboard The motherboard is the main system board in the Personal Computer that interconnects the other hardware components.

mount point In the UNIX and Linux operating systems, a mount point is a directory name that is associated with a partition.

MOVE A Microsoft DOS and Windows command used to move a file from one directory location to another.

MSDOS.SYS A DOS system hidden file.

multi-boot The capability of a computer to boot multiple operating systems.

multi-mode fiber-optic A type of fiber-optic that carries multiple light signals down the cable.

multi-processor An operating system kernel that can utilize several processors concurrently is called a multi-processor operating system.

multitasking A multitasking operating system kernel is one that appears to handle multiple tasks at the same time. A task is a program that is running. A task is also known as a job or process.

Multithread Operating systems supporting multithreaded applications allow threads to be executed concurrently. Also, see thread.

multi-user The capability of an operating system to support multiple users at the same time.

mv A UNIX/Linux command that allows you to move or rename files or directories.

My Computer An icon that is used to access hardware, software and files located on the computer.

My Documents icon A Microsoft Windows icon allowing you to quickly access the My Documents folder.

My Network Places A Microsoft Windows icon allowing you to view network computers. This icon is available only on Windows 2000 and XP. On Windows 98 and NT, the Network Neighborhood icon is used for the same purpose.

NET A Microsoft DOS and Windows command used for network trouble-shooting and configuration.

NetBIOS Enhanced User Interface (NetBEUI) NetBEUI is a non-routable network protocol.

NetWare The network operating system developed by Novell.

network A connection of two or more devices that can communicate.

Network Administrator A technical specialist responsible for all aspects of a network.

Network File System (NFS) A file system, developed by Sun Microsystems, allowing access to files and directories on remote computers across a network.

Network Interface Card (NIC) A hardware device allowing you to connect to other computers.

Network Neighborhood A Microsoft Windows icon allowing you to view network computers.

network number The portion of an IP address that represents which network segment a computer is on.

Network Operating System (NOS) An operating system that has network capabilities.

New Technology File System (NTFS) New Technology File System, provides performance and reliability not present in FAT. NTFS allows for disk quotas, compression, encryption, file level security, bad-cluster remapping, auditing and dynamic disks.

non-executable program A file such as a document, text file or graphics file that cannot be run, or executed, by the operating system.

Notepad A text processing Microsoft Windows application.

Novell NetWare A networking operating system developed by Novell.

NT Diagnostics A Microsoft Windows program that allows you to view configuration information about the computer's hardware, installed device drivers, and installed services.

NT Hardware Qualifier A Microsoft program that identifies what hardware is installed on a computer system.

NT Virtual DOS Machine (NTVDM) A software program that simulates a DOS environment inside NT.

NTBOOTDD.SYS A Microsoft Windows file used on a system with SCSI drives that have the SCSI BIOS disabled.

NTDETECT.COM A Microsoft Windows system command that detects the hardware on a PC.

NTLDR The Microsoft Windows operating system bootstrap loader.

NTOSKRNL.EXE The Microsoft Windows operating system kernel.

OK button A Microsoft Windows button that applies all changes when it is clicked.

Open Systems Interconnect (OSI) The OSI model is a standard for information transfer across the network.

Operating System An operating system (OS) is a set of software instructions that allows your computer system to operate. Operating systems are written in programming languages like application programs.

output devices A device that facilitates output from the computer system.

packet Data at the Network layer of the OSI model.

Packet Internet Groper (PING) A TCP/IP command that tests connectivity of a host computer.

page file The page file, or swap file, is the section of the hard disk used for virtual memory.

parent directory A directory that contains another directory or a file.

partial path A partial path is simply the file name without any reference to the root directory.

partition A section of a hard disk.

passwd A UNIX/Linux command that allows you to change your password.

path The path is the location of the file, or directory. The two types of paths are full and partial.

PATH command The DOS command that contains the directory path of available commands.

pattern matching A technique where symbols are used to represent characters in a file name. Also called wildcarding.

PAUSE A Microsoft DOS and Windows command allowing you to pause the execution of a batch file. You can enter the PAUSE command at the command prompt but it is typically used in a batch file. The user can press ENTER to continue when this statement is used.

peer-to-peer network (or peer network) In a peer-to-peer network, each computer acts as its own server.

Performance Monitor A Microsoft Windows program allowing you to monitor the performance of your computer.

Personal Computer (PC) A computer for personal use.

platform A platform is comprised of both the hardware and software that a given system runs on. For example, Microsoft Windows XP running on an Intel Pentium would be considered a platform.

Plug and Play (PnP) Plug and Play, developed by Intel, is a standard that allows a computer to automatically detect and configure the installed device. Both the operating system and the device must be Plug and Play compliant for this method to work.

plug-in An application designed to be used with a browser such as Macromedia Flash or Shockwave.

point and click A method whereby the mouse is used to navigate and manipulate Microsoft Windows.

Point-to-Point Protocol (PPP) PPP (Point-to- Point Protocol) is a connection-oriented, layer 2 protocol that encapsulates data for transmission over phone lines.

Port A receptacle on a hub or switch where devices connect.

portable This term is used to mean that an operating system can be fairly easily carried to a computer of another type and execute correctly.

positional parameters Data that can be passed (or given) to a batch file. You can pass up to 9 positional parameters. You can use the SHIFT command to allow more to be passed.

Power On Self Test (POST) Part of the boot sequence where components such as memory, the keyboard, the mouse and the disk drive are verified.

preemptive multitasking With preemptive multitasking, the operating system has the ability to take control of the computer system from an application.

primary partition The primary partition contains your operating system and is the first partition on a drive. In order to boot your operating system, you must have a primary partition and it must be marked as the active partition.

Process A running program. Also known as a task or job.

Process Identification (PID) number A process will be given a Process Identification number, called a PID, which is used by the operating system to reference the process.

program A software program is a collection of instructions that accomplish a task. Software programs are written in programming languages such as Pascal, C, C++, Java, Visual BASIC, COBOL, FORTRAN or Assembly.

Programmer Analyst The Programmer Analyst is the person responsible for analyzing the business needs of the user and writing the programs for users. This person must learn the programming language being used as well as the business processes so they can write programs effectively for users.

programming language Software programs are written in programming languages such as Pascal, C, C++, Java, Visual BASIC, COBOL, FORTRAN or Assembly. When a person writes a program, they write instructions to perform a certain function or task.

PROMPT A Microsoft DOS and Windows command used to set the command prompt.

protected mode In protected mode, the application software does not have direct access to the hardware. Also, multiple applications are assigned their own memory address space and are "protected" from one another. Microsoft Windows 98, NT, 2000, and XP boot into real mode and then run in protected mode. Protected mode is also called 32-bit mode.

protocol A set of rules that govern communication between devices.

pwd A UNIX and Linux command that displays your present working directory.

radio button A Microsoft Windows round circle where you can select an option.

Radio Frequency Interference (RFI) A specific type of EMI noise that occurs in the radio frequency range.

Random Access Memory (RAM) Random Access Memory, or RAM, is temporary. RAM is also shortened to "memory."

RD /S A Microsoft DOS and Windows command used to remove a tree, which includes files and directories.

RD A Microsoft DOS and Windows command allowing you to remove a directory.

RDISK A Microsoft utility used to create the Emergency Repair Disk (ERD).

Read-only attribute A Microsoft DOS and Windows attribute allowing you to only read the contents of a file. You cannot change the contents of the file if the read-only attribute is set.

read permission A UNIX/Linux file permission allowing you to read the contents of a file.

Readiness Analyzer A Microsoft Windows tool allowing you to check your system for hardware and software compatibility issues.

Read-Only Memory (ROM) Firmware is a set of instructions permanently stored on Read-Only Memory (ROM) chips. You cannot simply delete the contents of the ROM firmware by shutting down the computer.

real mode In real mode, the application software has direct (hence the term "real") access to the hardware (such as memory). In real mode, the failure of a single application can cause the whole computer to fail. Native DOS runs in real mode. Real mode is also known as 16-bit mode.

record Multiple fields together comprise a record. For example, a record could include these five fields: Employee ID, First Name, Last Name, Pay Rate, and Hours Worked.

Recovery Console A boot method for Microsoft Windows operating systems that allows you to boot to a command prompt. Here, you can start and stop services, repair the MBR with the FIXMBR command, repair boot sectors with the FIXBOOT command or format the disk drive with the FORMAT command. It takes up about 7 MB and is installed with the WINNT32 / CMDCONS command.

Recycle Bin An icon used to hold files and folders that have been deleted. Note that files and folders deleted from a floppy will not be held in the Recycle Bin.

Reduced Instruction Set Computer (RISC) A computer that contains few instructions in its instruction set and relies on hardware to handle the instructions not specifically contained within the instruction set.

REGEDIT A Microsoft Windows command to modify and view Registry components.

REGEDT32 A Microsoft Windows command to modify and view Registry components.

register A storage unit in the processor.

Registry The Registry is a hierarchical database that contains values for specific computer settings. Microsoft Windows 95, 98, NT, 2000 and XP all have a Registry.

relative path Another name for partial path.

REM A Microsoft DOS and Windows command used as a remark in a DOS batch file.

REN A Microsoft DOS and Windows command used to rename a file.

resource A resource is a hardware device, a software program, or a file needed by users.

rm A UNIX/Linux command that allows you to remove a file.

RMDIR The Microsoft DOS and Windows command that is the same as the RD command. See the RD command.

rmdir A UNIX/Linux command that allows you to remove a directory.

root directory The top level directory in the file system hierarchy.

root user The name of the UNIX and Linux user account that has full control of the operating system commands and files.

router A device that forwards a packet to a remote network.

rpm The Linux command to manage software applications.

Safe Mode A boot method for Microsoft Windows operating systems that loads the mouse, keyboard, CD-ROM and VGA drivers. This mode is useful for troubleshooting hardware problems.

Safe Recovery A Microsoft Windows feature allowing an installation to continue after a failure in the middle of the installation.

SCANDISK A Microsoft DOS and Windows command that checks the status of a disk.

SCANREG A command line based Registry Checking utility.

SCANREG.INI An initialization file for SCANREG.

SCANREGW A GUI line based Registry Checking utility.

screen saver A picture, color or pattern that displays when the computer is inactive.

Sector A sector is equal to 512 bytes.

Serial Line Internet Protocol (SLIP) An older protocol that was used with dial-up networking and was the predecessor to PPP.

server In a LAN, servers are computers that allow other computers to connect to the server's shared resources.

server-based network With a server-based network, computer users log in to a main computer called a server where they are authenticated (authorized to use the network).

service The Microsoft term for a running program.

Service Pack (SP) A software or set of software that fixes a problem. Microsoft uses the term SP.

Service Set Identifier (SSID) An SSID is a set of 32 alphanumeric characters used to differentiate between different wireless networks.

SETUP.EXE A Microsoft Windows software program allowing you to install the operating system.

shared resource A shared resource is a resource that is capable of being used on other computers. A printer that can be used by multiple users is an example of a shared resource. An application stored on a computer that is used by multiple users is another example of a shared resource.

shell In UNIX/Linux terminology, the command line interface is called a shell.

shell script A UNIX/Linux program that contains UNIX/Linux commands. A shell script should have the execute permission.

Shielded Twisted-Pair (STP) A type of twisted-pair cabling with additional insulation.

SHIFT A Microsoft DOS and Windows command used to shift positional parameters.

shortcut An icon that is used to quickly access to an application or file.

Simple Mail Transport Protocol (SMTP) SMTP is used for sending e-mail.

single-mode fiber-optic A type of fiber-optic that carries only one light beam down the cable.

Small Computer System Interface (SCSI) An interface standard that connects multiple small devices to the same adapter via a SCSI bus.

software Software is defined as a set of instructions that are processed by a computer system.

software interrupt An interrupt originating from software.

SORT A Microsoft DOS and Windows command used to sort, or arrange, data in a file.

sort A UNIX/Linux command that allows you sort, or arrange, data in a file.

spreadsheet Spreadsheet programs allow you to manage data in rows and columns.

Standard User The Standard User, or user, is the person who uses the system on a daily basis in support of their job. They use the system to access an application. The application they need depends upon their function in the organization.

star topology A topology where each device connects to a hub or switch.

Start button The button located in the lower left hand corner of a Microsoft Windows desktop used to launch application and utilities, find files and other computers, get help, and add/remove hardware and software.

Startup menu A Microsoft Windows menu that determines how the operating system will boot. Pressing the F8 key during the boot sequence will display the Startup menu.

subnet mask A 32-bit number that is used to determine which part of the IP address represents the network and which part represents the host.

subtree A Registry folder that contains hives.

Supervisor user A user with a great deal of access on a computer system.

swap file The swap file, or page file, is the section of the hard disk used for virtual memory.

switch A switch is an intelligent hub.

Symmetrical Multi-Processing (SMP) The ability of a computer system to use multiple processors that can execute different portions of a program is called Symmetrical Multi-Processing or SMP.

SYS A Microsoft DOS and Windows command used to transfer system files to a device.

SYSEDIT A Microsoft Windows program allowing you to modify system files.

System Administrator A person responsible for the operation of a computer system.

System Configuration Utility (MSCONFIG) A Microsoft Windows program allowing you to disable or enable entries in AUTOEXEC.BAT, CONFIG.SYS, SYSTEM.INI or WIN.INI.

System File Checker (SFC) A Microsoft Windows program that protects system files.

system files A set of Microsoft Windows operating system files.

system key A protection feature for Windows 2000 and XP passwords.

System Monitor A Microsoft Windows program that assists with monitoring system performance.

system partition Microsoft Windows NT, 2000, and XP use the terms system partition to refer to the location of the load files. The system partition must be marked as active in order for these operating systems to load.

system resources System resources are features that control how devices on a computer system work.

system software System software includes the core components of the system that must be present in order for the computer to operate. Examples of system software are the operating system kernel, process management, memory management, and device drivers.

System State A group of Microsoft Windows 2000 and XP files that consist of the Registry, system files, boot files and the COM+ Class Registration database.

system unit Also known as the computer system chassis. See chassis.

SYSTEM.DAT A Registry file that holds computer-specific hardware settings, PnP configurations and application settings.

T1 A type of WAN connection.

table A table is a collection of database data stored on disk.

task A running program. Also known as a job or process.

Task Manager A Microsoft Windows utility allowing you to manage processes.

taskbar The bar that run horizontally across the screen on Microsoft Windows operating systems.

telnet A TCP/IP related command that allows you to remotely connect to another computer.

Terabyte (TB) 1 trillion bytes.

terminals A terminal is a device that has no computing ability and is strictly dependent upon the processing power of another, such as the mainframe.

Terminate and Stay Resident (TSR) A TSR is a program that executes, terminates normally, and then stays resident in memory until needed by the operating system. Examples include

HIMEM.SYS, EMM386.EXE and COMMAND.COM.

text box A Microsoft Windows area where you can type a specific parameter (or value).

text mode A mode where characters are entered on the screen. DOS and Linux have a text mode.

The Cloud A graphical representation of the Internet.

thread A thread is a basic unit of instruction that is allocated processor time by the operating system. Think of a process as being made up of many threads. The threads of a process execute the process code. A thread can execute any part of the code, which includes portions that are currently run by another thread.

thread synchronization Threads synchronize with each other to coordinate resource access. This prevents one thread from interrupting another.

TIME A Microsoft DOS and Windows command used to display or change the time.

time slice A time slice is a unit of time allocated to a task

Title Bar The horizontal bar at the top of a window that indicates the name of the program that is running.

Token A packet of data passed between computers in a ring topology.

Token Ring A topology, developed by IBM, that is wired like a star topology but the computers are logically accessed in a ring where a "token" is passed to each computer. The computer with the "token" can use the network.

Topology The arrangement of devices in a network.

touch A UNIX/Linux command that allows you to update the modification time stamp on a file or directory.

tracert command A TCP/IP related command used to trace the route a packet takes from source to destination.

track A track is a concentric circle running around the center of the disk.

Transmission Control Protocol/Internet Protocol (TCP/IP) The protocol used to access the Internet.

TREE A Microsoft DOS and Windows command allowing you to view the file system hierarchy.

Tree A Linux command that displays the directory structure.

twisted-pair cable Cable used in a star topology.

TYPE A Microsoft DOS and Windows command allowing you to display the contents of a file.

Universal Disk Format (UDFS) Universal Disk Format, is primarily used for read-only DVD/CD-ROM media.

Uniform Resource Locator (URL) An address of a web page that is on the Internet.

Universal Naming Convention (UNC) A Microsoft naming convention for a shared resource. The UNC is comprised on the computer and shared device name.

UNIX A multi-user, multitasking operating system.

Unshielded Twisted-Pair (UTP) A type of twisted-pair cabling with no additional insulation.

upgrade The process of installing an operating system with a higher version on a computer with an operating system that contains a lower version operating system. For example, if Windows 2000 existed on a PC, you could upgrade it to Windows XP.

upper memory or Upper Memory Block (UMB) Upper memory is used for the system BIOS, video BIOS and other functions and is the 384 KB from 640 KB to 1 MB.

User The Standard User, or user, is the person who uses the system on a daily basis in support of their job. They use the system to access an application. The application they need depends upon their function in the organization.

user profile A Registry setting that contains a user's specific configuration settings.

USER.DAT A Registry file that holds user-specific settings such as logon name, desktop setting and Start button settings.

value entries Fields within the Registry that contain data.

VER A Microsoft DOS and Windows command used to view the operating system version.

vi A UNIX/Linux command that allows you to create, edit and modify text in a file.

virtual device driver (VxD) A device driver that operates in protected mode and can access hardware directly.

Virtual FAT (VFAT) Microsoft developed VFAT (Virtual FAT) with Microsoft Windows 95 to allow file names to exceed the 8.3 rule. VFAT allows Long File Names, or LFNs.

Virtual Memory Virtual memory uses RAM and a section of the hard disk to accommodate multitasking and multiple users.

Virtual Memory Manager (VMM) Software that manages virtual memory.

Virtual Private Networking (VPN) A VPN is a remote computer connecting to a remote network by securely "tunneling" over an intermediate network such as the Internet or a LAN.

virus A computer program that is designed to do harm.

wallpaper scheme A Microsoft Windows background picture, pattern, or color.

wasted space Unused space in a disk cluster caused by a file not completely occupying the cluster.

who A UNIX and Linux command that displays the current users logged on.

Wide Area Network (WAN) WANs connect devices in remote locations. When you connect to the Internet, you are using WAN technologies.

wildcard characters Characters, such as * and ?, which are used to represent (or match) other characters.

WIN.COM The DOS command that loads Microsoft Windows 3.1.

window Part of the screen that belongs to a specific application or utility.

Windows File Protection (WFP) A Microsoft Windows program that protects system files.

Windows Image Acquisition (WIA) Software allowing communication between applications and image-capturing devices.

Windows Internet Naming Service (WINS) A Microsoft program that resolves Microsoft computer names to IP addresses.

Windows On Windows (WOW) An environment simulator for 16-bit applications that runs inside of an NTVDM.

Windows Protection Error A Microsoft Windows error caused by a virtual device driver being loaded or unloaded.

Windows script A Microsoft Windows method of automating shortcuts for users and set or restrict access for the desktop, Start menu, and applications.

Windows Script Host (WSH) The Microsoft Windows scripting host computer.

WINIPCFG A Microsoft command that displays IP settings on Windows 95 and 98.

WINNT.EXE The Microsoft Windows NT installation program. It is used to install to a computer that currently has DOS, Windows 3.x, Windows 95 or Windows 98 installed.

WINNT32.EXE The Microsoft Windows NT installation program. It is used to upgrade from a previous version of Microsoft Windows NT.

WINVER A Microsoft DOS and Windows command that displays operating system version.

Wireless Equivalent Privacy (WEP) WEP encrypts data being transmitted.

word processing Word processing software allows you to create, modify, delete, save, and print documents that are office quality. They also have capabilities for spell checking, and include a dictionary and thesaurus.

write permission A UNIX/Linux file permission allowing you to write to a file.

WSCRIPT.EXE The Microsoft Windows scripting tool.

XCOPY A Microsoft DOS and Windows command used to copy a whole tree, including files and directories.

Index

Instructor's Answers

Chapter 1

1. B
2. A
3. B
4. C
5. D
6. A
7. C
8. B
9. C
10. A
11. D
12. D
13. D
14. C
15. C
16. C
17. A
18. D
19. A
20. B

Chapter 2

1. B
2. A
3. C
4. D
5. D
6. A
7. D
8. A
9. B
10. C
11. A
12. C
13. D
14. B
15. A
16. B

17. A
18. A
19. C
20. C

Chapter 3

1. A
2. C
3. B
4. A
5. D
6. B
7. B
8. D
9. D
10. A
11. C
12. B
13. C
14. D
15. A
16. D
17. B
18. B
19. A
20. C

Chapter 4

1. B
2. A
3. A
4. B
5. C
6. B
7. C
8. A
9. B
10. A
11. D
12. B

13. D
14. D
15. A
16. B
17. A
18. A
19. B
20. D

Chapter 5

1. C
2. A
3. A
4. B
5. C
6. A
7. C
8. B
9. B
10. D
11. D
12. B
13. B
14. B
15. D
16. A
17. A
18. C
19. B
20. A

Chapter 6

1. B
2. B
3. C
4. A
5. C
6. A
7. B

8. A
9. A
10. B
11. D
12. A
13. D
14. A
15. D
16. A
17. D
18. A
19. C
20. D

Chapter 7

1. D
2. C
3. D
4. A
5. B
6. A
7. D
8. C
9. B
10. C
11. B
12. A
13. C
14. A
15. B
16. C
17. D
18. B
19. A
20. D

Chapter 8

1. D
2. C and D
3. D
4. B
5. C
6. A
7. B
8. D
9. D
10. B
11. D
12. D
13. C
14. B
15. B
16. C
17. D
18. D
19. B
20. C

Chapter 9

1. B
2. A
3. C
4. D
5. B
6. C
7. C
8. A
9. C
10. A
11. B
12. B
13. A
14. B
15. D
16. D
17. C
18. C
19. D
20. C